Springer Theses

Recognizing Outstanding Ph.D. Research

Aims and Scope

The series "Springer Theses" brings together a selection of the very best Ph.D. theses from around the world and across the physical sciences. Nominated and endorsed by two recognized specialists, each published volume has been selected for its scientific excellence and the high impact of its contents for the pertinent field of research. For greater accessibility to non-specialists, the published versions include an extended introduction, as well as a foreword by the student's supervisor explaining the special relevance of the work for the field. As a whole, the series will provide a valuable resource both for newcomers to the research fields described, and for other scientists seeking detailed background information on special questions. Finally, it provides an accredited documentation of the valuable contributions made by today's younger generation of scientists.

Theses may be nominated for publication in this series by heads of department at internationally leading universities or institutes and should fulfill all of the following criteria

- They must be written in good English.
- The topic should fall within the confines of Chemistry, Physics, Earth Sciences, Engineering and related interdisciplinary fields such as Materials, Nanoscience, Chemical Engineering, Complex Systems and Biophysics.
- The work reported in the thesis must represent a significant scientific advance.
- If the thesis includes previously published material, permission to reproduce this must be gained from the respective copyright holder (a maximum 30% of the thesis should be a verbatim reproduction from the author's previous publications).
- They must have been examined and passed during the 12 months prior to nomination.
- Each thesis should include a foreword by the supervisor outlining the significance of its content.
- The theses should have a clearly defined structure including an introduction accessible to new PhD students and scientists not expert in the relevant field.

Indexed by zbMATH.

More information about this series at http://www.springer.com/series/8790

Fabien Niel

Classical and Quantum Description of Plasma and Radiation in Strong Fields

Doctoral Thesis accepted by
Sorbonne Université, Jussieu, Paris

Springer

Author
Fabien Niel
Laboratoire Utilisation Lasers Intenses
Sorbonne Université
Paris, France

Supervisor
Prof. Caterina Riconda
Laboratoire Utilisation Lasers Intenses
Sorbonne Université, Jussieu
Paris, France

ISSN 2190-5053 ISSN 2190-5061 (electronic)
Springer Theses
ISBN 978-3-030-73546-3 ISBN 978-3-030-73547-0 (eBook)
https://doi.org/10.1007/978-3-030-73547-0

This Springer imprint is published by the registered company Springer Nature Switzerland AG
The registered company address is: Gewerbestrasse 11, 6330 Cham, Switzerland

À mes parents et à ma soeur

Supervisor's Foreword

The study of quantum electrodynamics (QED) processes and strong-field QED processes, in particular, has long been considered the domain of elementary particle physics. Over the last decade, however, the study of QED processes in plasmas has become a very hot research topic, at the cross-road of various disciplines and as such approached differently by various scientific communities. The motivation behind the ongoing research effort has come, on the one hand, from the realization that QED processes are important in various extreme astrophysical environments dominated by relativistic plasma flows, such as active galactic nuclei, pulsars or gamma ray bursts. On the other hand, it is encouraged by the advent of new research facilities, such as the x-ray free-electron lasers and multi-petawatt laser facilities under construction worldwide, which will soon allow to create, in the laboratory, electromagnetic fields of unprecedented amplitude together with ultra-relativistic particles and/or plasma flows. Interestingly, the strong research activity in this field is also supported and in many ways driven by the development of high-performance kinetic simulation tools that more and more allow to account for additional QED processes in a fully consistent, relativistic framework for plasma modelling.

Fabien Niel's thesis is a contribution to this research effort, which focuses on the study of two processes: (i) the emission of high-energy gamma photons by ultra-relativistic electrons in strong electromagnetic fields and its back reaction on the electrons' dynamics; and (ii) the production of electron–positron pairs in the interaction of an ultra-high intensity laser beam carrying angular orbital momentum with a counter-propagating gamma-photon bunch.

The treatment of high-energy (gamma) photon emission by ultra-relativistic electrons in strong electromagnetic fields and its back reaction on the electron dynamics is a very old problem known as radiation reaction (RR). Its description and study in Fabien Niel's thesis have led to new results that are particularly relevant today. First, it presents in a self-consistent, mathematically clear and physically solid framework how classical RR—usually treated using a continuous radiation reaction force—emerges from a quantum description based on the statistical description of not a single electron, but an ensemble of electrons. By doing so, it is possible to understand and relate various treatments of RR whether in a theoretical framework (linear Boltzmann, Fokker–Planck or radiation reaction force descriptions) or in computer simulations (Monte Carlo and particle-in-cell simulations). The domains of validity of the various descriptions are also identified. In addition, the author presents an exhaustive analysis of the effects of RR on the distribution function of relativistic charged particles interacting with a strong background electromagnetic field, as well as on the emitted high-energy gamma-photon spectra. This second aspect has serious implications for future experiments on the forthcoming extreme light laser facilities such as the Apollon laser in construction on the Plateau de Saclay, South of Paris, and the European project Extreme Light Infrastructure (ELI) that aims at the construction of at least three multi-petawatt laser facilities in Central and Eastern Europe. Such laser facilities will indeed provide the scientific community with a new path towards a yet unexplored regime of laser–plasma interaction where the electron dynamics is not only ultra-relativistic but also of a quantum nature. The in-depth study of the resulting electron and photon distributions will prove an important asset in the study of RR on these forthcoming lasers.

The most exotic process and striking demonstration of QED envisioned on these extreme light lasers is the so-called nonlinear Breit–Wheeler process. It consists in the production of an electron–positron pair via the interaction of an energetic (gamma) photon with the strong electromagnetic field of the laser. A chapter in Fabien Niel's thesis is devoted to the study of this process in the interaction of a high-energy photon bunch colliding with a Laguerre–Gaussian beam carrying orbital angular momentum. This study clearly illustrates that QED processes in plasmas are not only interesting at the fundamental level, they can also help design new intense sources of radiation (gamma photons) and electron–positron pairs.

Clearly, the thesis by Fabien Niel has the potential to help the next generation of young scientists get acquainted with the emerging field of strong-field plasma interaction. Supported by an extensive literature, it answers many questions as well as poses new ones, thus paving the way to further studies in the subject. We are also confident that this thesis will make an excellent reference text for a much broader audience and a valuable textbook for both students or experienced researchers interested in the treatment of QED processes in plasmas. Indeed, beyond the presentation of original results in Part II and III of the thesis, Part I is intended as an advanced tutorial moving from classical electrodynamics, through the classical theory of RR, to quantum electrodynamics and quantum RR, while finally introducing the basics of the kinetic description of a system composed of a large number of particles at the basis of plasma physics. There the author presents in a concise, yet clear and self-contained framework the theoretical background necessary to a solid understanding of the inclusion of QED processes in plasmas. Beyond the pedagogical approach, we hope this thesis will offer the reader a glimpse of the excitement and new ideas blooming in this very active research field, and that it will motivate advances in basic physics problems past the near future applications of intense laser–plasma interaction.

Prof. Caterina Riconda

Acknowledgements

First of all, I would like to thank my thesis supervisor Caterina Riconda for her unwavering support during these three years from both a human and a scientific point of view. Thank you for always encouraging me, even during my periods of existential doubts and questions about my future. Thank you for your intellectual availability. You were always ready to discuss any scientific idea, whether it was in relation with the thesis or not. Finally thanks for having contributed to this friendly and relaxed working atmosphere. I will remember, in particular, the typical discussions about the latest movie you had seen, or other more "philosophical" questions (such as "who was nicer: Karl Popper or Ludwig Wittgenstein?" Or "who is the nice one in Death Note: L or Light?").

I would like to thank François Amiranoff for the trust you have placed in me during the "laser–plasma interaction at high flux" oral exam, at the end of which I told you I wished to pursue a thesis in theoretical physics and you proposed me this thesis without hesitation. Thank you for all the cakes, chocolates and other dishes you constantly gratified us, for the greatest pleasure of our palates and highest displeasure of our bellies! Finally thanks for all the "physics of everyday life" questions that you shared with us and that constitute for me an important source of inspiration for any physicist ("how does a diver feel the pressure of the water above him?").

Last but not least, I would like to thank Mickaël Grech. Thank you for your rigour, always searching for the devil in the details, or as you would say, "á e****** des mouches"! The sessions on the board to try to clarify what had been done have contributed without a doubt to the quality of this thesis. Thank you as well for all the discussions on physics ("what would happen to the twin paradox if we have a universe that folds in on itself?"), music (Leonard Cohen vs Bob Dylan) or the questions of existential nature that often started with "I'm going to grab a sandwich, let's go downstairs?".

I would also like to thank all the members of the jury, Pr. Mattias Marklund, Pr. Luis O. Silva, Pr. Antonino Di Piazza, Dr. Laurent Gremillet and Pr. Jean-Michel Raimond, for taking the time to read and evaluate this work, and also for the numerous questions and remarks. Thanks, in particular, to Pr. Mattias Marklund and Pr. Luis O. Silva for having accepted to be the rapporteurs of this thesis and to Jean-Michel Raimond for having accepted to chair the committee.

I would also like to thank Antonino Di Piazza to have welcomed me in his group during the summer of 2017. I think he is an outstanding scientist and I feel lucky to have had the chance to work with him. I would also like to thank him for his interest in my work that manifested itself in the several times he invited me for a seminar in his group and for the very interesting questions he always had about my work.

I would like to thank Pr. Francesco Pegoraro, a "giant" in plasma physics, for his interest in my work and the interesting discussions we had when he came to Paris and when he invited me to his group in Pisa.

Thanks Archana, Maitreyi, Daniel and Sergei for all the great moments we spent at Bishmarckplatz! I keep a wonderful memory of our parties at the castle! My time in Heidelberg would not have been the same without you guys!

I would also like to thank my current and past colleagues, without whom the work would not have been the same: Anna Grassi (despite her annoying tendency to want to keep the office temperature above 35 C, even during a heat wave), Marco Chiaramello, Andrea Sgattoni, Samuel Marini, Samuel Ongala, Paulo Angelo and Tomasso Vinci. Thank you to Fred Perez for our passionate discussions regarding "exotic" physics or history of science. I would also like to counter-thank Paula Kleij for keeping me from working and for falling asleep during my group meeting!

Merci à la team fusion, Noémie et Florian, pour nos fameuses soirées Gustino. Merci en particulier à Florian sans qui ces trois années n'auraient clairement pas été les mêmes. Je garde en particulier un très bon souvenir de notre premier été de thèse avec nos restos sur l'île Saint-Louis, ainsi que les week-ends au Boulois à réparer la 4L!

Merci à mes amis, les Pyrus, 8 ans d'amitié déjà , et oui on le dit tout le temps mais "ça passe trop vite !". Merci à mes amis de la maison de la Tunisie, sans qui cette première année de thèse n'aurait pas été la même. Merci William, toujours prèsent aprés tant d'annés. Même s'il est difficile d'évoquer tout le monde, j'aimerais citer en particulier: Maxime, Mathilde, Antoine, Dorine, Pauline, Salma, Anaïs, Houssem, Ghada, Rami, Alan, Sébastien, Paula.

Finalement je voudrais remercier mes parents, ma soeur et toute ma famille pour leur soutien depuis le début de mes études. Merci Papa et Maman pour m'avoir toujours encouragé et pour avoir su me donner les meilleures conditions possibles pour que tout cela soit possible. Merci Valentine pour ta présence et ton soutien, même quand je n'avais pas le moral. Pour finir, une pensée pour notre boule de poils préférée: Bonny.

Contents

List of Symbols

$\eta^{\mu\nu} = \mathrm{diag}(+,-,-,-)$	Flat Minkowski space-time metric
$\delta^\mu_\nu, \delta_{ij}$	Kronecker symbols
$\epsilon^{\mu\nu\rho\sigma}, \epsilon^{0123} = -\epsilon_{0123} = 1$	Four-dimensional Levi–Civita symbols
$\epsilon^{ijk}, \epsilon^{123} = 1$	Three-dimensional Levi–Civita symbols
$\boldsymbol{a}^i = (a^1, a^2, a^3)$	Three-vector (bold Latin letters, with Latin indices)
$a^\mu = (a^0, \boldsymbol{a})$	Four-vector (italic Latin letters, with Greek indices)
$(\boldsymbol{a} \times \boldsymbol{b})^i = \epsilon^{ijk} = \epsilon^{ijk} a^j b^k$	Cross-product
$\boldsymbol{a} \cdot \boldsymbol{b} = a^i b^i$	Scalar product of two three-vectors
$ab = a^\mu b_\mu = a^0 b^0 - \boldsymbol{a} \cdot \boldsymbol{b}$	Scalar product of two four-vectors
$x^\mu = (t, \boldsymbol{x}) = (t, x, y, z)$	Four-position vector
$\partial_t = \partial/\partial t = \partial/\partial x^0$	Partial derivative with respect to time
$\partial_i = \partial/\partial x^i$	Partial derivative with respect to the *i*th space component
$\boldsymbol{\nabla} = (\partial_i) = -(\partial^i)$	Space gradient
$\Box = \partial^2$	D'Alembert operator
$\partial_\mu = \partial/\partial x^\mu = (\partial_t, \boldsymbol{\nabla})$	Partial derivative with respect to the four positions
$f'(x)$	Derivative with respect to the argument
$[A, B] = AB - BA$	Commutator
$\{A, B\} = AB + BA$	Anticommutator
$T^{[\mu\nu]} = \frac{1}{2}(T^{\mu\nu} - T^{\nu\mu})$	Anti-symmetric part of a tensor
$T^{\{\mu\nu\}} = \frac{1}{2}(T^{\mu\nu} + T^{\nu\mu})$	Symmetric part of a tensor symmetric part of a tensor
$T^{*\mu\nu} = \frac{1}{2}\epsilon^{\mu\nu\rho\sigma} T_{\rho\sigma}$	Dual tensor
$A^\dagger, A^\mathrm{T}, A^{-1}$	Hermitian conjugate, transpose and inverse of an operator
$A^\dagger = A$	Hermitian operator
$A^\dagger = -A$	Anti-Hermitian operator
$A^\dagger = A^{-1}$	Unitary operator
γ^μ	Dirac gamma matrices (see Appendix B.1)
$\not{a} = \gamma^\mu a_\mu$	Feynman's slash notation

Ψ	Dirac spinor
$\bar{\Psi} = \Psi^{\dagger}\gamma^0$	Dirac conjugated spinor
$\bar{M} = \gamma^0 M^{\dagger}\gamma^0$	Dirac conjugated matrix

Conventions and Units

Natural Units

In Chapters 2 and 3, we use the so-called *natural* or *rationalized Gaussian c.g.s. units*. In this system of units, we have $\hbar = c = 1$ and the fine structure constant α is related to the electron charge e by

$$\alpha = \frac{e^2}{4\pi} \, . \tag{1}$$

Maxwell's equations then take the form

$$\nabla \cdot \boldsymbol{E} = \rho, \; \nabla \times \boldsymbol{B} = \partial_t \boldsymbol{E} + \boldsymbol{j} \, , \tag{2}$$

while the Coulomb field between two charges Q_1 and Q_2 takes the following form

$$V(r) = \frac{Q_1 Q_2}{4\pi r} \, . \tag{3}$$

Furthermore, the energy density of the electromagnetic field reads

$$\mathcal{E} = \frac{1}{2} \left(\boldsymbol{E}^2 + \boldsymbol{B}^2 \right) . \tag{4}$$

In this systems of units

$$[\text{velocity}] = \text{pure number} \tag{5}$$

$$[\text{energy}] = [\text{momentum}] = [\text{mass}] \, , \tag{6}$$

since $c = 1$, while

$$[\text{length}] = [\text{mass}]^{-1}\,, \tag{7}$$

since $\hbar/(mc)$ is a length. Therefore, all physical quantities can be expressed as powers of mass, or equivalently as powers of length.

SI Units

In Chapter 4 and Parts II and III, we use the *International System of Units* or *SI* units in short. In this system of units, the fine structure constant α is related to the electron charge e by

$$\alpha = \frac{e^2}{4\pi\epsilon_0\hbar c}\,. \tag{8}$$

Maxwell's equations then take the form

$$\boldsymbol{\nabla}\cdot\boldsymbol{E} = \frac{\rho}{\epsilon_0}\,, \boldsymbol{\nabla}\times\boldsymbol{B} = \mu_0\boldsymbol{j} + \frac{1}{c^2}\,\partial_t\boldsymbol{E}\,, \tag{9}$$

while the Coulomb field between two charges Q_1 and Q_2 takes the following form

$$V(r) = \frac{Q_1 Q_2}{4\pi\epsilon_0 r}\,. \tag{10}$$

Furthermore, the energy density of the electromagnetic field reads

$$\mathcal{E} = \frac{\epsilon_0}{2}\,\boldsymbol{E}^2 + \frac{1}{2\mu_0}\,\boldsymbol{B}^2\,. \tag{11}$$

Indices

Throughout this thesis, we use the flat Minkowski metric with signature $(+,-,-,-)$, i.e.

$$\eta^{\mu\nu} = \mathrm{diag}(+1,-1,-1,-1)\,. \tag{12}$$

Lorentz indices are written with Greek characters and take values $\mu = 0,1,2,3$, while we use Latin characters for spatial indices, $i,j,\ldots = 1,2,3$. Repeated upper and lower Lorentz indices are summed over, i.e. $A_\mu B^\mu = \sum_{\mu=0\ldots3} A_\mu B^\mu$. The Levi–Civita symbols $\epsilon^{\mu\nu\rho\sigma}$ are chosen so that $\epsilon^{0123} = +1$ (and therefore $\epsilon_{0123} = -1$).

Fourier Transform and Dirac δ Distribution

We adopt the following conventions. For four-dimensional Fourier transforms

$$f(x) = \int \frac{d^4k}{(2\pi)^4}\, e^{-ikx}\tilde{f}(k)\,, \tag{13a}$$

$$\tilde{f}(k) = \int d^4x\, e^{ikx} f(x)\,, \tag{13b}$$

and for three-dimensional Fourier transforms

$$f(\boldsymbol{x}) = \int \frac{d^3k}{(2\pi)^3}\, e^{+i\boldsymbol{k}\cdot\boldsymbol{x}} \tilde{f}(\boldsymbol{k})\,, \tag{14a}$$

$$\tilde{f}(\boldsymbol{k}) = \int d^3x\, e^{-i\boldsymbol{k}\cdot\boldsymbol{x}} f(\boldsymbol{x})\,. \tag{14b}$$

For any integer n, n-dimensional Dirac delta functions are defined so that

$$\int d^nx\, e^{ikx} = (2\pi)^n \delta(n)(k)\,. \tag{15}$$

Chapter 1
Introduction

In the fabric of space and in the nature of matter, as in a great work of art, there is, written small, the artist's signature

Carl Sagan (1985), in *Contact*

1.1 Strong-Field Regime of Quantum Electrodynamics

Modern physics describes nature using four fundamental forces: gravity, the electromagnetic force, the weak and the strong interactions. Excluding gravity, the three remaining forces are described through the *Standard Model of particle physics*, which describes all known elementary particles. It has been tested with an outstanding precision, and the last experimental evidence was recently given with the discovery at the LHC of the Higgs boson [1], which explains why elementary particles (except the photon and the gluon) are massive [2, 3]. Within this theory, quantum electrodynamics (QED) is the unbroken part of the spontaneously broken Glashow-Salan-Weinberg model of electro-weak interaction. More simply, QED is the quantum theory of the electromagnetic interaction. It is an extremely precise theory with an incredible agreement with experiment. For example, the fine structure constant α has been measured experimentally with a precision of more than 8 digits [4]. However, certain questions still remain unanswered. In particular, the Standard Model requires the experimental determination of 19 parameters (among which we find the masses and couplings) and for which the theory does not provide any explanation of why they have these precise values. Moreover, the simple fact that the Standard Model does not comprise gravity has lead physicists to think that it only represents the low-energy limit of a "grand unified theory" (GUT) that should include a quantum description of gravity at the Planck scale.

F. Niel, *Classical and Quantum Description of Plasma and Radiation in Strong Fields*, Springer Theses, https://doi.org/10.1007/978-3-030-73547-0_1

For all these reasons, it is desirable to test the Standard Model in new, unexplored regimes of interaction, in order to provide the theory a possibility to fail. In usual collider experiments used to confront the theory, computations rely on perturbation theory. A possible complementary approach can therefore be the investigation of highly nonlinear or nonperturbative phenomena such as color confinement in Quantum Chromodynamics (QCD). Another path is the study of QED in a strong background field such as a high-intensity laser field, for which the electron exhibits a nonperturbative coupling with the laser field. This is the field of Strong-field QED to which this thesis is devoted.

1.2 Schwinger-Sauter Mechanism

Among the nonperturbative features that QED acquires in the presence of a background field, stands the so-called Schwinger-Sauter mechanism. This process was first discovered by Sauter [5] as an explanation for the Klein paradox [6] and later formalized by Schwinger in the context of QED [7]. Because of the Heisenberg uncertainty principle and of the Einstein mass-energy equivalence, an electron-positron pair with energy $\epsilon \gtrsim mc^2$ can spontaneously appear in vacuum for a short time $t \lesssim \hbar/\epsilon$. Pictorially speaking, we say that vacuum is full of virtual electron-positron pairs. In the presence of a strong external electric field, electrons (and positrons) will be accelerated as $\epsilon = ecEt$ where E is the external electric field. If this field has an amplitude $E \geq E_S = m^2c^3/(e\hbar)$, it will provide a work of mc^2 over a Compton wavelength $\lambda_c = \hbar/mc$ to the virtual pairs and will therefore bring them on shell. We talk about spontaneous pair creation from vacuum. This QED critical field is called *Schwinger field* and has the numerical value $E_S \simeq 1.3 \times 10^{18}$ V/m. Unfortunately, fields corresponding to such an intensity $I_S \simeq 4.6 \times 10^{29}$ W/cm^2 are not attainable in the near future in experiments. Several catalyzing mechanisms are however possible, among which we find the so-called assisted Schwinger mechanism [8, 9], where a strong, low frequency (for example a high-intensity optical laser) and a weak, high frequency field (for example, a XFEL) are superimposed. Intuitively speaking, the effective tunnelling barrier is reduced by absorbing several photons from the high-frequency field. This configuration has also been proposed in order to observe other pure vacuum QED effects such as the vacuum birefringence [10]. Alternatively, the Schwinger field can be reached in a boosted Lorentz frame. In the case of a photon beam with energy $\gtrsim 2\,mc^2$, counterpropagating with a laser field with amplitude E, pair creation occurs if $\chi_\gamma = 2\gamma_\gamma E/E_S \gtrsim 1$. This quantity χ_γ is called *nonlinear photon quantum parameter* and controls the process of pair production by a high energy photon in a strong background field. This process is called *nonlinear Breit-Wheeler pair production* and will be studied in this work.

1.3 Strong-Field QED Processes

In addition to the nonlinear Breit-Wheeler process, many other strong-field QED processes exist, that take advantage of this Lorentz boost factor. Among these, we find the well known *nonlinear Compton scattering*. This process consists in the emission, by an electron in a strong external field, of a high-energy photon, after the absorption of n external field photons: $e^- + n\hbar\omega_0 \rightarrow e^- + \gamma$. It is controlled by the nonlinear electron quantum parameter χ, which measures the ratio of the electric field seen by the electron in its proper frame and the Schwinger field. For a counterpropagating configuration, it is given by $\chi = 2\gamma E/E_S$, similarly to the photon case. In matter, there exist analogous processes, where the external field is the Coulombian field of a heavy nucleus. The decomposition of a gamma photon into an electron-positron pair at proximity of a heavy nucleus is called *Bethe-Heitler* process: $\gamma + Z \rightarrow e^- + e^+ + Z$. Similarly, the emission of a high-energy photon by an electron deflected at proximity of a heavy nucleus is called *Bremsstrahlung*: $e^- + Z \rightarrow e^- + \gamma + Z$. Finally, because of crossing symmetry, all the inverse processes are also possible, such as the photon absorption by an electron (or positron): $\gamma + e^{-/+} \rightarrow e^{-/+}$ or the recombination of an electron-positron pair into a gamma photon $e^- + e^+ \rightarrow \gamma$.

In this work, we will focus our attention on the nonlinear Compton scattering and the nonlinear Breit-Wheeler process which are dominant in vacuum.

1.4 Sources of Strong Fields and High-Energy Particles in Nature and in the Laboratory

All the previously described processes occur in a strong external background field and involve high energy particles. These can be found in a variety of environnements, from astrophysics to the laboratory:

- Pulsars are highly magnetized rotating neutron stars. They emit collimated electromagnetic radiation and are believed to be the source of ultra-high-energy cosmic rays. The magnetic field at their surface is of the order of 10^8 T, i.e. almost of the order of the Schwinger magnetic field ($B_S \simeq 4.4 \times 10^9$ T).
- Magnetars are a type of neutron star that have an extremely powerful magnetic field, going from 100 to 1000 times that of a usual pulsar ($\sim 10^{10} - 10^{11}$ T), that is, from 10 to 100 times the Schwinger field.
- Heavy nuclei with very high charge have an ultra-strong Coulomb field at their vicinity. Using the estimate[1] $r \sim \lambda_c$, nuclei with a Z bigger than $Z \sim 4\pi\epsilon_0\hbar c/e^2 = 1/\alpha \simeq 137$ would therefore be unstable and create pairs. This explanation is sometimes invoked to justify why the periodic table has less than 137 elements.

[1]Electron-positron pairs cannot be created in the Euler-Heisenberg theory, which is valid for wavelengths greater than $\sim \lambda_c$. Pair production would therefore begin for Z large enough so that $E_S \sim eZ/(4\pi\epsilon_0\lambda_c^2)$.

- Ultra-High Intensity (UHI) optical lasers are now considered an interesting platform to test strong-field theories. The next generation of high-power lasers should reach intensities of the order of 10^{24-25} W/cm^2 (Fig. 1.1). They play a central role in the processes/scenarios studied in this thesis.
- X-ray Free Electron Lasers (XFELs). Although their output peak power does not exceed the GW level (to be compared to the PW levels offered by UHI optical lasers), their short wavelength allows for much tighter diffraction limited focal spots, opening the possibility for very high peak fields. In particular, the European XFEL aims at reaching field strengths of the order of 10^{17} V/m [11].
- Linear or circular accelerators. Nowadays, the biggest circular accelerator is the LHC at CERN, which is able to accelerate protons to energies up to 8 TeV. The previous world record was the Tevatron at Fermilab which was able to produce proton-proton collisions with energies of up to 1.6 TeV. The RF linear accelerator (LinAc) of the Linac Coherent Light Source (LCLS) at SLAC accelerates electrons and positrons up to 50 GeV. It was used in the Burke experiment [12] where it was collided with a terawatt pulse from a Nd:glass laser at 527 nm. Other accelerators are today being constructed [13].
- Plasma accelerators are able to support fields of the order of 100 GV/m (to be compared to the RF linac limitation which is currently around 100 MV/m, partly due to the breakdown that occurs on the walls of the structure). Currently, laser wakefield accelerators (LWFA) are able to accelerate electrons up to the GeV level on the centimeter scale [14].

1.5 Laser-Plasma Interaction at Ultra-High Intensity

The introduction, in the mid-1980's by Strickland & Mourou [15], of the Chirped-Pulse Amplification (CPA) technique in the optic domain has since allowed for the generation of short, ultra-intense laser pulses. Today, ultra-short (from few to tens of femtosecondes) laser pulses with central wavelengths $\sim$0.8 μm (for Ti-Sapphire laser) or $\sim$10 μm routinely reach intensities beyond 10^{18} W/cm^2. Under such a large intensity, the electron in the laser electromagnetic field quickly (in less that an optical cycle) reaches relativistic velocities. This is the field of so-called ultra-high intensity (UHI) physics, a field that opens a wide range of applications, from particle acceleration to new sources of light and laboratory astrophysics studies. This paragraph briefly reviews some of these applications, focusing on those most interesting for my host-laboratory, and provides references for the interested reader.

- Laser wakefield acceleration (LWFA): LWFA of electrons was first envisioned by Tajima & Dawson in the late 1970's [16]. It consists in exciting an electron plasma wave in the wake of an UHI laser pulse propagating through a low density (undercritical) plasma. Electrons can be efficiently accelerated in this longitudinal wave, and gain relativistic velocities over few mm to few cm of propagation. In 2004, the so-called *Dream Beam* issue in Nature presented the work of three experimental teams demonstrating the feasibility of creating high-energy electron bunches with low energy spread and divergence. Mangles et al. [17] and Geddes et al. [18] demonstrated the production of 70–80 MeV electrons, with small ($\lesssim 2$ MeV) energy spread and small (few mrad) divergence using 40–55 fs pulses at intensity of a few 10^{18} W/cm^2, for a total laser energy of about 500 mJ. Faure et al. [19] demonstrated the production of an electron beam with an energy of 170 MeV at the price of a larger energy spread ($\sim$20 MeV) and divergence ($\sim$10 mrad), but that also carried an important charge, of the order of 0.5 nC. Since then, LWFA has been reproduced and is routinely used in various laboratories worldwide, in particular in the strongly nonlinear, so-called bubble regime [20]. Today, electron beams with energy of 4.2 GeV have been produced at the BELLA facility in Berkeley, US [21]. At Apollon, electron beams with energies of the order 10 GeV are envisioned [22], and will play a significant role in experiments testing the physics studies in this thesis. Recent papers also investigate the possibility of LWFA of positrons [23] with the (long-term) goal to build-up laser-plasma-based electron-positron colliders [24].

- Laser-plasma acceleration of ions has also been a strong focus of the UHI community. As ions are much more massive than electrons, accelerating them is more tricky. Today, most of the experiments aiming at ion acceleration rely on the so-called Target Normal Sheath Acceleration (TNSA) process [25]. This mechanism relies on the irradiation by an UHI laser pulse of a solid density target, and proceeds as follows: first, electrons at the laser-plasma interaction surface are strongly heated by the UHI laser. Some of these hot electrons travel through the target, and as they cross its rear-side build up a strong (of the order the TV/m) electrostatic field. It is this electrostatic field that accelerates the ions (mainly protons present in pollutants at the target rear-side). As put forward by Mora in his seminal paper [26] ion acceleration then proceeds as the expansion of a (hot) plasma in a vacuum. This mechanism is routinely reproduced in today's experiments, and laser-plasma accelerated protons are already used in the lab as a diagnostic: the so-called proton radiography indeed allows to probe with interesting time (ps) and spatial (few μm) resolutions, the electromagnetic fields developing in various experiments [27]. Yet, this mechanism comes with strong limitations. In particular, the resulting accelerated ions have a very broad (thermal) energy distribution. Also, today's experiments face difficulties in reaching very high proton energies, and 100 MeV/nucleon stands as one limit yet to overcome. This is particularly constraining for some envisioned applications of laser-accelerated ions that range from hadron-therapy [28, 29] to the abundant production of neutrons, as proposed e.g. by Julien Fuchs in his Genesis ERC project.[2] To break this limit, different acceleration schemes have been proposed. While it would be too long to explain all of them, one should note the possibility to accelerate ions using the strong radiation pressure associated with UHI lasers [30]. Recent experiments have claimed evidence of this process using circularly-polarized light pulses [31], and future

[2] https://www.polytechnique.edu/en/content/julien-fuchs-recipient-2018-erc-advanced-grants-his-genesis-project.

experiments on the Apollon laser will be devoted to this regime of ion acceleration. More details on laser-plasma acceleration of ions the reader are available in the review article [32].

- Laser-plasma interaction in the UHI regime also offers new opportunities for the generation of novel light sources, ranging from the THz [33] to XUV [34] and gamma-rays [35]. On the experimental side, XUV and X-ray sources have attracted a special attention. In the review [36], the authors discuss the generation of femtosecond X-ray pulses from betatron, synchrotron or nonlinear Thomson scattering radiation from LWFA electrons. These sources are particularly interesting for a wide range of applications, from ultra-fast highly resolved radiography to biological or medical applications [37]. XUV radiations, and in particular attosecond pulses of light, can also be generated by irradiating a solid-density plasma target. There exist various mechanisms responsible for the generation of ultra-short XUV light pulses [38]. Among the most popular is the relativistic oscillating mirror where, due to the strongly nonlinear dynamics of the target electrons interacting with the UHI laser, the irradiated target surface acts as a relativistic oscillating mirror and leads to the generation of high-harmonics of the reflected laser. This mechanism has been well established in various experiments worldwide, and has recently been brought forward as a potential path toward the generation of extremely short, extremely intense light pulses, potentially up to the Schwinger limit [39, 40].
- The possibility to bring matter under extreme conditions of temperature and pressure or to drive fast plasma flows with UHI laser can also be used to recreate, in the laboratory, conditions relevant to space plasmas or astrophysics studies. While these studies are usually conducted using more energetic but less intense lasers [41], recent theoretical works [42–45] have brought forward the possibility of recreating collisionless shocks of paramount interest for the astrophysics community as held responsible for accelerating the most energetic particles—so-called cosmic rays—in the Universe. In [42, 43], it was for instance shown that fast quasi-neutral flows can be launched into an overdense target by the strong laser radiation pressure, as a result of the fast flows interaction with the background plasma of the target, strongly nonlinear effects develop that lead to the formation of a collisionless shock. In these studies, however, most of the nonlinear effects at play were driven by the hot electron population, easily obtained with UHI lasers, but absent in most of the astrophysical scenarios of interest. Grassi et al. [45] showed that by carefully tuning the laser-target interaction conditions (in particular the incidence angle and laser polarization), it was possible to strongly mitigate electron heating, so that shock formation follows from the nonlinear development of the ion Weibel instability, as expected in various astrophysical environments such as supernovae remnants or gamma ray bursts. The required laser energy however stays beyond the reach of current lasers. Similarly, Lobet et al. [44] have studied the possibility of driving such shocks in the interpenetration of two dense electron-positron plasmas produced by irradiating two dense targets with two extreme light lasers. These two lasers, with duration of typically 60 fs and intensity of the order of 10^{24} W/cm^2 drive the Breit-Wheeler pair production process earlier introduced. Recent works,

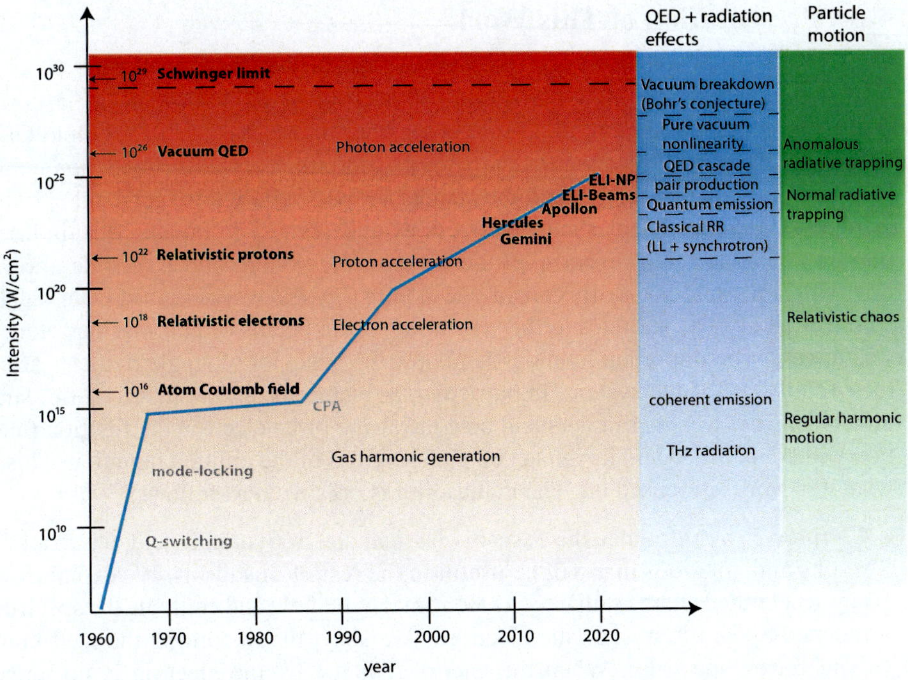

Fig. 1.1 Maximum reachable intensity as a function of the year. The accessible physics at each given intensity is also described in the red part, in addition to the names of the facilities at which these intensities are accessible. In the blue part, the different physical phenomena related to radiation and QED effects are listed. Finally, the green part describes the dynamics of particles in each given field.

that build on the possibility to create dense electron-positron flows in dense high-Z targets [46, 47], have also started considering the interaction of electron-positron pair jets colliding with a background electron-proton plasma [48]. These on-going efforts shall help shade a new light on various processes taking place in the most violent and energetic environments in the Universe such as Pulsar Wind Nebulae, Gamma Ray Bursts or at the vicinity of Active Galactic Nuclei.

- The advent of multi-petawatt laser systems such as CoReLS (4PW demonstrated, in South-Korea) [49], Apollon in France [50] or ELI [51] in Czech Republic, Hungary and Romania, offer the possibility to probe laser-matter under extreme light conditions (Fig. 1.1). This PhD work is largely devoted to this new physics, that bridges relativistic plasma physics and QED. These extreme light facilities also provide a wide range of new opportunities in fields as exotic as relativistic atomic dynamics and nuclear physics in strong electromagnetic fields, vacuum polarization effects, pair production beyond electron-positron pairs (e.g. muon-antimuon or pion-antipion pairs), particle physics within and beyond the standard model, etc. All these exciting studies are presented in the review article [52].

1.6 Presentation of This Work

With the advent of the new generation of petawatt and multi-petawatt lasers, it will become more and more important in the near future to study strong-field QED (SFQED). As previously mentioned, SFQED can either serve as a new nonperturbative regime for physics beyond the standard model, allow to understand extreme astrophysics events (such as magnetars), or even become a dominant effect in laser-plasma interaction. However, in all these cases, the system under consideration is not a single particle as usually considered in pure SFQED works [53–55], but will be a collection of such particles (either electron/photon beams or pair plasmas etc.). It will therefore be important to understand how the behavior of single particles affect the overall state of the system. In particular, as electrons are the first to suffer such effects, one has to better understand how the shape of its electron distribution function will be modified and its impact on the spectrum of the emitted radiation. This is what this work is focused on. The manuscript is organized as follows:

- **Chapter 2**: we introduce the basis of classical electrodynamics and derive all the results and notations that will be useful in the rest of this thesis. We explain why any accelerated charge radiates an electromagnetic field and compute the spectrum radiated by an ultrarelativistic electron.[3] We show that it follows the well-know synchrotron spectrum. When the energy radiated by the electron is no longer negligible, the emitted radiation will counteract on the trajectory of the particle itself. This is the so-called *radiation reaction*. We derive the Lorentz-Abraham-Dirac (LAD) equation, that describes the motion of an electron, taking into account radiation reaction (RR). We show that this equation presents unphysical solutions and deduce the Landau-Lifshitz (LL) equation, that we will use to describe RR in the classical regime in the rest of this work. We then consider the solution of the LL equation in simplified fields such as plane-waves or constant uniform magnetic fields. The classical radiation dominated regime (CRDR) is described and the limit of validity of the classical description analyzed.
- **Chapter 3**: when the electron quantum parameter is of the order of unity, emitted photons can have an energy close to that of the emitting electron. In this case, radiation reaction can no longer be treated classically. In this chapter, we present the basis of quantum electrodynamics (QED), which is the framework in which such quantum effects can be computed. We derive the Volkov states that take into account exactly the nonperturbative coupling between the electron and the strong background field. These fields are used in the so-called Furry picture in order to compute the different QED processes such as the nonlinear Compton scattering or the nonlinear Breit-Wheeler process. The cross-sections for these two processes are then analyzed.
- **Chapter 4**: so far, the description of radiation concerned only single particles. Here we introduce the Vlasov equation that describes the evolution of the func-

[3] All the results remain valid for any charged particle but from now on, we will use the generic term of "electron" for simplicity.

tion distribution of a set a particles. We then see how to modify this equation in order to take into account classical and quantum RR. The numerical resolution of this equation, together with Maxwell's equations is then described, in particular through the well-known PIC loop. We describe how to modify this classical PIC loop in order to include classical and quantum RR, in particular through the use of a Monte-Carlo module.

- **Chapter 5**: after a brief reminder of the Landau-Lifshitz (LL) equation, which describes radiation reaction (RR) in classical electrodynamics (CED) as a deterministic force in the particle momentum equation, we recall the emission properties of a quantum radiating electron. We then turn our attention to the linear Boltzmann equation which is at the center of the kinetic description of RR explored in this thesis. Our analysis relies on the local constant field approximation (LCFA) and concerns ultrarelativistic radiating charges. An expansion for small ratios of the energy of the emitted photon over the electron's energy to second order allows to simplify the linear Boltzmann (lB) equation to a Fokker-Planck (FP) equation. By considering the expansion at first order, we find the classical deterministic description corrected by the quantum Gaunt factor. We then consider the equation of evolution of the successive moments of the electron distribution as described by the full linear Boltzmann (lB) equation, the second order expansion [referred to as the Fokker-Planck (FP) model] and the first order expansion [referred to as the corrected Landau-Lifshitz (cLL) model]. The equation of evolution of the mean electron energy is found to be formally the same in the three descriptions. The effective discrepancy in the average energy in the three models is computed and shown to be smaller than a few % for $\chi \lesssim 1$. The equation of evolution of the electron energy spread is shown to be formally the same in the lB and FP descriptions but different from the cLL description. In the cLL case, the electron distribution function can only cool down because of RR while in the FP and lB case, we compute analytically the threshold between the cooling and heating domains. Finally, the equation of evolution of the third order moment in energy of the electron distribution function is different in the cLL, FP and lB models. This third order moment μ_3 can either increase or decrease in the lB model and the threshold between these two domains is computed analytically. The decrease of this moment yielding to a negative μ_3 is identified as a signature of quantum RR. Another interesting consequence of this study is that, for physical problems in which only the average electron energy is important, the cLL description of RR is sufficient. If one needs only the energy average and spread, the FP description is enough. If finally the exact shape of the electron distribution function is important, one needs to use the full lB description.
- **Chapter 6**: it is usually admitted that the classical LL equation is valid for small values of the electron quantum parameter χ. Based on the kinetic description of RR, we derive more rigorous limits of validity of the different models depending not only on the average χ parameter, but also on the electron distribution itself. We find that quantum effects can be important in some cases for relatively small values of χ and vice versa the cLL description can be sufficient in some cases for quantum parameters close to 1. We support these claims with numerical simulations of the

three models with initially narrow electron beams as well as broad Maxwell-Jüttner distributions, interacting with linearly or circularly polarized plane-waves and constant uniform magnetic fields. These simulations also allow us to verify the results derived in the previous chapter, such as the discrepancy in the average energy in the three models, the maximum electron energy spread as well as the evolution of these quantities and of the third order moment.

- **Chapter 7**: we then turn our attention to the photon spectrum and link its properties to the electron distribution function described in the previous chapter. The instantaneous and time integrated spectra are investigated using numerical simulations of the three models. We find that the predictions of the FP and lB model are in good agreement even for χ close to one and different than the cLL model. In particular, the lB and FP models exhibit a hardening of the photon spectrum. This phenomenon is observed for initially narrow electron beams as well as broad Maxwell-Jüttner distributions, interacting with linearly or circularly polarized plane-waves and constant uniform magnetic fields. The discrepancies between the three models are much smaller than on the electron distribution functions which makes us conclude that the electron distribution function is a better candidate to observe the transition between the classical and quantum regime.
- **Chapter 8**: in the previous chapters, the external fields that were considered (linearly or circularly polarized plane-waves and constant uniform magnetic fields) had in common that *they did not increase the mean energy of the electron distribution*. Here we consider the superposition of two circularly polarized plane waves. The resulting field is a stationary wave, which, in the magnetic nodes, reduces to a pure rotating electric field. We study the evolution of an electron population initially at rest or of an initially hot Maxwell-Jüttner distribution function, initially in these magnetic nodes. In this case, the motion is 2D and the electrons remain infinitely in the magnetic node plane. We find that after a few rotations of the electric field, the electron distribution function reaches a stationary state. All of its moments are therefore constant. We then apply the previous kinetic equations, taking into account the new source term, that comes from the fact that the external field now brings energy to the particles. The average asymptotic energy is computed numerically in the cLL and lB models and the discrepancy between these two values is deduced analytically from the kinetic equations. We show that the source term in the equation of evolution of the energy spread is negligible for $\chi \lesssim 1$ and the asymptotic energy spread in this configuration is therefore the same as the threshold of the energy spread computed in Chap. 6. For $\chi \lesssim 1$, the asymptotic electron distribution function is shown to be a gaussian in energy. For $\chi > 1$, nonlinear Breit-Wheeler pair production (NBWPP) is no longer negligible. We generalize the previous approach to this case and derive the kinetic equations for the electron, positron and photon distributions and the corresponding equations of evolutions of the successive energy moments in this case.
- **Chapter 9**: in this third part, we turn our study to the so-called Laguerre-Gaussian (LG) beams. These beams have recently received a great attention from the physics community. They indeed carry orbital angular momentum (OAM) which is believed to play a role in some astrophysical phenomena, such as the radiation

from the accretion disk around Kerr black holes or the radiation by rotating pulsars or quasars. Here we derive them from the paraxial equation and review their main properties such as their helical phase and orbital angular momentum.

- **Chapter 10**: we then examine the possibility of producing pairs by colliding such a LG beam with a high-energy photon beam (so called *shower*). It is found that the number of created pairs increases with the orbital angular momentum of the LG beam (at fixed energy). This effect is found to be geometrical and linked to the effective transverse size of the beam. In particular, it increases when increasing the orbital angular momentum of the external field. We confirm this effect by comparing the pair yields of gaussian beams with different transverse sizes. We therefore obtain the counter-intuitive conclusion that the number of created pairs increases with decreasing intensity at fixed energy (as long as the photon quantum parameter remains larger than a threshold value).
- **Chapter 11**: we finally conclude on our findings and present perspectives for future works.

References

1. Chatrchyan S, Khachatryan V, Sirunyan AM, Tumasyan A, Adam W, Aguilo E, Bergauer T, Dragicevic M, Erö J, Fabjan C, Friedl M (2012) Observation of a new boson at a mass of 125 GeV with the CMS experiment at the LHC. Phys Lett B 716(1):30–61
2. Englert F, Brout R (1964) Broken symmetry and the mass of gauge vector mesons. Phys Rev Lett 13:321–323
3. Higgs PW (1964) Broken symmetries and the masses of gauge bosons. Phys Rev Lett 13:508–509
4. Aoyama T, Hayakawa M, Kinoshita T, Nio M (2012) Tenth-order QED contribution to the electron $g-2$ and an improved value of the fine structure constant. Phys Rev Lett 109:111807
5. Sauter F (1931) Uber das Verhalten eines Elektrons im homogenen elektrischen Feld nach der relativistischen Theorie Diracs. Z Phys 69:742
6. Klein O (1929) Die Reflexion von Elektronen an einem Potentialsprung nach der relativistischen Dynamik von Dirac. Zeitschrift fur Physik 53:157–165
7. Schwinger J (1951) On Gauge Invariance and Vacuuwn Polarization. Phys Rev 82(5):664
8. Schützhold R, Gies H, Dunne G (2008) Dynamically assisted Schwinger mechanism. Phys Rev Lett 101:130404
9. Dunne GV, Gies H, Schützhold R (2009) Catalysis of Schwinger vacuum pair production. Phys Rev D 80:111301
10. Schlenvoigt H-P, Heinzl T, Schramm U, Cowan TE, Sauerbrey R (2016) Detecting vacuum birefringence with X-ray free electron lasers and high-power optical lasers: a feasibility study. Physica Scripta 91(2):023010
11. Ringwald A (2001) Pair production from vacuum at the focus of an X-ray free electron laser. Phys Lett B 510(1):107–116
12. Burke DL, Field RC, Horton-Smith G, Spencer JE, Walz D, Berridge SC, Bugg WM, Shmakov K, Weidemann AW, Bula C, McDonald KT, Prebys EJ, Bamber C, Boege SJ, Koffas T, Kotseroglou T, Melissinos AC, Meyerhofer DD, Reis DA, Ragg W (1997) Positron production in multiphoton light-by-light scattering. Phys Rev Lett 79:1626–1629
13. Facility for Advanced Accelerator Experimental Tests (ELI). https://facet.slac.stanford.edu

14. Leemans WP, Nagler B, Gonsalves AJ, Tóth C, Nakamura K, Geddes CG, Esarey ES, Schroeder CB, Hooker SM (2006) GeV electron beams from a centimetre-scale accelerator. Nat Phys 2:696
15. Strickland D, Mourou G (1985) Compression of amplified chirped optical pulses. Opt Commun 56(3):219–221
16. Tajima T, Dawson JM (1979) Laser electron accelerator. Phys Rev Lett 43:267–270
17. Mangles SPD, Murphy CD, Najmudin Z, Thomas AGR, Collier JL, Dangor AE, Divall EJ, Foster PS, Gallacher JG, Hooker CJ, Jaroszynski DA, Langley AJ, Mori WB, Norreys PA, Tsung FS, Viskup R, Walton BR, Krushelnick K (2004) Monoenergetic beams of relativistic electrons from intense laser–plasma interactions. Nature, 431:535–538
18. Geddes CGR, Toth C, van Tilborg J, Esarey E, Schroeder CB, Bruhwiler D, Nieter C, Cary J, Leemans WP (2004) High-quality electron beams from a laser wakefield accelerator using plasma-channel guiding. Nature 431:538
19. Faure J, Glinec Y, Pukhov A, Kiselev S, Gordienko S, Lefebvre E, Rousseau JP, Burgy F, Malka V (2004) A laser–plasma accelerator producing monoenergetic electron beams. Nature 431:541
20. Pukhov A, Meyer-ter Vehn J (2002) Laser wake field acceleration: the highly non-linear broken-wave regime. Appl Phys B 74(4):355–361
21. Leemans WP, Gonsalves AJ, Mao H-S, Nakamura K, Benedetti C, Schroeder CB, Tóth C, Daniels J, Mittelberger DE, Bulanov SS, Vay JL, Geddes CGR, Esarey E (2014) Multi-GeV electron beams from capillary-discharge-guided subpetawatt laser pulses in the self-trapping regime. Phys Rev Lett 113:245002
22. Cros B, Paradkar B, Davoine X, Chancé A, Desforges F, Dobosz-Dufrénoy S, Delerue N, Ju J, Audet T, Maynard G, Lobet M, Gremillet L, Mora P, Schwindling J, Delferrière O, Bruni C, Rimbault C, Vinatier T, Piazza AD, Grech M, Riconda C, Marquès J, Beck A, Specka A, Martin P, Monot P, Normand D, Mathieu F, Audebert P, Amiranoff F (2014) Laser plasma acceleration of electrons with multi-pw laser beams in the frame of CILEX. Nucl Instrum Methods Phys Res Sect A 740:27–33. Proceedings of the first European Advanced Accelerator Concepts Workshop 2013
23. Vieira J, Mendonça JT (2014) Nonlinear laser driven donut wakefields for positron and electron acceleration. Phys Rev Lett 112:215001
24. Leemans W, Esarey E (2009) Laser-driven plasma-wave electron accelerators. Phys Today 62:44
25. Wilks SC, Langdon AB, Cowan TE, Roth M, Singh M, Hatchett S, Key MH, Pennington D, MacKinnon A, Snavely RA (2001) Energetic proton generation in ultra-intense laser? Solid interactions. Phys Plasmas 8(2):542–549
26. Mora P (2003) Plasma expansion into a vacuum. Phys. Rev. Lett. 90:185002
27. Borghesi M, Schiavi A, Campbell DH, Haines MG, Willi O, MacKinnon AJ, Gizzi LA, Galimberti M, Clarke RJ, Ruhl H (2001) Proton imaging: a diagnostic for inertial confinement fusion/fast ignitor studies. Plasma Phys Controlled Fusion 43(12A):A267–A276
28. Tajima T, Habs D, Yan X (2009) Laser acceleration of ions for radiation therapy. Rev Accelerator Sci Technol 02(01):201–228
29. Ledingham, LWD, Bolton PR, Shikazono N, Ma CM (2014) Towards laser driven hadron cancer radiotherapy: a review of progress. arXiv:1405.2657
30. Esirkepov T, Borghesi M, Bulanov SV, Mourou G, Tajima T (2004) Highly efficient relativistic-ion generation in the laser-piston regime. Phys Rev Lett 92:175003175003
31. Henig A, Steinke S, Schnürer M, Sokollik T, Hörlein R, KieferD, Jung D, Schreiber J, Hegelich BM, Yan XQ, Meyer-ter Vehn J, Tajima T, Nickles PV, Sandner W, Habs D (2009) Radiation-pressure acceleration of ion beams driven by circularly polarized laser pulses. Phys Rev Lett 103:245003
32. Macchi A, Borghesi M, Passoni M (2013) Ion acceleration by superintense laser-plasma interaction. Rev Mod Phys 85:751–793
33. Déchard J, Debayle A, Davoine X, Gremillet L, Bergé L (2018) Terahertz pulse generation in underdense relativistic plasmas: from photoionization-induced radiation to coherent transition radiation. Phys Rev Lett 120:144801

34. Thaury C, George H, Quéré F, Loch R, Geindre JP, Monot P, Martin P (2008) Coherent dynamics of plasma mirrors. Nat Phys 4:631
35. Kalmykov SY, Davoine X, Ghebregziabher I, Shadwick BA (2018) Optically controlled laser-plasma electron accelerator for compact gamma-ray sources. New J Phys 20:023047
36. Corde S, Ta Phuoc K, Lambert G, Fitour R, Malka V, Rousse A, Beck A, Lefebvre E (2013) Femtosecond X rays from laser-plasma accelerators. Rev Mod Phys 85:1–48
37. Albert F, Thomas AGR (2016) Applications of laser wakefield accelerator-based light sources. Plasma Phys Controlled Fusion 58(10):103001
38. Thaury C, Quéré F (2010) High-order harmonic and attosecond pulse generation on plasma mirrors: basic mechanisms. J Phys B At Mol Opt Phys 43(21):213001
39. Baumann C, Pukhov A, Nerush EN, Kostyukov IY (2018) Probing fully non-perturbative QED with electron-laser collisions. arXiv:1811.03990
40. Vincenti H (2018) Achieving extreme light intensities using relativistic plasma mirrors. arXiv:1812.05357
41. Remington BA, Drake RP, Ryutov DD (2006) Experimental astrophysics with high power lasers and z pinches. Rev Mod Phys 78:755–807
42. Fiuza F, Fonseca RA, Tonge J, Mori WB, Silva LO (2012) Weibel-instability-mediated collisionless shocks in the laboratory with ultraintense lasers. Phys Rev Lett 108:235004
43. Ruyet C (2014) Kinetic instabilities in plasmas: from electromagnetic fluctuations to collisionless shocks. Ph.D. thesis, Universite de Paris XI
44. Lobet M, Ruyer C, Debayle A, d'Humières E, Grech M, Lemoine M, Gremillet L (2015) Ultrafast synchrotron-enhanced thermalization of laser-driven colliding pair plasmas. Phys Rev Lett 115:215003215003
45. Grassi A (2017) Collisionless shocks in the context of Laboratory Astrophysics. Ph.D. thesis, Paris 6
46. Chen H, Wilks SC, Bonlie JD, Liang EP, Myatt J, Price DF, Meyerhofer DD, Beiersdorfer P (2009) Relativistic positron creation using ultraintense short pulse lasers. Phys Rev Lett 102:105001
47. Sarri G, Poder K, Cole JM, Schumaker W, Di Piazza A, Reville B, Dzelzainis T, Doria D, Gizzi LA, Grittani G, Kar S, Keitel CH, Krushelnick K, Kuschel S, Mangles SPD, Najmudin Z, Shukla N, Silva LO, Symes D, Thomas AGR, Vargas M, Vieira J, Zepf M (2015) Generation of neutral and high-density electron-positron pair plasmas in the laboratory. Nat Commun 6:6747
48. Dieckmann ME, Alejo A, Sarri G, Folini D, Walder R (2018) One-dimensional thermal pressure-driven expansion of a pair cloud into an electron-proton plasma. Phys Plasmas 25(6):064502
49. Nam CH, Sung JH, Lee HW, Youn JW, Lee SK (2018) Performance of the 20 fs, 4 PW Ti: sapphire laser at CoReLS. In: Conference on lasers and electro-optics, p. STu4O.3. Optical Society of America
50. Cros B, Paradkar B, Davoine X, Chancé A, Desforges F, Dobosz-Dufrénoy S, Delerue N, Ju J, Audet T, Maynard G, Lobet M, Gremillet L, Mora P, Schwindling J, Delferrière O, Bruni C, Rimbault C, Vinatier T, Piazza AD, Grech M, Riconda C, Marquès J, Beck A, Specka A, Martin P, Monot P, Normand D, Mathieu F, Audebert P, Amiranoff F (2014b) Laser plasma acceleration of electrons with multi-PW laser beams in the frame of CILEX. Nucl Instrum Methods Phys Res Sect A 740:27–33 Proceedings of the first European Advanced Accelerator Concepts Workshop 2013
51. Extreme light infrastructure (ELI). https://eli-laser.eu/
52. Di Piazza A, Müller C, Hatsagortsyan KZ, Keitel CH (2012) Extremely high-intensity laser interactions with fundamental quantum systems. Rev Mod Phys 84:1177–1228
53. Ritus VI (1979) Quantum effects of the interaction of elementary particles with an intense electromagnetic field. Ph.D. thesis
54. Harvey C, Heinzl T, Ilderton A (2009) Signatures of high-intensity Compton scattering. Phys Rev A 79:063407
55. Piazza Di A, Müller C, Hatsagortsyan K, Keitel CH (2012) Extremely high-intensity laser interactions with fundamental quantum systems. Rev Mod Phys 84:1177

Part I
Theoretical Background

Chapter 2
Classical Electrodynamics

The views of space and time which I wish to lay before you have sprung from the soil of experimental physics, and therein lies their strength. They are radical. Henceforth, space by itself, and time by itself, are doomed to fade away into mere shadows, and only a kind of union of the two will preserve an independent reality.

Hermann Minkowski (Sep 21, 1908), to the 80th Assembly of German Natural Scientists and Physicians.

2.1 Outline

This first chapter focuses on classical electrodynamics (CED), that is, the classical and relativistic theory of charged particles[1] and electromagnetic field interaction. It aims at introducing the principal notions and physical quantities that will be used throughout this thesis, and it is presented in a logico-deductive order so that the reader will be able to draw some lines between this classical description and the quantum description of electrodynamics presented in the next chapter. In order to maintain the presentation both logical and compact, not all derivations presented here are straightforward: some results may require quite some algebra before being derived. I have however tried to state clearly all assumptions used in the derivations, and to provide physically intuitive pictures when possible in order to get a qualitative idea before going to the computational part. The reader interested in the details of the derivations is referred to the seminal books on CED by Jackson [1] and Landau & Lifshitz [2]. Moreover, during this work, I have benefited a lot from the book *Classical charged particles* by Rohrlich [3] which I have found illuminating on many aspects of CED.

[1]Throughout this chapter, we will in particular focus on the electron.

F. Niel, *Classical and Quantum Description of Plasma and Radiation in Strong Fields*, Springer Theses, https://doi.org/10.1007/978-3-030-73547-0_2

This chapter is structured as follows. Section 2.2 introduces some general results on Classical Field Theory (CFT). The notions of action, Lagrangian and classical equations of motion are introduced. One then presents Noether's theorem which by allowing to derive conservation laws from symmetries plays a central role in theoretical physics. Introducing Poincaré's group of symmetry, Noether's theorem is used to derive the energy-momentum as conserved quantity. In Sect. 2.3, one then applies these results to CED explicitly defining its Lagrangian. Equations of motion and conserved quantities are then derived considering either a free electron, a free electromagnetic field or the coupled system. We then use these results to derive in Sect. 2.4 important results on the radiation by moving charges, which is central to this work. The basic properties (power, angular and frequency spectra) of the emitted radiation are derived. Section 2.5 then introduces the problem of (classical) radiation reaction which is an other central notion to this PhD work. This long standing problem of CED arises when one wants to describe the dynamics of a radiating charge in a prescribed (background) electromagnetic field. Here radiation reaction is presented considering a single electron, and will lead us to introduce some limit of validity for CED itself.

2.2 Classical Field Theory

2.2.1 Euler-Lagrange Equations

Let us consider a physical system described by a set of fields $[\phi_i(x)]_{i=1..n}$ that depend on the space-time coordinate[2] x^μ. We assume that for this system, there exists a scalar function $\mathcal{L}(\phi_i, \partial_\mu\phi_i)$, called Lagrangian density, from which we build the action [2]

$$S = \int d^4x\, \mathcal{L}(\phi_i, \partial_\mu\phi_i). \tag{2.1}$$

The principle of least action then states that the classical dynamics is found when the action is stationary: $\delta S = 0$. This condition can be shown to be equivalent to the following set of partial differential equations [4]

$$\boxed{\partial_\mu \frac{\partial\mathcal{L}}{\partial(\partial_\mu\phi_i)} - \frac{\partial\mathcal{L}}{\partial\phi_i} = 0} \tag{2.2}$$

which are called the Euler-Lagrange equations (ELE) [5] and will also be referred to as the *classical equations of motion*.

[2] Throughout this work, we assume a flat Minkowski metric with signature $(+, -, -, -)$. Furthermore, greek indices take values $\mu = 0, ..., 3$ while spatial indices are denoted by Latin letters $i = 1, 2, 3$.

2.2.2 Noether's Theorem

In field theories, the link between symmetries and conserved quantities is given by Noether's theorem [6]. Basically, this theorem states that to any continuous symmetry of the action of a system, there exists a conserved quantity. More precisely, let us consider a set of infinitesimal space-time (or external) and intrinsic (or internal) transformations:

$$T = \begin{cases} x^\mu & \rightarrow \quad x'^\mu = x^\mu + \epsilon^a A_a^\mu(x) \\ \phi_i(x) & \rightarrow \quad \phi'_i(x') = \phi_i(x) + \epsilon'^a F_{i,a}(\phi_j, \partial\phi_j) \end{cases} \tag{2.3}$$

where $(\epsilon^a)_{a=1..4N}$ and $(\epsilon'^a)_{a=1..4N}$ are two sets of $4N$ infinitesimal parameters and $[A_a^\mu(x)]_{a=1..4N}$ (resp. $[F_{i,a}(\phi, \partial\phi)]_{a=1..4N}$) are given functions of x (resp. ϕ_j and $\partial\phi_j$). Noether's theorem states that, if T is a symmetry of the system (i.e. if T leaves the action S invariant), then there are $4N$ currents $j_a^\mu(\phi)$ that are conserved on a classical solution ϕ_{cl} of the equations of motion [2, 7]

$$\partial_\mu j_a^\mu(\phi_{\mathrm{cl}}) = 0, \tag{2.4}$$

with

$$\boxed{j_a^\mu = \frac{\partial\mathcal{L}}{\partial(\partial_\mu\phi_i)}\left[A_a^\nu\partial_\nu\phi_i - F_{i,a}(\phi_j, \partial\phi_j)\right] - A_a^\mu\mathcal{L}} \tag{2.5}$$

In this work, a will be[3] a set of N Minkowski indices $(\mu_i)_{i=1...N}$ and $j_a^\mu = j^\mu_{\mu_1...\mu_N}$ a rank-(N+1) tensor (see for example Sect. 2.2.4). Integrating (2.4) over all space yields

$$\partial_0 Q_a = \int d^3x\, \partial_0 j_a^0 = -\int d^3x\, \partial_i j_a^i = 0, \tag{2.6}$$

since this is the integral of a total divergence (and assuming that the fields decrease sufficiently fast at infinity). We therefore deduce that the charge Q_a is conserved. Finally note that, since Noether's current is derived as the quantity satisfying Eq. (2.4), it is defined up to a divergenceless tensor [8]. This fact will be used in Sect. 2.2.4 to define "refined" Noether's currents [9].

2.2.3 Poincaré Invariance

In practice, Lagrangians will have symmetries and more precisely, they will be invariant by Poincaré transformations. Poincaré invariance is a fundamental symmetry in physics and any relativistic field theory must have a Poincaré-invariant action

[3] At the moment, a is just an indice.

[10–12]. Poincaré transformations form a Lie group called the *Poincaré group* (or *inhomogeneous Lorentz group*) $ISO(3,1)$. This group is composed of two subgroups: the group of space-time translations $\mathbb{R}^{1,3}$ and the Lorentz group[4] $O(3,1)$ [13].

Space-time translations form a four dimensional group (one dimension for time translations and the three others for ordinary space translations). Using the same notations as in Eq. (2.3), the index "a" will be a Minkowski index "μ" and infinitesimal space-time translations will be described by Eq. (2.3) with[5] $A^\mu_\nu = \delta^\mu_\nu$ and $F_{i,a} = 0$, that is to say by

$$T = \begin{cases} x^\mu & \rightarrow \quad x'^\mu = x^\mu + a^\mu, \\ \phi_i(x) & \rightarrow \quad \phi'_i(x') = \phi_i(x). \end{cases} \tag{2.7}$$

The Lorentz group is a six dimensional group composed of spatial rotations and Lorentz boosts (three dimensions for rotations and three dimensions for Lorentz boosts). More formally, it is the group of linear coordinate transformations

$$x^\mu \rightarrow \quad x'^\mu = \Lambda^\mu{}_\nu \, x^\nu, \tag{2.8}$$

which leave the quantity

$$\eta^{\mu\nu} \, x_\mu \, x_\nu = x_0^2 - x_1^2 - x_2^2 - x_3^2, \tag{2.9}$$

invariant. For this condition to hold, we must have

$$\eta_{\mu\nu} = \eta_{\rho\sigma} \, \Lambda^\rho{}_\mu \, \Lambda^\sigma{}_\nu. \tag{2.10}$$

Using the same notation as in Eq. (2.3), the index "a" will be written as two Minkowski indices "$\nu\rho$" and infinitesimal Lorentz transformations will be described by $A^\mu_{(\nu\rho)} = -\frac{i}{2}(J_{\nu\rho})^\mu{}_\sigma \, x^\sigma$ and $F_{\eta,(\nu\rho)} = -\frac{i}{2}(J_{\nu\rho})^\alpha{}_\eta \, \phi_\alpha(x)$, that is to say

$$T = \begin{cases} x^\mu & \rightarrow \quad x'^\mu = x^\mu - \frac{i}{2} \, \omega^{\nu\rho} (J_{\nu\rho})^\mu{}_\sigma \, x^\sigma, \\ \phi^\eta(x) & \rightarrow \quad \phi'^\eta(x') = \phi^\eta(x) - \frac{i}{2} \, \omega^{\nu\rho} (J_{\nu\rho})^\eta{}_\alpha \, \phi^\alpha(x). \end{cases} \tag{2.11}$$

where $(J_{\nu\rho})^\eta{}_\alpha$ is the *generator* of Lorentz transformations on 4-vectors.

2.2.4 Energy-Momentum Tensor

As stated before, all relativistic field theories are Poincaré invariant, so in particular, they will be invariant by space-time translations. Using Eq. (2.7), we deduce the

---

[4]More precisely, we consider the proper orthochronous Lorentz group $SO^+(3,1)$.

[5]Here the index a of Eq. (2.3) becomes a Minkowski index ν. Moreover, $F_{i,a} = 0$ since all fields are scalar under translation.

associated Noether currents $j^{\mu}_{(\nu)} \equiv \Theta^{\mu}_{\nu}$ (i.e. "a" is here a Minkowski index that we call "ν"). They form a rank-2 Lorentz tensor called the *energy-momentum tensor*

$$\boxed{\Theta^{\mu\nu} = \frac{\partial \mathcal{L}}{\partial(\partial_\mu \phi_i)} \partial^\nu \phi_i - \eta^{\mu\nu} \mathcal{L}} \tag{2.12}$$

which is conserved according to Noether's theorem [6]

$$\partial_\mu \Theta^{\mu\nu} = 0. \tag{2.13}$$

Let us define the Noether's charge associated to Noether's current (2.12)

$$P^\mu = \int d^3x \, \Theta^{0\mu}, \tag{2.14}$$

which is a 4-vector called 4-momentum (and where the integral is over all 3-space). The $0th$ component P^0 will therefore be interpreted as the energy W and the ith component P^i as the ith component of the momentum. The remaining components Θ^{ij} are interpreted by integrating the i component of Eq. (2.13) over a finite volume V around the system. This gives rise to the following continuity equation

$$d_t P^i_V = -\int_V d^3x \, \partial_j \Theta^{ji} = \int_{\partial V} dS \, \hat{n}_j \, \Theta^{ji}, \tag{2.15}$$

where $P^i_V = \int_V d^3x \, \Theta^{0\mu}$ is the momentum contained in the volume V and $\hat{\mathbf{n}}$ is a unitary vector normal to the surface ∂V around V and directed outward. This equation allows us to interpret Θ^{ij} as the flux of the component i of the momentum P^i through a closed surface ∂V around the volume V. Doing the same with the 0th component of Eq. (2.14) yields

$$d_t W_V = -\int_V d^3x \, \partial_i P^i = \int_{\partial V} dS \, \hat{n}_i \, P^i, \tag{2.16}$$

where $W_V = \int_V d^3x \, \Theta^{00}$ is interpreted as the energy contained in the volume V. This allows us to interpret P as the flux of energy through a closed surface ∂V around the volume V. We finally define the energy density w and momentum density p by

$$w = \Theta^{00}, \tag{2.17a}$$

$$p^i = \Theta^{0i}. \tag{2.17b}$$

Note that it is possible to obtain the angular momentum tensor following the same procedure using the rotational invariance of the action. However, we will only use the energy-momentum tensor in this work

2.3 Classical Electrodynamics

Classical electrodynamics (CED) is described by the following action [2]

$$S_{\text{CED}} = \underbrace{-m \int d\tau}_{S_{\text{matter}}} \underbrace{- \int d^4x \, \frac{1}{4} F^{\mu\nu} F_{\mu\nu}}_{S_{\text{field}}} \underbrace{- \int d^4x \, A^\mu j_\mu}_{S_{\text{int}}} .$$

All the terms in this action will now be discussed one by one using the theoretical tools described in the previous section.

2.3.1 Lagrangian

Matter in the absence of external field is described classically by the following action

$$S_{\text{matter}} = -m \int d\tau, \tag{2.18}$$

where m is the so-called *bare mass*[6] of the particle. The corresponding Lagrangian density can be written as

$$\mathcal{L}_{\text{matter}} = -\frac{1}{2} m \, \eta_{\mu\nu} u^\mu u^\nu, \tag{2.19}$$

where $u^\mu = \dot{x}^\mu = dx^\mu / d\tau$ is the particle four-velocity, $x^\mu = (t, \mathbf{x})$ its four-position and τ its proper time.

The free electromagnetic field action already takes the form of a four-dimensional integral. Its Lagrangian density is therefore readily deduced as

$$\mathcal{L}_{\text{field}} = -\frac{1}{4} F^{\mu\nu} F_{\mu\nu}, \tag{2.20}$$

where

$$F^{\mu\nu} = \partial^\mu A^\nu - \partial^\nu A^\mu, \tag{2.21}$$

is the antisymmetric electromagnetic strength tensor, derived from the four-potential $A^\mu = (V, \mathbf{A})$ [1]. Defining the dual electromagnetic strength tensor as $F^{*\mu\nu} = \frac{1}{2} \epsilon^{\mu\nu\rho\sigma} F_{\rho\sigma}$, we deduce the two following Lorentz invariants

$$\zeta_1 = -\frac{1}{4} F^{\mu\nu} F_{\mu\nu}, \tag{2.22a}$$

---

[6]See Sect. 2.5 for a discussion about the bare mass in CED.

$$\zeta_2 = -\frac{1}{4} F^{*\mu\nu} F_{\mu\nu}, \tag{2.22b}$$

where $\epsilon^{\mu\nu\rho\sigma}$ is the four dimensionnal Levi-Civita antisymmetric symbol. The usual electric and magnetic fields **E** and **B** are linked to these tensors as

$$F^{0i} = -E^i, \tag{2.23a}$$
$$F^{ij} = -\epsilon^{ijk} B^k, \tag{2.23b}$$
$$F^{*0i} = -B^i, \tag{2.23c}$$
$$F^{*ij} = \epsilon^{ijk} E^k, \tag{2.23d}$$

and thus the two previous invariants read $\zeta_1 = \frac{1}{2}(\mathbf{E}^2 - \mathbf{B}^2)$, $\zeta_2 = \mathbf{E} \cdot \mathbf{B}$.

Finally, the interaction of a charged particle with an external electromagnetic field is described by

$$\mathcal{L}_{\text{int}} = -A^\mu j_\mu. \tag{2.24}$$

where j^μ is the charged particle current density. Note that, in order for the action to remain gauge invariant[7] when the interaction term is considered, we must have

$$\partial_\mu j^\mu = 0. \tag{2.26}$$

In other words, in order to have a gauge invariant Lagrangian, the interaction has to take place with a conserved current.[8] In the same way as for Noether's charge, we can define the Lorentz invariant charge

$$Q = \int_V d^3x \, j^0, \tag{2.27}$$

which will satisfy the continuity equation

$$d_t Q = -\int_{\partial V} dS \, \hat{n}_i \, j^i, \tag{2.28}$$

[7]The action indeed transforms as

$$S' - S = -\int d^4x \, j^\mu \partial_\mu \epsilon = -\int d^4x \, \partial_\mu (j^\mu \epsilon) + \int d^4x \, \partial_\mu j^\mu \epsilon \,. \tag{2.25}$$

under the gauge transformation $A^\mu \to A^\mu + \partial^\mu \epsilon(x)$.

[8]Note that, contrarily to Noether's theorem, there is no need for the classical equations of motion to be verified for this conservation law to hold.

where j^i therefore denotes the flux of charge, called 3-current density. The Lagrangian density of CED finally reads

$$\boxed{\mathcal{L}_{\text{CED}} = -\frac{1}{2}m\,\eta_{\mu\nu}u^\mu u^\nu - \frac{1}{4}\,F^{\mu\nu}F_{\mu\nu} - A^\mu j_\mu} \tag{2.29}$$

2.3.2 Equations of Motion

Application of the ELE for the 4-coordinate x^μ to $\mathcal{L}_{\text{CED}}$ yields the Newton-Lorentz equation

$$\boxed{ma^\mu = q\,F^{\mu\nu}u_\nu} \tag{2.30}$$

where $a^\mu = du^\mu/d\tau$ is the particle four-acceleration and q the *bare charge*[9] of the particle. On a classical equation of motion, the four-velocity satisfies the so-called *on-shell condition* $u^2 = 1$. Differentiating this identity with respect to the proper time τ, we see that the 4-acceleration a and the 4-velocity u are orthogonal[10] to each other, that is

$$au = 0. \tag{2.31}$$

Further differentiation of this identity yields[11]

$$\dot{a}u + a^2 = 0, \tag{2.32a}$$

$$2a\dot{a} + \dot{a}^2 + u\ddot{a} = 0. \tag{2.32b}$$

The same procedure for the 4-potential A^μ yields the usual inhomogeneous Maxwell's equations [14]

$$\partial_\mu F^{\mu\nu} = j^\nu \tag{2.33a}$$

$$\partial_\mu F^{*\mu\nu} = 0 \tag{2.33b}$$

In 3-vector notation, Eq. (2.33a) gives the first pair of Maxwell's equations, namely Poisson and Ampere's equations

$$\nabla\cdot\mathbf{E} = \rho, \tag{2.34a}$$

$$\nabla\times\mathbf{B} = \mathbf{j} + \partial_t\mathbf{E}, \tag{2.34b}$$

[9] See Sect. 2.5 for a discussion about the bare charge in CED.

[10] This can also be seen by multiplying Eq. (2.30) by u_μ and by antisymmetry of F.

[11] These relations will be useful for the derivation of the regularized self-field in Sect. 2.5.

while Eq. (2.33b) yields the second pair, namely the free-divergence magnetic field and Faraday's equations

$$\nabla \cdot \mathbf{B} = 0, \tag{2.35a}$$

$$\nabla \times \mathbf{E} = -\partial_t \mathbf{B}. \tag{2.35b}$$

2.3.3 *Energy-Momentum*

Applying Noether's theorem for space-time translations (2.7) to the Lagrangian of the free electromagnetic field (3.25), we get the energy-momentum tensor for the electromagnetic field

$$\boxed{\Theta^{\mu\nu}_{\text{field}} = -F^{\mu\rho}\partial^{\nu}A_{\rho} + \frac{1}{4}\eta^{\mu\nu}F^2} \tag{2.36}$$

which satisfies $\partial_\mu \Theta^{\mu\nu}_{\text{field}} = 0$. We will call this energy-momentum tensor the *canonical energy-momentum tensor*. This tensor has a number of deficiencies [1]. In particular, it is manifestly not gauge invariant, is not symmetric and not traceless.[12] Let us therefore follow the procedure described in Sect. 2.2.2 with $A^{\rho\mu\nu} = F^{\mu\rho}\partial^{\nu}\theta$ and change $\Theta^{\mu\nu}_{\text{field}}$ as $T^{\mu\nu}_{\text{field}} = \Theta^{\mu\nu}_{\text{field}} + \partial_\rho(F^{\mu\rho}\partial^{\nu}\theta)$. This yields

$$\boxed{T^{\mu\nu}_{\text{field}} = F^{\mu\rho}F^{\nu}_{\ \rho} + \frac{1}{4}\eta^{\mu\nu}F^2.} \tag{2.37}$$

which we call the *symmetrized (or Belinfante-Rosenfeld) energy-momentum tensor* [9, 15, 16]. Of course this energy-momentum tensor is still conserved

$$\partial_\mu T^{\mu\nu}_{\text{field}} = 0, \tag{2.38}$$

and it is now manifestly gauge-invariant. Moreover, it is obviously symmetric and traceless [1, 2]. Its different components read

$$W = \int d^3x\, T^{00} = \frac{1}{2}\int d^3x\, (\mathbf{E}^2 + \mathbf{B}^2), \tag{2.39a}$$

$$P^i = \int d^3x\, T^{0i} = \int d^3x\, \epsilon^{ijk} E_j B_k, \tag{2.39b}$$

$$T^{ij}_M = -T^{ij} = E^i E^j + B^i B^j - \frac{1}{2}\delta^{ij}(\mathbf{E}^2 + \mathbf{B}^2). \tag{2.39c}$$

---

[12] Which should be the case since the photon is massless.

Integrating the ith component of Eq. (2.38) over a finite volume V around the system gives rise to the two following continuity equations

$$d_t W = \int_V d^3x\, \partial_i P^i = -\int_{\partial V} dS\, \hat{n}_i\, P^i, \tag{2.40a}$$

$$d_t P^i = \int_V d^3x\, \partial_j T_{\mathrm{M}}^{ji} = -\int_{\partial V} dS\, \hat{n}_j\, T_{\mathrm{M}}^{ji}, \tag{2.40b}$$

where P^i is the momentum contained in the volume V and $\hat{\mathbf{n}}$ is a unitary vector normal to the surface ∂V around V and directed outward. This equation allows us to interpret T_{M} as the flux of momentum and P^i as the flux of energy through a surface.

Similarly, one can derive the conserved energy-momentum tensor for a free charge

$$\boxed{T^{\mu\nu}_{\mathrm{matter}} = m u^\mu u^\nu \frac{d\tau}{dt} \delta^{(3)}(\mathbf{x} - \mathbf{r}(\tau))} \tag{2.41}$$

where $\mathbf{x}$ is the position at which this tensor is evaluated and $\mathbf{r}(\tau)$ the world-line of the charge [17]. The associated Noether's charge yields the energy-momentum 4-vector

$$P^\mu_{\mathrm{matter}} = \int d^3x\, T^{0\mu}_{\mathrm{matter}} = m u^\mu, \tag{2.42}$$

where $u^\mu = (\gamma, \mathbf{u}) = \gamma(1, \boldsymbol{\beta})$, $\gamma = 1/\sqrt{1 - \boldsymbol{\beta}^2}$ is the so-called Lorentz gamma factor and $\boldsymbol{\beta}$ the three-velocitiy of the particle. In 3-vector notation, this reads

$$W_{\mathrm{matter}} = \int d^3x\, T^{00}_{\mathrm{matter}} = m\gamma, \tag{2.43a}$$

$$P^i_{\mathrm{matter}} = \int d^3x\, T^{0i}_{\mathrm{matter}} = m\gamma\beta^i. \tag{2.43b}$$

Finally, for a system of interacting charges and fields, the full energy-momentum tensor is simply the sum of the two previous free energy-momentum tensor

$$T^{\mu\nu} = T^{\mu\nu}_{\mathrm{field}} + T^{\mu\nu}_{\mathrm{matter}}, \tag{2.44}$$

and the corresponding conservation law reads

$$\partial_\mu T^{\mu\nu} = \partial_\mu \left(T^{\mu\nu}_{\mathrm{field}} + T^{\mu\nu}_{\mathrm{matter}}\right) = 0. \tag{2.45}$$

2.4 Radiation by Moving Charges

2.4.1 Physical Picture

Let us first note that, because of the equivalence of all inertial frames, a uniformly moving particle does not radiate. Let us now assume that, in an inertial frame in which the particle is at rest at position A, it accelerates to a velocity Δv in a short time Δt, to reach position B. Then the information about this change in velocity propagates at a the speed of light[13] c and reaches a radius $R = ct$ after time t (see Fig. 2.1). Therefore, for radiuses $r < R$, there exists a static Coulomb field pointing to B. However for $r > R + c\Delta t$, the information about the acceleration of the charge has not arrived yet and there exists a static Coulomb field pointing to the former charge's position A. Because of the continuity of the electromagnetic field, one thus deduces that, in the region $R < r < R + c\Delta t$, there exists a transverse field, that is to say, a pulse of electromagnetic radiation.

Let us evaluate the magnitude of this transverse field. As can be seen in Fig. 2.1, the ratio of the perpendicular to the parallel field is given by

$$\frac{E_\perp}{E_\parallel} = \frac{t\Delta v \sin\theta}{c\Delta t}, \tag{2.46}$$

where $E_\parallel$ is the usual Coulomb field given by $E_\parallel = q/(4\pi\epsilon_0 r^2)$ with $r = ct$. We then find

$$E_\perp = \frac{qa\sin\theta}{4\pi\epsilon_0 c^2 r}, \tag{2.47}$$

where $a = \Delta v/\Delta t$. We can then deduce the Poynting vector magnitude as

$$S = \frac{q^2 a^2 \sin^2\theta}{16\pi^2\epsilon_0 c^3 r^2} \tag{2.48}$$

which after integration over a spherical surface surrounding the radiating particle yields the well-know *Larmor power*

$$P_{\text{Larmor}} = \frac{q^2 a^2}{6\pi\epsilon_0 c^3}. \tag{2.49}$$

---

[13]While the theoretical calculations are done in natural units, we use SI units for all the numerical applications and physical pictures.

Fig. 2.1 Intuitive explanation of the radiation of accelerated charges

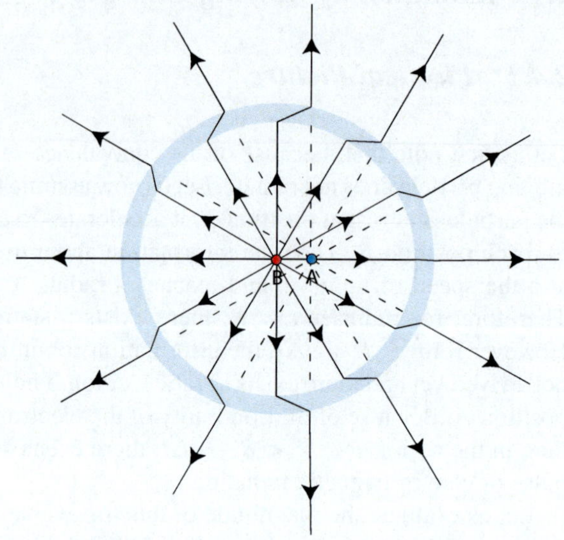

(a) We consider a charge initially in A and accelerating to a speed Δv in an interval of time Δt. It is then in B at time t. The blue circle represents the place where the information about the velocity change arrives at time t (this circle has a radius $R = ct$). For $r < R$ (inside the circle), the electric field is a pure Coulomb field pointing to B. Outside of the blue circle ($r > R$), the information about the velocity change did not arrive yet and the electric field is a pure Coulomb field pointing to A (black dashed lines). In the blue area, the field is a burst of electromagnetic radiation. We will detail this zone in the figure below.

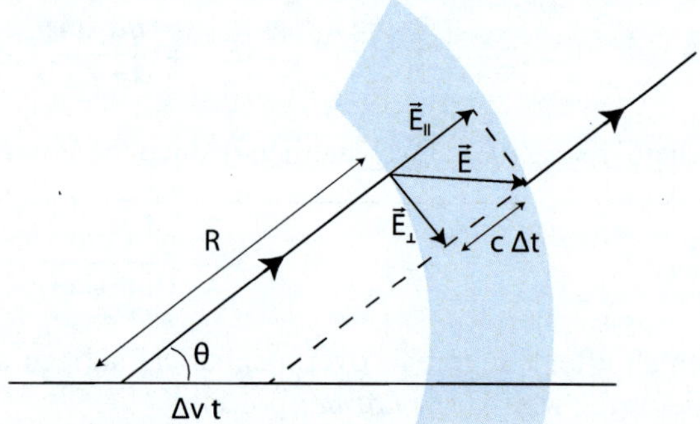

(b) In the interval $R < r < R + c\Delta t$, the electromagnetic field will have a perpendicular component because of the particle's acceleration. It is possible to link this perpendicular component to the parallel Coulomb field [Eq. (2.46)] from which we deduce the Larmor power [Eq. (2.49)].

This explanation is attributed to Thomson[14] and is useful to understand qualitatively was happens when a particle radiates because of its acceleration. We will now rederive this result in a more rigorous way by direct use of the so-called *Liénard-Wiechert potentials*.

---

[14]And was reproduced here based on [18].

2.4.2 Retarded and Advanced Propagators

As previously explained, accelerated charges emit electromagnetic radiation [1, 3]. To compute the radiated field in terms of the particle trajectory, we need to solve Maxwell's Eq. (2.33a) knowing the trajectory of the particle (and therefore j^μ). This can be done using Green's function method. Let us express Eq. (2.33a) in terms of A^μ in the Lorenz gauge $(\partial A) = 0$, we get (introducing $\Box = \partial_\mu \partial^\mu$)

$$\Box A^\mu = j^\mu, \tag{2.50}$$

for which there are two Green's functions,

$$D_r(x - x') = \frac{\theta(x_0 - x_0')}{4\pi|\mathbf{x} - \mathbf{x}'|}\delta(x_0 - x_0' - |\mathbf{x} - \mathbf{x}'|), \tag{2.51a}$$

$$D_a(x - x') = \frac{\theta(x_0' - x_0)}{4\pi|\mathbf{x} - \mathbf{x}'|}\delta(x_0 - x_0' + |\mathbf{x} - \mathbf{x}'|), \tag{2.51b}$$

which are called, respectively, the *retarded* and the *advanced* Green's functions. They can be rewritten in a covariant way as[15]

$$D_r(x - x') = \frac{1}{2\pi}\,\theta(x_0 - x_0')\,\delta[(x - x')^2], \tag{2.52a}$$

$$D_a(x - x') = \frac{1}{2\pi}\,\theta(x_0' - x_0)\,\delta[(x - x')^2]. \tag{2.52b}$$

The solutions of the wave equation (2.50) now read

$$A_r^\mu(x) = A_{\text{in}}^\mu(x) + \int d^4x'\, D_r(x - x') j^\mu(x'), \tag{2.53a}$$

$$A_a^\mu(x) = A_{\text{out}}^\mu(x) + \int d^4x'\, D_a(x - x') j^\mu(x'), \tag{2.53b}$$

where A_r^μ and A_a^μ are the so-called retarded and advanced fields while A_{in}^μ and A_{out}^μ are the *incoming* and *outgoing* fields.[16]

[15] The theta functions, apparently noninvariant are actually Lorentz invariant when constrained by the delta functions.

[16] See [19] of their interpretation.

2.4.3 Liénard-Wiechert Potentials

Considering the potential created by a single particle, j^μ is given by

$$j^\mu(x) = q \int d\tau\, \delta^4 \left[x - r(\tau)\right] u^\mu(\tau). \tag{2.54}$$

Inserting this expression in Eq. (2.53) yields the so-called Liénard-Wiechert potentials [3]

$$\boxed{A^\mu_{\substack{r\\a}}(x) = \pm \left.\frac{q u^\mu}{4\pi\rho}\right|_{\substack{\tau=\tau_{\mathrm{ret}}\\ \tau=\tau_{\mathrm{adv}}}}} \tag{2.55}$$

where $\tau_{\substack{\mathrm{ret}\\ \mathrm{adv}}} = r^0(\tau) \mp |\mathbf{x} - \mathbf{r}(\tau)|$ and $\rho = u\,(x - r)$. In 3-vector notation, this reads

$$\phi_{\substack{r\\a}}(x) = \left.\frac{q}{4\pi R\,(1 \mp \boldsymbol{\beta}\cdot\mathbf{n})}\right|_{\substack{\tau=\tau_{\mathrm{ret}}\\ \tau=\tau_{\mathrm{adv}}}}, \tag{2.56a}$$

$$\mathbf{A}_{\substack{r\\a}}(x) = \left.\frac{q\boldsymbol{\beta}}{4\pi R\,(1 \mp \boldsymbol{\beta}\cdot\mathbf{n})}\right|_{\substack{\tau=\tau_{\mathrm{ret}}\\ \tau=\tau_{\mathrm{adv}}}}, \tag{2.56b}$$

where $R = x^0 - r^0(\tau) = |\mathbf{x} - \mathbf{r}(\tau)|$, $u^\mu(\tau) = \gamma(1, \boldsymbol{\beta}(\tau))$ and $\mathbf{n} = \mathbf{r}/R$. Note that the factor $(1 \mp \boldsymbol{\beta}\cdot\mathbf{n})$ is a purely relativistic term. For nonrelativistic motions, $|\boldsymbol{\beta}| \ll 1$ and we find the well-known nonrelativistic form of the potentials.

2.4.4 Radiated Electromagnetic Field

Using the definition of the electromagnetic field tensor (2.21) and the previous expression of the Liénard-Wiechert potentials, we get the expression of the electromagnetic field created by a single accelerated particle

$$\boxed{F^{\mu\nu}_{\substack{r\\a}}(x) = \pm \frac{q}{2\pi\rho}\frac{d}{d\tau}\left.\left(\frac{R^{[\mu}u^{\nu]}}{\rho}\right)\right|_{\substack{\tau=\tau_{\mathrm{ret}}\\ \tau=\tau_{\mathrm{adv}}}}} \tag{2.57}$$

where $R^\mu = x^\mu - r^\mu$ and $a^{[\mu}b^{\nu]} = \frac{1}{2}\left(a^\mu b^\nu - b^\mu a^\nu\right)$. Because of relativistic causality, from now on, we will only consider the retarded fields. More explicitly (and dropping the "ret" subscripts), Eq. (2.57) reads, in 3-vector notation

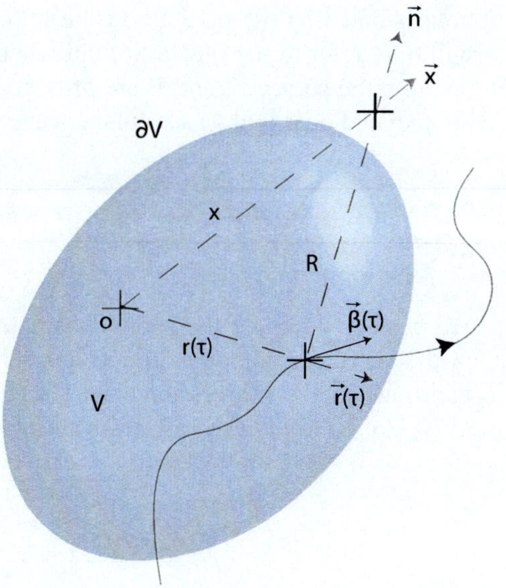

Fig. 2.2 Schematic representation of the different vectors and coordinates involved in the computation of Liénard-Wiechert potentials (2.56) and fields (2.57). The particle trajectory is represented by the black line and is parametrized by the vector $\mathbf{r}(\tau)$ which is at a distance $r(\tau)$ from the origin o. Its normalized velocity $\boldsymbol{\beta}(\tau)$ is tangential to the trajectory. We consider an observation point at a distance x from the origin o in the direction $\mathbf{x}/x$. The unitary vector $\mathbf{n}$ points from the particle's position to the observation point. For observers far from the particle, $x \gg r(\tau)$, we can approximate the distance from the observer to the particle by its distance to the origin $x \sim R$ and the direction from the particle to the observer by the direction from the origin to the observer $\mathbf{n} \sim \mathbf{x}/x$. We talk about *radiation zone*. The blue surface represents a fictitious surface around the particle that illustrates the integration volumes that we use in the derivation of the radiated power

$$\mathbf{E}(x) = \left[\frac{q}{4\pi} \frac{\mathbf{n} - \boldsymbol{\beta}}{\gamma^2 (1 - \boldsymbol{\beta} \cdot \mathbf{n})^3 R^2} + \frac{q}{4\pi} \frac{\mathbf{n} \times \left[(\mathbf{n} - \boldsymbol{\beta}) \times \dot{\boldsymbol{\beta}} \right]}{(1 - \boldsymbol{\beta} \cdot \mathbf{n})^3 R} \right], \tag{2.58a}$$

$$\mathbf{B}(x) = \mathbf{n} \times \mathbf{E}(x). \tag{2.58b}$$

We immediately see that $\mathbf{E}$ is composed of two physically different fields: a "velocity field" that does not depend on the acceleration of the particle and that falls off as R^{-2}. This field represents the static field of the particle. The second part of $\mathbf{E}$ is an "acceleration field" which is electromagnetic and falls off as R^{-1}. This part is the field radiated by the particle. In the following, except if otherwise mentioned, we will only refer to the radiation field when referring to **E**. Moreover, we notice that far from the source, **E** and **B** are orthogonal to each other and to **n**. In other words, the radiated field looks like a plane wave far from the radiating particle.

2.4.5 Larmor Formula

We want to compute the energy-momentum tensor of the radiating field in terms of the particle's quantities like its acceleration. Plugging Eq. (2.57) into Eq. (2.37) yields the energy-momentum tensor for the radiated field [3] from which we deduce, using Eq. (2.38), the so-called *Larmor formula*

$$\boxed{\frac{dP^{\mu}_{\text{rad}}}{d\tau} = \frac{2}{3}q^2 a^2 u^{\mu}} \tag{2.59}$$

which represents the rate at which the charge loses 4-momentum in terms of the particle's proper time τ. As stated in Sect. (2.4.4), the radiated 4-momentum corresponds to the radiation field (which is in $1/\rho$) which is the only one which "detaches" itself from the particle, contrarily to the velocity field which remains attached to the particle and has a vanishing 4-momentum far from the charge. Integrating over $d\tau$ and using Eq. (2.30), we get

$$\Delta P^{\mu}_{\text{rad}} = P_0 \int \eta^2 \, dx^{\mu}, \tag{2.60}$$

where $P_0 = \frac{2}{3}\frac{m}{\tau_e}$, $\tau_e = q^2/m$ and

$$\boxed{\eta = \left| \frac{F^{\mu\nu}}{E_{\text{cr}}} \frac{p_{\nu}}{m} \right| = \frac{\gamma}{E_{\text{cr}}} \sqrt{(\mathbf{E}^2 + \boldsymbol{\beta} \times \mathbf{B})^2 - (\boldsymbol{\beta} \cdot \mathbf{E})^2}} \tag{2.61}$$

which is manifestly Lorentz and gauge invariant and where $E_{\text{cr}} = m^2/q^3$. Moreover, it will be useful in the rest of this work to define the quantity

$$P_{\text{cl}} = P_0 \, \eta^2, \tag{2.62}$$

that we call the *classical radiated power*.

In order to interpret these newly introduced quantities, let us briefly use SI units. We can then rewrite $\tau_e = r_e/c = q^2/(4\pi\epsilon_0 mc^3) \simeq 9.37 \times 10^{-23}$ s where $r_e = q^2/(4\pi\epsilon_0 mc^2) \simeq 2.82 \times 10^{-15}$ m is the so-called *classical electron radius*, defined as the radius of the sphere in which the Coulomb potential energy is equal to the relativistic electron rest-energy mc^2. This represents the characteristic length of CED. The quantity τ_e therefore represents the time for light to travel the classical electron radius and represents a characteristic time for CED. From this characteristic time it is possible to construct a characteristic power $P_0 = \frac{2}{3}mc^2/\tau_e \simeq 36.4\,\text{keV/fs}$ which is the characteristic power of CED. Finally, $E_{\text{cr}} = 4\pi\epsilon_0 m^2c^4/q^3 \simeq 1.8 \times 10^{20}$ V/m is the so-called *critical field of CED* and represents the field that produces a work of mc^2 over r_e. It corresponds to an intensity of $I_{\text{cr}} \simeq 3.18 \times 10^{31}$ W/cm^2. In order to interpret η, we evaluate it in the particle's proper frame. We obtain

$$\eta = \frac{E}{E_{\text{cr}}}, \tag{2.63}$$

which therefore represents the magnitude of the electric field in the particle's proper frame compared to the critical field of CED E_{cr}. In the particular case of a plane-wave of amplitude E_0, we have

$$\eta = \gamma\frac{E_0}{E_{\text{cr}}}(1 - \beta_{\parallel}), \tag{2.64}$$

where $\beta_{\parallel}$ is the component of $\boldsymbol{\beta}$ parallel to the direction of propagation of the plane-wave. This expression shows that, not only the amplitude of the field plays a role in the value of η (and therefore in the total radiated power) but also the geometric configuration. In particular, we see that, for a given E_0, η (and therefore in the total radiated power) is maximum when the electron counter-propagates against the plane-wave. On the contrary, for an ultra-relativistic charge propagating in the direction of propagation of the plane-wave, $\eta \to 0$ whatever the value of E_0.

2.4.6 Radiated Spectrum

Plugging Eq. (2.58) into (2.39b) and (2.40a) yields

$$\partial_t W = -\int_{\partial V} \underbrace{d\Omega\,|R\mathbf{E}|^2}_{dP_{\text{rad}}}, \tag{2.65}$$

in spherical coordinates where $d\Omega$ is an infinitesimal solid angle element. This equation means that the energy leaving a closed surface surrounding the radiating charge (blue surface on Fig. 2.2) is carried away by a radiating field with an infinitesimal radiating power dP_{rad}. We therefore deduce that

$$\frac{dP_{\text{rad}}(t)}{d\Omega} = |R\mathbf{E}(t)|^2, \tag{2.66}$$

Using Parseval's theorem, we deduce the energy radiated per unit solid angle per unit frequency as

$$\frac{d^2 I}{d\omega\, d\Omega} = 2\,|\boldsymbol{C}(\omega)|^2, \tag{2.67}$$

where $\boldsymbol{C}(\omega)$ is the Fourier transform[17] of $\boldsymbol{C}(t)$ and is given by

$$\boldsymbol{C}(\omega) = \frac{e}{4\pi}\int_{-\infty}^{+\infty} e^{i\omega t}\left[\frac{\mathbf{n}\times\left[(\mathbf{n}-\boldsymbol{\beta})\times\dot{\boldsymbol{\beta}}\right]}{(1-\boldsymbol{\beta}\cdot\mathbf{n})^3 R}\right]_{\text{ret}} dt. \tag{2.68}$$

Assuming that the observation point is far from the radiating particle, we have

$$|\mathbf{x}-\mathbf{r}(t)| \simeq |\mathbf{x}| - \mathbf{n}\cdot\mathbf{r}(t). \tag{2.69}$$

Using this approximation, we finally get the energy radiated per unit solid angle per unit frequency

$$\boxed{\frac{d^2 I}{d\omega\, d\Omega} = \frac{e^2\omega^2}{4\pi^2}\left|\int_{-\infty}^{+\infty} \mathbf{n}\times(\mathbf{n}\times\boldsymbol{\beta})\, e^{i\omega(t-\mathbf{n}\cdot\mathbf{r}(t))}\, dt\right|^2} \tag{2.70}$$

This formula can be rewritten in 4-vector notation as

$$\frac{d^2 I}{d\omega\, d\Omega} = \frac{e^2\omega^2}{4\pi^2}\left|\int_{-\infty}^{+\infty} dt\, \frac{p^\mu(t)}{\epsilon(t)} e^{ikr}\right|^2, \tag{2.71}$$

where $k = \omega(1, \mathbf{n})$, $r = (t, \mathbf{r}(t))$ and $\epsilon(t) = p^0(t)$.

2.4.7 *Charged Particle in a Plane-Wave Field*

General Solution

The simplest possible electromagnetic field is that of a plane monochromatic wave. Such a field depends only on the Lorentz invariant parameter $\phi = kx$ and can be described by the following four-potential

$$A^\mu(\phi) = A_1^\mu \psi_1(\phi) + A_2^\mu \psi_2(\phi), \tag{2.72}$$

---

[17] and where the factor 2 comes from the fact that negative frequencies are unphysical.

where ψ_i are arbitrary scalar functions and A_i^μ are the two polarization directions satisfying $A_1 A_2 = 0$ and $kA_i = 0$. Defining $a^\mu(\phi) = qA^\mu(\phi)/m$, we have

$$a^\mu(\phi) = a_1^\mu \psi_1(\phi) + a_2^\mu \psi_2(\phi), \tag{2.73}$$

where $a_1 a_2 = 0$ and $ka_i = 0$. From this four-potential, we deduce the normalized electromagnetic strength tensor $f^{\mu\nu}(\phi) = qF^{\mu\nu}(\phi)/m$

$$f^{\mu\nu}(\phi) = \mathfrak{f}_1^{\mu\nu} \psi_1'(\phi) + \mathfrak{f}_2^{\mu\nu} \psi_2'(\phi), \tag{2.74}$$

where

$$\mathfrak{f}_i^{\mu\nu} = k^\mu a_i^\nu - k^\nu a_i^\mu, \quad \mathfrak{f}_i^{\mu\rho} \mathfrak{f}_{j,\rho}^\nu = -k^\mu k^\nu a_i^2 \delta_{ij}, \quad k_\mu \mathfrak{f}_i^{\mu\nu} = 0. \tag{2.75}$$

Knowing that $kp = kp_0 = cte$ and changing the proper time derivative to a derivative with respect to ϕ, with $\frac{d\phi}{d\tau} = (kp_0)/m = (ku_0)$, the solution is readily obtained as [20, 21]

$$\begin{aligned} p^\mu(\phi) = p_0^\mu &- q\left[A^\mu(\phi) - A^\mu(\phi_0)\right] + q\,\frac{[A(\phi) - A(\phi_0)]\,p_0}{kp_0}\,k^\mu \\ &- q^2\,\frac{\left[A^2(\phi) - A^2(\phi_0)\right]}{2kp_0}\,k^\mu. \end{aligned} \tag{2.76}$$

Defining the integrated electromagnetic strength tensor as

$$\mathcal{F}^{\mu\nu}(\phi, \phi_0) = \int_{\phi_0}^{\phi} d\phi' f^{\mu\nu}(\phi') = \sum_{i=1,2} f_i^{\mu\nu} \left[\psi_i(\phi) - \psi_i(\phi_0)\right], \tag{2.77}$$

we can rewrite the solution in a more compact way

$$\boxed{p^\mu(\phi) = p_0^\mu + \frac{\mathcal{F}^{\mu\nu}(\phi, \phi_0)\,p_{0,\nu}}{ku_0} + \frac{\mathcal{F}^{2\mu\nu}(\phi, \phi_0)\,p_{0,\nu}}{2(ku_0)^2}} \tag{2.78}$$

More explicitly, for the 4-velocity, we have

$$\begin{aligned} u^\mu(\phi) = u_0^\mu &+ \frac{1}{ku_0} \sum_{i=1,2} \left[\psi_i(\phi) - \psi_i(\phi_0)\right] \left[(a_i u_0) k^\mu - (ku_0) a_i^\mu\right] \\ &- \frac{1}{2(ku_0)} \sum_{i=1,2} a_i^2 \left[\psi_i(\phi) - \psi_i(\phi_0)\right]^2 k^\mu, \end{aligned} \tag{2.79}$$

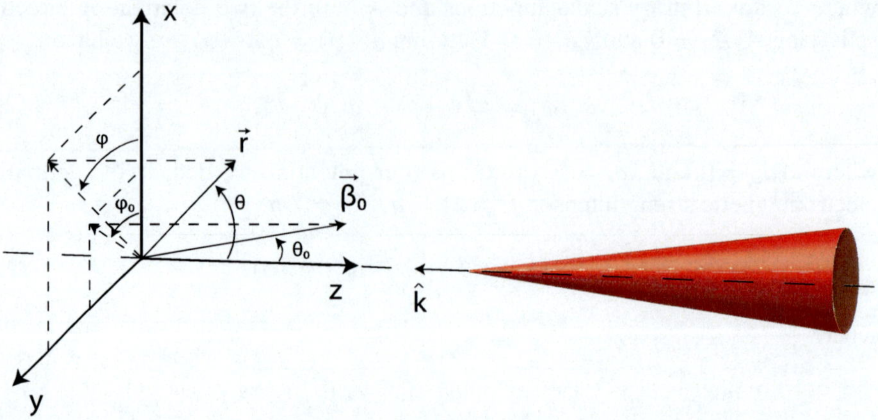

Fig. 2.3 Schematic of the different vectors and angles involved in the solution (2.82) and in (2.81). The initial electron velocity is represented by $\boldsymbol{\beta}_0$ and forms an angle θ_0 with the z axis and an angle φ_0 from the x axis in the (x, y) plane. The position $\mathbf{r}$ forms an angle θ with respect to the z axis and an angle φ from the x axis in the (x, y) plane. The plane-wave propagates in the direction $\hat{\mathbf{k}} = -\hat{\mathbf{z}}$. In the head-on configuration (2.84) and (2.88), we have $\theta_0 = \varphi_0 = 0$

from which we deduce the classical trajectory

$$\boxed{x^\mu(\phi) = x_0^\mu + \int_{\phi_0}^{\phi} d\phi' \frac{u^\mu(\phi')}{ku_0}} \tag{2.80}$$

where $x_0^\mu = x^\mu(\phi_0)$ is the initial four-position. Taking

$$\begin{aligned} a^\mu(\phi) &= a_0\Big(0, \delta\cos(\phi), \sqrt{1-\delta^2}\sin(\phi), 0\Big), \quad u^\mu = \gamma(1, \boldsymbol{\beta}\Big), \\ u_0^\mu &= \gamma_0(1, \boldsymbol{\beta}_0), \quad k^\mu = \omega_0(1, \hat{\mathbf{k}}), \quad x^\mu = (t, \mathbf{x}), \\ \phi &= \omega_0(t - \hat{\mathbf{k}}\cdot\mathbf{x}), \quad ku_0 = \gamma_0\omega_0(1 - \hat{\mathbf{k}}\cdot\boldsymbol{\beta}_0), \end{aligned} \tag{2.81}$$

we obtain in 3-vector notation,

$$\gamma(\phi) = \gamma_0 - \frac{[\mathbf{a}(\phi) - \mathbf{a}(\phi_0)]\cdot\boldsymbol{\beta}_0}{1 - \hat{\mathbf{k}}\cdot\boldsymbol{\beta}_0} + \frac{\big[\mathbf{a}^2(\phi) - \mathbf{a}^2(\phi_0)\big]}{2\gamma_0(1 - \hat{\mathbf{k}}\cdot\boldsymbol{\beta}_0)}, \tag{2.82a}$$

$$\begin{aligned} \mathbf{u}(\phi) = \mathbf{u}_0 - [\mathbf{a}(\phi) - \mathbf{a}(\phi_0)] - \frac{[\mathbf{a}(\phi) - \mathbf{a}(\phi_0)]\cdot\boldsymbol{\beta}_0}{1 - \hat{\mathbf{k}}\cdot\boldsymbol{\beta}_0}\hat{\mathbf{k}} \\ + \frac{\big[\mathbf{a}(\phi)^2 - \mathbf{a}(\phi_0)^2\big]}{2\gamma_0(1 - \hat{\mathbf{k}}\cdot\boldsymbol{\beta}_0)}\hat{\mathbf{k}}, \end{aligned} \tag{2.82b}$$

from which we deduce (Fig. 2.3)

$$\mathbf{x}(\phi) = \mathbf{x}_0 + \int_{\phi_0}^{\phi} d\phi' \, \frac{\mathbf{u}_0 - \left[\mathbf{a}(\phi') - \mathbf{a}(\phi_0)\right]}{\gamma_0\omega_0(1 - \hat{\mathbf{k}} \cdot \boldsymbol{\beta}_0)} - \int_{\phi_0}^{\phi} d\phi' \, \frac{\left[\mathbf{a}(\phi') - \mathbf{a}(\phi_0)\right] \cdot \boldsymbol{\beta}_0}{\gamma_0\omega_0(1 - \hat{\mathbf{k}} \cdot \boldsymbol{\beta}_0)^2} \hat{\mathbf{k}}$$
$$+ \int_{\phi_0}^{\phi} d\phi' \, \frac{\left[\mathbf{a}(\phi')^2 - \mathbf{a}(\phi_0)^2\right]}{2\gamma_0^2\omega_0(1 - \hat{\mathbf{k}} \cdot \boldsymbol{\beta}_0)^2} \hat{\mathbf{k}}. \tag{2.83}$$

Head-on Collision

In the following, we will mainly consider the situation of an initially ultra-relativistic electron[18] counter-propagating with the plane-wave propagating in the direction $\hat{\mathbf{k}} = -\hat{\mathbf{z}}$. In this case, $\beta_0 \simeq 1 - \frac{1}{2\gamma_0^2}$ and we write $ku_0 = \omega_0 h_0$ with $h_0 = \gamma_0(1 + \beta_0)$ In the case of the linearly polarized plane-wave ($\delta = 1$), we get

$$u_x(\phi) = a_0 \cos\phi, \tag{2.84a}$$
$$u_y(\phi) = 0, \tag{2.84b}$$
$$u_z(\phi) = \frac{h_0}{2}\left[1 - \frac{1}{h_0^2}\left[1 + \frac{a_0^2}{2}(1 + \cos 2\phi)\right]\right], \tag{2.84c}$$

from which we deduce the position of the particle

$$x(\phi) = x_0 + x_1 \sin\phi, \tag{2.85a}$$
$$y(\phi) = y_0, \tag{2.85b}$$
$$z(\phi) = z_0 + z_1\,\phi + z_2 \sin 2\phi, \tag{2.85c}$$

where $\mathbf{x}$ is normalized in ω_0, x_0, y_0 and z_0 are the initial positions normalized in ω_0 and

$$x_1 = \frac{a_0}{h_0}, \tag{2.86a}$$
$$z_1 = \frac{1}{2}\left[1 - \frac{1}{h_0^2}\left(1 + \frac{a_0^2}{2}\right)\right], \tag{2.86b}$$
$$z_2 = -\frac{a_0^2}{8h_0^2}. \tag{2.86c}$$

From Eqs. (2.85), we see that the electron drifts at a speed $h_0 z_1$ in the z direction, oscillates in the direction of the laser polarization and follows a figure-8 trajectory (Lissajou curve) with a longitudinal (transverse) extension of the order of a_0^2 (a_0) (see Fig. 2.4). The parameter a_0 is linked to the potential and electric field amplitudes A_0 and E_0 by

$$a_0 = \frac{e\sqrt{-A^2}}{m} = \frac{eE_0}{m\omega_0}. \tag{2.87}$$

[18] i.e. we take in the preceding equations $q = -e$.

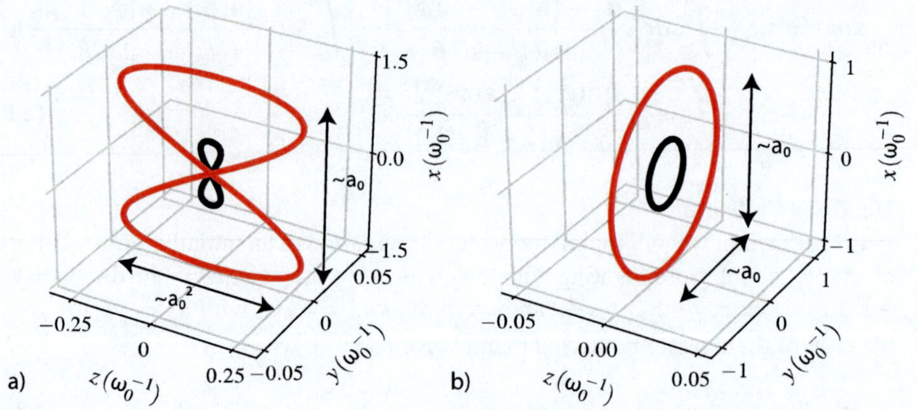

Fig. 2.4 Trajectory of an electron initially at rest a) in a linearly polarized plane-wave (LPPW) with $a_0 = 0.5$ and $a_0 = 1.5$ in black and red line (resp.) and b) in a circularly polarized plane-wave (CPPW) with $a_0 = 0.5$ and $a_0 = 1.5$ in black and red line (resp.). The trajectories are represented in the average rest frame which is defined as the frame in which the average drift velocity is equal to 0. The distances are normalized to ω/c. In the case of the LPPW, the trajectory has a figure-8 shape with a longitudinal (transverse) extension of the order of a_0^2 (a_0), while in the CPPW case, the trajectory is a circle with radius $\sim a_0$ in the (x, y) plane

It is a Lorentz and gauge invariant parameter. Going back temporarily to SI units, we can rewrite $a_0 = eE_0\lambda_0/(2\pi mc^2)$ where $\lambda_0 = 2\pi c/\omega_0$ is the laser wavelength in the laboratory frame. The parameter a_0 thus represents the work of the laser over one wavelength λ_0. If $a_0 \sim 1$, the electron becomes relativistic over one wavelength and we therefore talk about *relativistic regime*. In this regime, the magnetic component of the Lorentz force becomes of the same order of magnitude as the electric part and the electron motion becomes nonlinear. We therefore sometimes refer to a_0 as the *classical nonlinear parameter*.

In the nonrelativistic regime ($a_0 \ll 1$), the equation of motion is linear and the electron oscillates sinusoidally in the external plane-wave. In terms of radiation, we therefore expect the electron to radiate at the same frequency as the laser frequency in its rest frame. On the contrary, for $a_0 \gg 1$, the electron motion becomes nonlinear and the trajectory sharpens at the instants where the velocity along the laser polarization reverses. In terms of radiation, we expect the electron to radiate harmonics of the laser frequency in its rest frame. Finally, the solution (2.85) indicates that in the relativistic regime, the electron acquires a drift in the longitudinal direction, which is proportional to a_0^2.

Let us now consider the case of a circularly polarized plane-wave ($\delta = 1/\sqrt{2}$), we get

$$u_x(\phi) = \frac{a_0}{\sqrt{2}} \cos\phi, \tag{2.88a}$$

$$u_y(\phi) = \frac{a_0}{\sqrt{2}} \sin\phi, \tag{2.88b}$$

$$u_z(\phi) = \frac{h_0}{2}\left[1 - \frac{1}{h_0^2}\left(1 + \frac{a_0^2}{2}\right)\right], \tag{2.88c}$$

The position of the particle is then given by

$$x(\phi) = x_0 + x_2 \sin\phi, \tag{2.89a}$$

$$y(\phi) = y_0 + y_2 \cos\phi, \tag{2.89b}$$

$$z(\phi) = z_0 + z_1 \phi, \tag{2.89c}$$

where $\mathbf{x}$ is normalized in ω_0, x_0, y_0 and z_0 are the initial positions normalized in ω_0 and

$$x_2 = \frac{a_0}{\sqrt{2}h_0}, \tag{2.90a}$$

$$y_2 = -\frac{a_0}{\sqrt{2}h_0}, \tag{2.90b}$$

$$z_1 = \frac{1}{2}\left[1 - \frac{1}{h_0^2}\left(1 + \frac{a_0^2}{2}\right)\right]. \tag{2.90c}$$

We draw the same conclusions in the case of the linearly polarized plane-wave except that the trajectory is now a circle with radius proportional to a_0 and a drift proportional to a_0^2 (see Fig. 2.4).

2.4.8 *Linear and Nonlinear Thomson Scattering*

In order to evaluate Eq. (2.70) explicitly, we need to compute its vector part $\mathbf{n} \times (\mathbf{n} \times \boldsymbol{\beta})$ and its phase $\psi(\phi) = \omega(t - \mathbf{n} \cdot \mathbf{r}(\phi))$. To do so, we choose the observation direction $\mathbf{n} = \mathbf{E}_r$ and we take $\mathbf{r} = (x, y, z)$ and $\boldsymbol{\beta} = (\beta_x, \beta_y, \beta_z)$. The coordinates of $\boldsymbol{\beta}$ in spherical coordinates are then given by

$$\beta_r = \beta_x \sin\theta \cos\varphi + \beta_y \sin\theta \sin\varphi + \beta_z \cos\theta, \tag{2.91a}$$

$$\beta_\theta = \beta_x \cos\theta \cos\varphi + \beta_y \cos\theta \sin\varphi - \beta_z \sin\theta, \tag{2.91b}$$

$$\beta_\varphi = -\beta_x \sin\varphi + \beta_y \cos\varphi. \tag{2.91c}$$

We then have the vector part and phase given by

$$\omega(t - \mathbf{n}\cdot\mathbf{r}(\phi)) = \frac{\omega}{\omega_0}[\phi - \tilde{x}(\phi)\sin\theta\cos\varphi - \tilde{y}(\phi)\sin\theta\sin\varphi - \tilde{z}(\phi)(1+\cos\theta)], \tag{2.92}$$

$$\mathbf{n}\times(\mathbf{n}\times\boldsymbol{\beta}) = -(\beta_x\cos\theta\cos\varphi + \beta_y\cos\theta\sin\varphi - \beta_z\sin\theta)\mathbf{E}_\theta + (\beta_x\sin\varphi - \beta_y\cos\varphi)\mathbf{E}_\varphi, \tag{2.93}$$

where $\tilde{\mathbf{r}}(t) = (\tilde{x}, \tilde{y}, \tilde{z}) = \omega_0(x, y, z)$. The evaluation of Eq. (2.70) now amounts to evaluate the integrals

$$\mathbf{K} = \int_{-\eta}^{\eta} d\phi\, \frac{d\tilde{\mathbf{r}}(\phi)}{d\phi}\, e^{i\psi(\phi)}, \tag{2.94}$$

where η is an arbitrary phase. The scattered radiation will be polarized in the direction of $\mathbf{n}\times(\mathbf{n}\times\boldsymbol{\beta})$. We can thus decompose (2.70) as

$$\frac{d^2I}{d\omega\, d\Omega} = \frac{d^2I_\theta}{d\omega\, d\Omega} + \frac{d^2I_\phi}{d\omega\, d\Omega}, \tag{2.95}$$

where I_θ and I_φ are the energies radiated with polarization in the $\mathbf{E}_\theta$ and $\mathbf{E}_\varphi$ directions. Using (2.93), we get

$$\frac{d^2I_\theta}{d\omega\, d\Omega} = \frac{e^2\omega^2}{4\pi^2}\left|\cos\theta\cos\varphi\, K_x + \cos\theta\sin\varphi\, K_y - \sin\theta\, K_z\right|^2, \tag{2.96a}$$

$$\frac{d^2I_\varphi}{d\omega\, d\Omega} = \frac{e^2\omega^2}{4\pi^2}\left|\sin\varphi\, K_x - \cos\varphi\, K_y\right|^2. \tag{2.96b}$$

Circularly Polarized Plane-Wave

Let us evaluate explicitly Eqs. (2.96) in the case of a circularly polarized plane-wave. To do so, we use the results of Sect. 2.4.7 and compute the phase and vector part of Eq. (2.94). Using the Bessel function identities developed in Appendix A.1 and in particular the Jacobi-Anger expansion, we get [22, 23]

$$\left.\frac{d^2I}{d\omega\, d\Omega}\right|_{\text{CP}} = \frac{e^2\omega^2\eta^2}{\pi^2}\sum_{n=1}^{+\infty} R(\omega, n\omega_0)^2\left[\left(\frac{\cos\theta - z_1(1+\cos\theta)}{\sin\theta}\right)^2 J_n(\tilde{\psi}_2)^2 + x_2^2\, J_n'(\tilde{\psi}_2)^2\right],$$

where

$$\tilde{\psi}_2 = \frac{n\, x_2\, \sin\theta}{1 - z_1(1+\cos\theta)}, \tag{2.97a}$$

$$z_1 = \frac{1}{2}\left[1 - \frac{1}{h_0^2}\left(1 + \frac{a_0^2}{2}\right)\right], \tag{2.97b}$$

$$x_2 = \frac{a_0}{\sqrt{2}h_0}, \tag{2.97c}$$

$$R(\omega, n\omega_0) = \frac{1}{2\eta}\int_{-\eta}^{\eta} d\phi\, e^{i[\psi_1 - n]\phi} = \frac{\sin[(\psi_1 - n)\eta]}{(\psi_1 - n)\eta}. \tag{2.97d}$$

The resonance function $R(\omega, n\omega_0)$ is strongly peaked at $\psi_1 = n$. We therefore have a spectrum of harmonics at

$$\omega_n = \frac{n\omega_0}{1 - z_1(1 + \cos\theta)}. \tag{2.98}$$

with width $\Delta\omega_n/\omega_n = 1/(nN_0)$, where N_0 is the number of wavelengths of the pulse. In the low intensity case ($a_0 \ll 1$), the spectrum is strongly peaked at the fundamental frequency ω_0. On the contrary, for high intensity lasers ($a_0 \gg 1$), the spectrum is composed of many closely spaced harmonics that appear broadband. Assuming that $n \gg 1$, and using Eqs. (A.9a) to (A.10), we deduce that the spectrum has the following asymptotic form

$$\left.\frac{d^2 I}{d\omega\, d\Omega}\right|_{\mathrm{CP}} \sim \exp\left[-\frac{n}{3}\left(1 - \hat{\psi}_2^2\right)^{3/2}\right],$$

where

$$\hat{\psi}_2 = \frac{x_2 \sin\theta}{1 - z_1(1 + \cos\theta)}. \tag{2.99}$$

It therefore extends up to a critical frequency corresponding to $n_c = 3(1 - \hat{\psi}_{2,\max}^2)^{-3/2}$. This critical frequency is reached (at fixed a_0) for

$$\cos\theta_0 = \frac{z_1}{1 - z_1}, \tag{2.100}$$

and corresponds to

$$n_c = \frac{3a_0^3}{2\sqrt{2}}. \tag{2.101}$$

Finally, in the ultra-relativistic case ($a_0 \gg 1$), $n \gg 1$ and the radiation is confined around the angle θ_0. Therefore, taking $\theta = \theta_0 + \delta\theta$ with $\delta\theta \ll 1$ in Eq. (2.97) and using Eqs. (A.9a), we get

$$\left.\frac{d^2 I}{d\omega\, d\Omega}\right|_{\mathrm{CP}} \simeq N_0 \frac{3e^2}{\pi^2}\frac{\gamma^2\xi^2}{1 + \gamma^2\delta\theta^2}\left[\frac{\gamma^2\delta\theta^2}{1 + \gamma^2\delta\theta^2}K_{1/3}^2(\xi) + K_{2/3}^2(\xi)\right], \tag{2.102}$$

where

$$\xi = \frac{\omega}{\omega_c}\left(1+\gamma^2\delta\theta^2\right)^{3/2}, \tag{2.103}$$

$$\omega_c = n_c\,\frac{1-z_1}{1-2z_1}\,\omega_0. \tag{2.104}$$

Integrating this equation over all frequencies and angles, we get the intensity distribution in angles and frequencies (respectively)

$$\left.\frac{dI}{d\Omega}\right|_{\rm CP} \simeq N_0\,\frac{7e^2}{48}\,\frac{\omega_c\gamma^2}{(1+\gamma^2\delta\theta^2)^{5/2}}\left[1+\frac{5}{7}\,\frac{\gamma^2\delta\theta^2}{(1+\gamma^2\delta\theta^2)}\right], \tag{2.105}$$

$$\left.\frac{dI}{d\omega}\right|_{\rm CP} \simeq N_0\,2\sqrt{3}\,e^2\gamma\,\frac{\omega}{\omega_c}\int_{2\omega/\omega_c}^{+\infty} d\xi\;K_{5/3}(\xi). \tag{2.106}$$

Linearly Polarized Plane-Wave

Let us now study the case of a linearly polarized plane-wave. We follow the same steps as for the case of the circularly polarized plane-wave and get (see Appendix A.6 for more details)

$$\left.\frac{d^2I}{d\omega\,d\Omega}\right|_{\rm LP} = \frac{e^2\omega^2}{4\pi^2}\sum_{n=1}^{+\infty} R(\omega,n\omega_0)^2\left[(1-\sin^2\theta\cos^2\varphi)\tilde{K}_x^2 - \sin^2\theta\cos\varphi\,\tilde{K}_x\tilde{K}_z + \sin^2\theta\,\tilde{K}_z^2\right],$$

where we used the same notations as in Sect. 2.4.7 and where

$$\tilde{K}_x = x_1\eta\sum_{m=-\infty}^{+\infty}(-1)^m\,J_m(\tilde{\psi}_3)\,[J_{n-2m-1}(\tilde{\psi}_2) + J_{n-2m+1}(\tilde{\psi}_2)], \tag{2.107}$$

$$\tilde{K}_z = 2\eta\sum_{m=-\infty}^{+\infty}(-1)^m\,J_m(\tilde{\psi}_3)\left[z_1\,J_{n-2m}(\tilde{\psi}_2) + z_2\left(J_{n-2m-2}(\tilde{\psi}_2)+J_{n-2m+2}(\tilde{\psi}_2)\right)\right], \tag{2.108}$$

$$R(\omega,n\omega_0) = \frac{1}{2\eta}\int_{-\eta}^{\eta} d\phi\; e^{i[\psi_1-n]\phi} = \frac{\sin\left[(\psi_1-n)\eta\right]}{(\psi_1-n)\eta}, \tag{2.109}$$

$$\psi_1 = \frac{\omega}{\omega_0}\left[1-z_1(1+\cos\theta)\right], \tag{2.110}$$

$$\tilde{\psi}_2 = \frac{nx_1\sin\theta\cos\varphi}{1-z_1(1+\cos\theta)}, \tag{2.111}$$

$$\tilde{\psi}_3 = \frac{-nz_2(1+\cos\theta)}{1-z_1(1+\cos\theta)}. \tag{2.112}$$

Similarly to the circularly polarized case, the resonance function $R(\omega, n\omega_0)$ is sharply peaked in $\psi_1 = n$. We therefore have a spectrum of harmonics at frequencies

$$\omega_n = \frac{n\omega_0}{1 - z_1(1 + \cos\theta)}. \tag{2.113}$$

For $a_0 \lesssim 1$ the radiation is mainly in the fundamental mode $n = 1$ (as was already predicted from the electron motion). On the contrary, when $a_0 \gg 1$, the radiation is dominated by high harmonics and appears like a continuum spectrum. Considering the case where $\varphi = \pi/2$, and in the limits where $a_0 \gg 1$, $n \gg 1$ and the radiation is mainly backscattered. We therefore assume that $\theta \ll 1$ and using Eqs. (A.9a) to (A.10) in a completely analogous way to the circularly polarized case, we get

$$\left.\frac{d^2 I}{d\omega\, d\Omega}\right|_{LP} \simeq N_0 \frac{12e^2}{\pi^2} \frac{\hat{\gamma}^2\xi^2}{1 + \hat{\gamma}^2\theta^2} \left[\frac{\hat{\gamma}^2\theta^2}{1 + \hat{\gamma}^2\theta^2} K_{1/3}^2(\xi) + K_{2/3}^2(\xi)\right], \tag{2.114}$$

where

$$\xi = \frac{\omega}{\omega_c}\left(1 + \hat{\gamma}^2\theta^2\right)^{3/2}, \tag{2.115a}$$

$$\hat{\gamma} = h_0/2, \tag{2.115b}$$

$$\omega_c = \frac{n_c}{1 - 2z_1}\omega_0, \tag{2.115c}$$

$$n_c = 3a_0^3/4, \tag{2.115d}$$

2.4.9 Radiation by an Ultra-Relativistic Particle in Arbitrary Motion

We saw previously that in the ultra-relativistic limit, the spectrum radiated by a charged particle in a plane wave field is the same in linear as in circular polarization. We will see that this result is actually true for any particle[19] in ultra-relativistic motion [1, 24]. First we will analyze qualitatively the main features of the radiation to then compute its full analytical expression from Eq. (2.70).

In the nonrelativistic limit, it is well known that the angular distribution of radiated energy is $dI/d\Omega \sim \sin^2\theta_a$ (dipole-like radiation [1]), where θ_a is the angle between the acceleration and the direction of observation. For latter convenience, we define the angle θ as $\theta = \pi/2 - \theta_a$ (Fig. 2.5a). Because of relativistic aberrations, the angle θ, in which a ray of light is seen by a non-moving observer, will be seen by an observer in inertial motion with speed β at an angle θ' satisfying

[19] We don't consider here pure longitudinal accelerations.

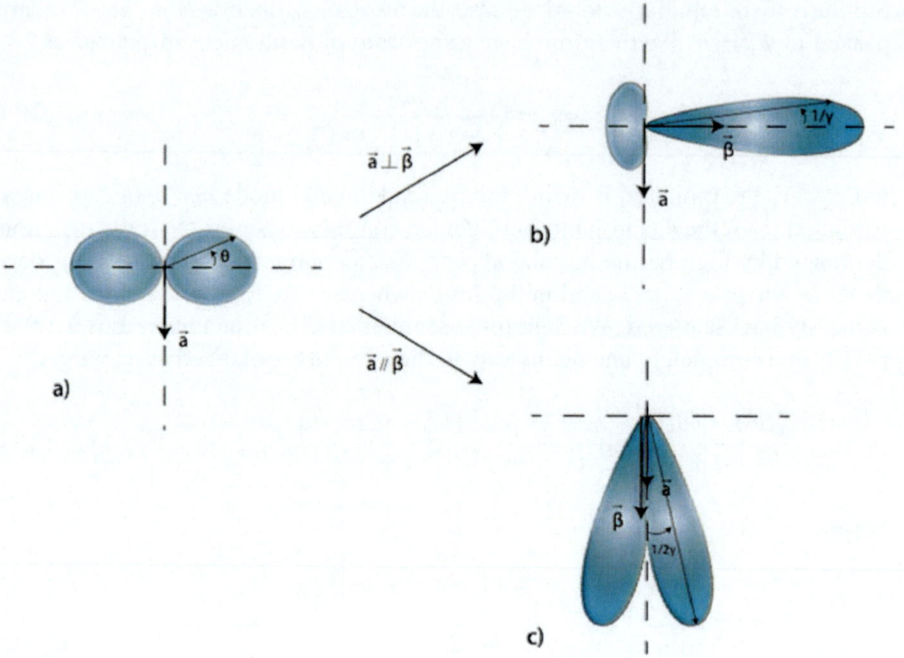

Fig. 2.5 a) In the nonrelativistic regime, the angular distribution of radiated energy is that of a dipole: $dI/d\Omega \sim \sin^2\theta_a$. Because of the relativistic aberration (2.116), if b) the particle moves at a speed $\boldsymbol{\beta}$ (with $\gamma \gg 1$) orthogonal to its acceleration $\boldsymbol{a}$, most of its radiation is emitted within an angle $1/\gamma$ around its direction of propagation. If c) the particle moves at a speed $\boldsymbol{\beta}$ (with $\gamma \gg 1$) parrallel to its acceleration $\boldsymbol{a}$, there is no radiation in the direction of propagation and all the radiation is confined in two lobes separated by an angle of $1/\gamma$

$$\tan\frac{\theta'}{2} = \sqrt{\frac{1-\beta}{1+\beta}}\,\tan\frac{\theta}{2}. \tag{2.116}$$

For an ultrarelativistic motion, light that would be emitted at $\theta = \pi/2$ in the rest frame of the emitter would then be seen at

$$\tan\frac{\theta'}{2} \sim \frac{1}{2\gamma} \Longrightarrow \theta' \sim \frac{1}{\gamma}, \tag{2.117}$$

by an observer moving at $\beta \simeq 1 - 1/(2\gamma^2)$ with respect to the emitter. For an emitter with its velocity $\boldsymbol{\beta}$ perpendicular to its acceleration $\boldsymbol{a}$, most of the radiation is emitted in an angle $1/\gamma$ in the direction of its velocity (Fig. 2.5b), while in the case of an acceleration parallel to its velocity, there is no radiation in the direction of propagation and all the radiation is confined in two lobes separated by an angle of $1/\gamma$ (Fig. 2.5c).

It is possible to show that the radiation due to a charge accelerated by an external force (with comparable parallel and perpendicular components) is mainly due to the perpendicular component of the force (more precisely, it is γ^2 larger than due to the parallel motion). The radiation due to a particle in ultra-relativistic motion can therefore be approximated by the radiation due a particle in an instantaneous circular path of radius of curvature $\rho \sim 1/\dot{v}_\perp$ where $\dot{v}_\perp$ denotes the perpendicular acceleration of the particle (see Fig. 2.6). Let us evaluate the duration Δt of the beam of radiation as seen by a fixed observer at a distance R of point B of the trajectory of the radiating particle. As was previously shown, the radiation is mainly emitted in an angle of $1/\gamma$. The segment of the charge's orbit that contributes to the radiation seen by the observer therefore comes from a segment of length $L = \rho/\gamma$ of the charge's trajectory (highlighted in purple in Fig. 2.6). However, the duration of the radiation seen by the observer is not simply L/β but much less since for an ultra-relativistic charge, the radiation emitted at the trailing edge of the purple area in Fig. 2.6 almost catches up the radiation emitted at the leading edge. Approximating this arc of circle by a straight line, the radiation emitted at the trailing edge takes $L + R$ to reach the observer, while the one emitted at the leading edge takes $\frac{L}{\beta} + R$. The duration of the beam of radiation seen by the observer has therefore a duration[20] of $\Delta t = L/\beta + R - L - R = L(1-\beta)/\beta \simeq L/(2\gamma^2)$. Using the uncertainty relation for time and angular frequency, we find that the higher frequency emitted by the charge is of the order of

$$\omega_c \simeq \frac{2\gamma^3}{\rho}. \tag{2.118}$$

Let us now evaluate quantitatively the spectrum of the emitted radiation. Following the same steps as in the previous section, we are led to evaluate the vector part and the phase part of Eq. (2.70). Choosing the coordinates so that the particle is at the origin at $t = 0$, we approximate the trajectory by a circle of radius of curvature ρ for $t \ll 1$. We get

$$\omega(t - \mathbf{n}\cdot\mathbf{r}(t)) = \omega\left[t - \rho\sin\left(\frac{\beta t}{\rho}\right)\cos\theta\right], \tag{2.119a}$$

$$\mathbf{n}\times(\mathbf{n}\times\boldsymbol{\beta}) = \beta\left[-\boldsymbol{\epsilon}_\parallel\sin\left(\frac{\beta t}{\rho}\right) + \boldsymbol{\epsilon}_\perp\cos\left(\frac{\beta t}{\rho}\right)\sin\theta\right], \tag{2.119b}$$

where $\boldsymbol{\epsilon}_\parallel$ is a unitary vector in the y direction corresponding to the polarization in the plane of the orbit, while $\boldsymbol{\epsilon}_\perp = \mathbf{n}\times\boldsymbol{\epsilon}_\parallel$ is the orthogonal polarization and corresponds approximately to polarization perpendicular to the plane of the orbit for $\theta \ll 1$ (see Fig. 2.6). As mentioned previously, the radiation perceived by a fixed observer is a coherent sum of the radiation emitted on an arc of circle of length $\sim \rho/\gamma$. We are thus interested in the part of Eq. (2.119) for which $\theta \ll 1$ and around $t = 0$. We get the following approximate form

[20] Note that this factor is already present in the expression of Lineard-Wiechert potentials (2.56).

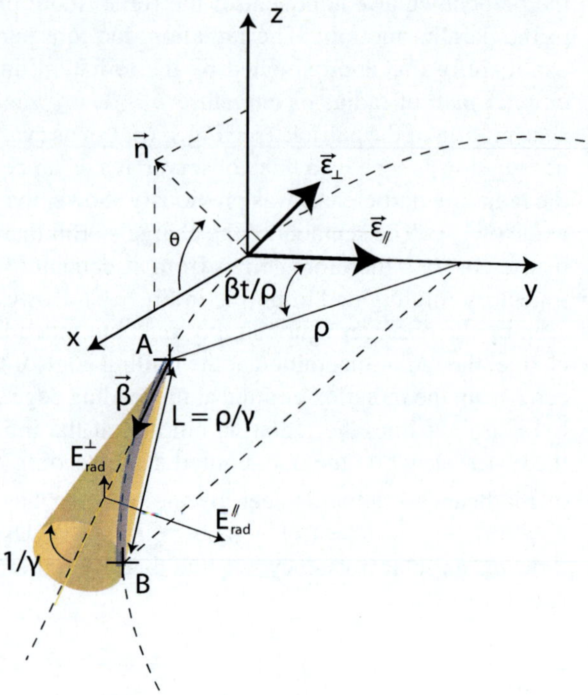

Fig. 2.6 It is possible to approximate the trajectory of any ultra-relativistic particle as locally circular with radius of curvature $\rho = 1/\dot{v}_\perp$, where $\dot{v}_\perp$ is the perpendicular acceleration. This trajectory is represented in black dotted line. The direction of observation **n** forms an angle θ with the x axis. According to Eq. (2.116), the radiation is emitted in an angle $\sim 1/\gamma$ in the direction of propagation (represented as a yellow cone). The contribution to the radiation, as seen by a distant observer, therefore comes from a portion of trajectory of length $L = \rho/\gamma$ and represented in purple. The polarization of the emitted radiation is computed using the vector $\boldsymbol{\epsilon}_\parallel$ (resp. $\boldsymbol{\epsilon}_\perp$) which is in the (x, y) and orthogonal to the particle's trajectory (in the (x, z) plane and orthogonal to the direction of observation **n**.)

$$\omega(t - \mathbf{n} \cdot \mathbf{r}(t)) \simeq \frac{\omega}{2}\left[\left(\frac{1}{\gamma^2} + \theta^2\right)t + \frac{1}{3\rho^2}t^3\right], \tag{2.120a}$$

$$\mathbf{n} \times (\mathbf{n} \times \boldsymbol{\beta}) \simeq -\boldsymbol{\epsilon}_\parallel \frac{t}{\rho} + \boldsymbol{\epsilon}_\perp \theta. \tag{2.120b}$$

Inserting these equations into (2.70) yields

$$\frac{d^2 I}{d\omega\, d\Omega} = \frac{e^2\omega^2}{4\pi^2}\left|-\boldsymbol{\epsilon}_\parallel K_\parallel + \boldsymbol{\epsilon}_\perp K_\perp\right|^2, \tag{2.121}$$

where

$$K_\parallel \simeq \frac{1}{\rho}\int_{-\infty}^{\infty} dt\, t\, e^{i\frac{\omega}{2}\left[\left(\frac{1}{\gamma^2}+\theta^2\right)t+\frac{t^3}{3\rho^2}\right]}, \tag{2.122a}$$

$$K_\perp \simeq \theta \int_{-\infty}^{\infty} dt\, e^{i\frac{\omega}{2}\left[\left(\frac{1}{\gamma^2}+\theta^2\right)t+\frac{t^3}{3\rho^2}\right]}, \tag{2.122b}$$

correspond, respectively, to the parallel and perpendicular polarizations of the radiation and where we recognize the Airy function and its derivative (See Appendix A.2). Expressing them in terms of modified Bessel functions of the second kind thanks to Eqs. (A.15), we finally get

$$\frac{d^2I}{d\omega\, d\Omega} = \frac{e^2}{3\pi^2}\,\omega^2\rho^2\left(\frac{1}{\gamma^2}+\theta^2\right)^2\left[\frac{\theta^2}{(1/\gamma^2)+\theta^2}\,K_{1/3}^2(\xi)+K_{2/3}^2(\xi)\right], \tag{2.123}$$

where $\xi = \frac{2}{3}\frac{\omega}{\omega_c}\,\theta_\gamma^3$, where ω_c is given by Eq. (2.118) and $\theta_\gamma = (1+\theta^2\gamma^2)$.

Integrating this equation over all frequencies yields the distribution of energy in angle

$$\frac{dI}{d\Omega} = \int_0^{+\infty} d\omega\,\frac{d^2I}{d\omega\, d\Omega} = \frac{7}{16}\frac{e^2}{\rho}\frac{1}{(\theta^2+1/\gamma^2)^{5/2}}\left[1+\frac{5}{7}\frac{\theta^2}{\theta^2+1/\gamma^2}\right]. \tag{2.124}$$

Finally, integration over all angles of Eq. (2.123) gives the emitted power distribution as a function of the frequency ω of the emitted photons [1]

$$\boxed{\frac{dI}{d\omega} = \frac{9\sqrt{3}}{8\pi}\frac{P_{\rm cl}}{\omega_c}\frac{\omega}{\omega_c}\int_{\omega/\omega_c}^{+\infty} dy\, \mathrm{K}_{5/3}(y)} \tag{2.125}$$

where $\omega_c = 3\,\gamma\eta/(2\tau_e)$ is the critical frequency for synchrotron emission, and where the instantaneous power radiated away is given by Eq. (2.62). This distribution has a maximum for photon frequencies $\omega \sim 0.29\,\omega_c$ and admits the two following limiting forms

$$\frac{dI}{d\omega} \sim \begin{cases} (\omega/\omega_c)^{1/3}, & \omega < \omega_c \\ \sqrt{\omega/\omega_c}\,\exp(-\omega/\omega_c), & \omega > \omega_c \end{cases} \tag{2.126}$$

2.5 Classical Radiation-Reaction

So far we have either considered the motion of charged particles in prescribed external electromagnetic fields, or the electromagnetic field generated by charged particles moving on prescribed trajectories. However, in the most general case, these two problems need to be solved consistently since an accelerated charge moving in external fields $F^{\mu\nu}_{\rm ext}$ radiates and therefore will then interact not only with the initial prescribed external electromagnetic field but with the sum of the external field and its own radiated field $F^{\mu\nu}_{\rm ext} + F^{\mu\nu}_{\rm rad}$. We are thus led to solve consistently the system (Fig. 2.7)

$$m\frac{du^\mu}{d\tau} = q\left(F^{\mu\nu}_{\text{ext}} + F^{\mu\nu}_{\text{rad}}\right)u_\nu, \tag{2.127a}$$

$$\partial_\mu F^{\mu\nu}_{\text{rad}} = q\int d\tau\, \delta[x - r(\tau)]\, u^\mu(\tau). \tag{2.127b}$$

Accounting for the back-reaction of radiation emission on the electron dynamics has been a long standing problem of classical electrodynamics (CED) [1, 3]. As such, it has been the focus of many studies, and various equations of motion of a radiating charge in an external field have been proposed [25]. In this Section, we derive the so-called LAD (Lorentz-Abraham-Dirac) equation of motion [26] and deduce its limiting case, valid in CED, the LL (Landau-Lifschitz) equation [2]. We then analyze the motion of a charged particle with RR in some simple external fields (such as a constant-uniform magnetic field or a plane-wave field) and compare it to the corresponding case without RR.

2.5.1 The Lorentz-Abraham-Dirac Equation

The most natural way to solve the problem (2.127) is to solve Eq. (2.127b) using the Green function (2.53) and insert it into Eq. (2.127a). However, in the same way as the stationary (Coulomb) field of a particle diverges at the particle's position, the full field (A.62) will diverge at the particle's position. This type of divergence is ubiquitous in quantum field theory and we will solve it using the *renormalization theory* [19, 27, 28].

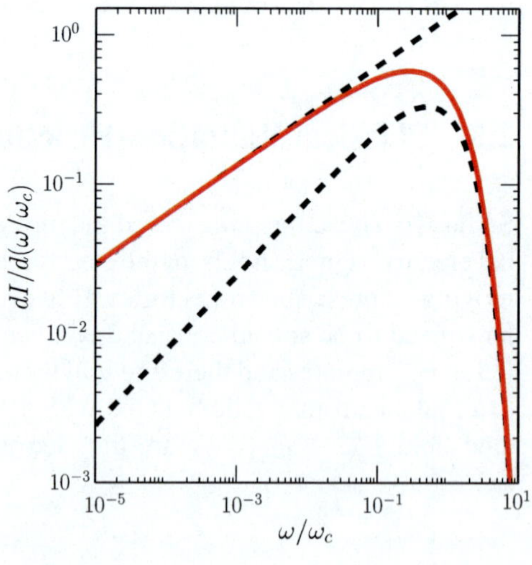

Fig. 2.7 Spectrum radiated by an ultra-relativistic electron as given by Eq. (2.125) (red line) and its asymptotic expressions (black dotted lines) as given by Eqs. (2.126)

When we introduced the Lagrangian of CED (2.29), we did not discuss the meaning of the parameters m and q. These parameters are called *bare mass* and *bare charge* (we shall write them m_0 and q_0 in this Section) and are different from the physical mass and charge m_{phys} and q_{phys}. Indeed, current theories cannot predict such quantities such as the mass or charge of elementary particles and these quantities need to be measured experimentally and then plugged into the theory. The parameters m_0 and e_0 must therefore be "tuned" so that the theory reproduces a particle with physical mass and charge m_{phys} and q_{phys}. We will carry this identification on CED after having introduce the *regularized self-field.*

- Regularized self-field:
  To better study how the field diverges near the position of the particle, we introduce a cutoff parameter ϵ. Basically, this means that length-scales that are under ϵ will not be taken into account in the "regularized" theory.
  We first compute the regularized potential [19, 27]

$$A^\mu_\epsilon(x) = \frac{q}{2\pi}\int_{-\infty}^{+\infty} d\tau\, \theta(x_0 - x_0')\, \delta\left[(x - x')^2 - \epsilon^2\right] u^\mu(\tau), \qquad (2.128)$$

  from which we deduce the regularized self-field at first order in ϵ (see Appendix A.7)

$$F^{\mu\nu}[r(\tau)] = \frac{q}{2\pi}\left[-\frac{1}{2\epsilon}a^{[\mu}u^{\nu]} + \frac{2}{3}\dot{a}^{[\mu}u^{\nu]}\right] + O(\epsilon)\,. \qquad (2.129)$$

  Inserting this field into Eq. (2.127a) yields

$$m_0 a^\mu = q_0 F^{\mu\nu}_{\text{ext}} u_\nu - \frac{1}{2\epsilon}\left(\frac{q_0^2}{4\pi}\right) a^\mu + \frac{2}{3}\left(\frac{q_0^2}{4\pi}\right)\left[\dot{a}^\mu + a^2 u^\mu\right] + O(\epsilon). \qquad (2.130)$$

- Mass renormalization:
  We notice that when $\epsilon \to 0$, $\frac{1}{2\epsilon}\left(\frac{q_0^2}{4\pi}\right) a^\mu$ diverges, which seems problematic. However, as mentioned above, this comes from the fact that m_0 is not the physical mass. Going back to the definition of the inertial mass, we consider the situation of a very slightly accelerated particle. Equation (2.130) then reduces to

$$\left[m_0 + \frac{1}{2\epsilon}\left(\frac{q_0^2}{4\pi}\right)\right] a^\mu = q_0 F^{\mu\nu}_{\text{ext}} u_\nu\,. \qquad (2.131)$$

  The coefficient multiplying the acceleration is clearly the physical mass of the particle, and we therefore identify $m_{\text{phys}} \equiv m_0 + \frac{1}{2\epsilon}\left(\frac{q_0^2}{4\pi}\right)$. In other words, we tune $m_0(\epsilon)$ (which depends on the cutoff ϵ), so that m_{phys} remains unchanged and equal to the measured particle's mass.

- Charge renormalization:
  Doing the same for the particle's charge, let us now consider Eq. (2.127b) in which we set $A^\mu = (V, \mathbf{0})$, we get

$$\Delta V = q_0\, \delta^{(3)}(\mathbf{x}). \tag{2.132}$$

  The coefficient q_0 is therefore obviously the electric charge and we make the identification $q_{\text{phys}} \equiv q_0$[21].

- Equation of motion:
  Taking the limit $\epsilon \to 0$, all lengthscales are now included in the theory and the final equation of motion reads

$$\boxed{a^\mu = f^{\mu\nu} u_\nu + \tau_0 \left(\dot{a}^\mu + a^2\, u^\mu\right)} \tag{2.133}$$

  where $\tau_0 = 2/3\tau_e$, $\tau_e = q^2/(4\pi m)$, $f^{\mu\nu} = (q/m)F^{\mu\nu}$ and where we dropped the subscript "phys" in the physical mass and charge, and "ext" in the external field for clarity. This equation is the *Lorentz-Abraham-Dirac (LAD) equation* [26].

The LAD equation is known to be plagued with physical inconsistencies [25]. We will illustrate these unphysical behaviours. In the case of a free particle, multiplying the LAD equation by a_μ gives

$$a^2 = \tau_0 \frac{d}{d\tau} a^2 \implies a^2 = C e^{\tau/\tau_0}, \tag{2.134}$$

where we have a runaway solution for $C \neq 0$, that is to say, the particle accelerates to the speed of light in the absence of external force. Of course, we could choose $C = 0$ to avoid such runaway solutions. Let us examine when we impose such condition. Let us assume that the field F is weak and that the solution of (2.133) reads $x^\mu = x_0^\mu + \delta x^\mu$ where x_0^μ is the solution in absence of field, and therefore the corresponding velocity u_0^μ is constant. The total velocity therefore reads $u^\mu = u_0^\mu + \delta u^\mu$ and its corresponding acceleration $a^\mu = \delta a^\mu$. At first order, Eq. (2.133) therefore reads

$$\delta a^\mu(\tau) = f^\mu(\tau) + \tau_0\, \delta\dot{a}^\mu(\tau), \tag{2.135}$$

where $f^\mu(\tau) = f^{\mu\nu}[x_0(\tau)]u_{0,\nu}$. Assuming that there is no acceleration at infinity, Eq. (2.135) integrates into

[21] There is therefore no charge renormalization in CED, contrarily to QED.

$$a^{\mu}(\tau) = e^{\frac{\tau}{\tau_0}} \int_{\tau}^{+\infty} \frac{d\tau'}{\tau_0} f^{\mu}(\tau') e^{-\frac{\tau'}{\tau_0}}. \tag{2.136}$$

Changing the integration variable to $s = (\tau - \tau')/\tau_0$, we get

$$a^{\mu}(\tau) = \int_{0}^{+\infty} ds\, f^{\mu}(\tau + s\tau_0)\, e^{-s}. \tag{2.137}$$

This solution violates causality since the acceleration at τ depends only on times later than τ.

2.5.2 *The Landau-Lifshitz Equation*

In order to better understand the issue with the LAD equation, let us recall that, in its derivation, we neglected the terms in $O(\epsilon)$ in the regularized self-field expansion (See Sect. 2.5.1 and Appendix A.7). These terms were of the form $\epsilon^n \frac{d^n a^{\mu}}{d\tau^n}$. So the approximation that led to the LAD equation is that the motions that we consider do not vary on proper-time scales smaller than the cut-off. More precisely, we require that

$$\forall n, \quad \left| \epsilon^n \frac{d^n a^{\mu}}{d\tau^n} \right| \ll |a^{\mu}|, \tag{2.138}$$

and we call such motions *admissible motions* [3, 29]. Going back to the LAD equation, let us evaluate τ_e, the time for light to cross the classical radius of the electron. To do so, we briefly go back to SI units and get $\tau_e = r_e/c = e^2/(4\pi\epsilon_0 m_e c^3) \simeq 9.36 \times 10^{-24}$ s $\ll \tau_c = \alpha\tau_e$. In other words, τ_e is much smaller than the scale at which quantum effects start to appear, i.e. well in the quantum regime. We therefore don't expect the LAD equation, which is a classical equation, to work at such scale. Choosing the cutoff ϵ to be τ_e, we therefore require that

$$|\tau_0 \dot{a}^{\mu}| \ll |a^{\mu}|. \tag{2.139}$$

This allows us to do a reduction of order in the LAD equation to obtain

$$\boxed{a^{\mu} = f^{\mu\nu} u_{\nu} + \tau_0 \left[\partial_{\eta} f^{\mu\nu} u_{\nu} u^{\eta} - f^{\mu\nu} f_{\eta\nu} u^{\eta} + (f^{\nu\eta} u_{\eta})(f_{\nu\alpha} u^{\alpha}) u^{\mu} \right]}$$

This equation is the *Landau-Lifshitz (LL) equation* [2]. It is valid for admissible motions (2.138) and is free of all the inconstancies that plagued the LAD equation. The fact that the LL equation is derived from the LAD equation which exhibits unphysical behaviours has led people to question its validity. However, it has been shown by [30] that the subspace of physical solutions of the LAD equation coincides with the solutions of the LL equation.

In 3-vector notation, this equation reads

$$\begin{aligned} m\,\frac{d\gamma}{dt} &= q\,\boldsymbol{\beta}\cdot\mathbf{E} + q\tau_0\gamma\,\dot{\mathbf{E}}\cdot\boldsymbol{\beta} \\ &+ \tau_0\,\mathbf{E}\cdot(\mathbf{E}+\boldsymbol{\beta}\times\mathbf{B}) \\ &- \tau_0\,\gamma^2\left[(\mathbf{E}+\boldsymbol{\beta}\times\mathbf{B})^2-(\boldsymbol{\beta}\cdot\mathbf{E})^2\right], \end{aligned} \tag{2.140}$$

$$\begin{aligned} \frac{d\mathbf{p}}{dt} &= q\,(\mathbf{E}+\boldsymbol{\beta}\times\mathbf{B}) + q\tau_0\,\gamma\,\left(\dot{\mathbf{E}}+\boldsymbol{\beta}\times\dot{\mathbf{B}}\right) \\ &+ \tau_0\,\left[(\boldsymbol{\beta}\cdot\mathbf{E})\,\mathbf{E}-\mathbf{B}\times(\mathbf{E}+\boldsymbol{\beta}\times\mathbf{B})\right] \\ &- \tau_0\,\gamma^2\left[(\mathbf{E}+\boldsymbol{\beta}\times\mathbf{B})^2-(\boldsymbol{\beta}\cdot\mathbf{E})^2\right]\boldsymbol{\beta}. \end{aligned} \tag{2.141}$$

We see that the second and last terms are respectively γ times and γ^2 times bigger than the first one. For ultra-relativistic particles, the last term will therefore be dominant and we can approximate Eq. (2.140) as

$$m\,\frac{d\gamma}{dt} = q\,\boldsymbol{\beta}\cdot\mathbf{E} - P_{\mathrm{cl}}, \tag{2.142}$$

$$\frac{d\mathbf{p}}{dt} = q\,(\mathbf{E}+\boldsymbol{\beta}\times\mathbf{B}) - P_{\mathrm{cl}}\,\boldsymbol{\beta}, \tag{2.143}$$

where

$$P_{\mathrm{cl}} = P_0\,\eta^2. \tag{2.144}$$

We shall actually see in Sect. 5.2 a more careful way to approximate Eq. (2.140) for an ultra-relativistic charge which takes into account the conservation of the on-shell condition and the fact that the radiation-reaction force should be a 4-force.

2.5.3 Charged Particle in a Constant-Uniform Magnetic Field with RR

The LL equation for a charged particle in a constant-uniform magnetic field reads in 3-vector notation

$$\frac{d\mathbf{u}}{d\tau} = \frac{q}{m}(\mathbf{u}\times\mathbf{B}) - \frac{\tau_0}{m}\mathbf{B}\times(\boldsymbol{\beta}\times\mathbf{B}) - \frac{\tau_0}{m}(\mathbf{u}\times\mathbf{B})^2\,\mathbf{u}. \tag{2.145}$$

We decompose the 4-velocity into its transverse component to the magnetic field $\mathbf{u}_\perp = (u_x, u_y)$ and its longitudinal component u_z, we get

$$\frac{d\mathbf{u}_\perp}{d\tau} = \frac{q}{m}(\mathbf{u}_\perp\times\mathbf{B}) - \frac{\tau_0}{m}B^2\,(1+\mathbf{u}_\perp^2)\,\mathbf{u}_\perp, \tag{2.146a}$$

$$\frac{du_z}{d\tau} = -\frac{\tau_0}{m} B^2 \mathbf{u}_\perp^2 u_z. \tag{2.146b}$$

Going to polar coordinates in the plane orthogonal to the magnetic field, we write $(u_x, u_y) = u_\perp(\cos\varphi, \sin\varphi)$ where $\varphi = \Omega\tau$ and $\Omega = qB/m$ is the rotation frequency in the particle's rest frame. Equation (2.146) then takes the form

$$\frac{du_\perp}{d\varphi} = -K\,(1 + u_\perp^2)\,u_\perp, \tag{2.147a}$$

$$\frac{du_z}{d\varphi} = -K\,u_\perp^2\,u_z, \tag{2.147b}$$

where $K = \tau_0 B^2/(\mathrm{m}\Omega)$. Equation (2.147a) admits a separation of variables and integrates into

$$u_\perp(\varphi) = \frac{u_{\perp 0}}{\sqrt{(1 + u_{\perp 0}^2)\,e^{2K(\varphi-\varphi_0)} - u_{\perp 0}^2}}. \tag{2.148}$$

Injecting this solution into Eq. (2.147b), the obtained equation admits a separation of variables and integrates into

$$u_z(\varphi) = \frac{u_{z0}\,e^{K(\varphi-\varphi_0)}}{\sqrt{(1 + u_{\perp 0}^2)\,e^{2K(\varphi-\varphi_0)} - u_{\perp 0}^2}}. \tag{2.149}$$

The 4-velocity thus finally reads

$$\boxed{\mathbf{u}(\varphi) = \frac{1}{\sqrt{(1 + u_{\perp 0}^2)\,e^{2K(\varphi-\varphi_0)} - u_{\perp 0}^2}} \left(u_{\perp 0}\cos(\varphi - \varphi_0),\, u_{\perp 0}\sin(\varphi - \varphi_0),\, u_{z0}\,e^{K(\varphi-\varphi_0)}\right)}$$

Direct integration of $\mathbf{u}$ with respect to φ yields

$$\begin{aligned} \begin{bmatrix} x(\varphi) - x_0 \\ y(\varphi) - y_0 \end{bmatrix} = \begin{bmatrix} \Re \\ \Im \end{bmatrix} \Bigg[\frac{i}{u_{\perp 0}} \sqrt{(1 + u_{\perp 0}^2)\,e^{2K(\varphi'-\varphi_0)} - u_{\perp 0}^2} \\ \times\ {}_2F_1\left(1, \frac{1}{2} + \frac{i}{2K}, 1 + \frac{i}{2K}, \frac{(1 + u_{\perp 0}^2)}{u_{\perp 0}^2} e^{2K(\varphi'-\varphi_0)}\right)\Bigg]\Bigg|_{\varphi_0}^{\varphi}. \end{aligned} \tag{2.150}$$

Equation (2.150) is exact, but not easy to interpret. Let us derive an approximate solution for the electron trajectory that will be transparent to interpret. The equation of evolution of the electron Lorentz factor reads

$$\frac{d\gamma}{dt} = K(1 - \gamma^2), \tag{2.151}$$

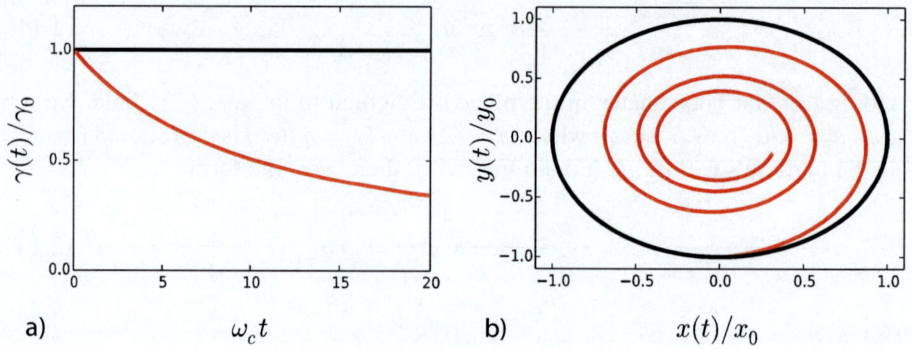

Fig. 2.8 Normalized energy a) and trajectory b) with (red line) and without (black line) RR for an electron in a constant uniform magnetic field with quantum parameter $\chi = 10^{-2}$

with $K = AB^2/B_{\rm cr}^2$ and $A = \frac{2}{3}\frac{\alpha^2}{\tau_0}$. This easily integrates into

$$\gamma(t) = \tanh\left(\operatorname{artanh}\gamma_0 + Kt\right), \tag{2.152}$$

which approximates to

$$\gamma(t) \simeq \frac{\gamma_0}{1 + \gamma_0 K t}, \tag{2.153}$$

for $\gamma_0 \gg 1$ where $\gamma_0 = \gamma(0)$. Assuming that the electron motion with RR is close enough to its trajectory without RR over one period, we can write

$$\mathbf{r}(t) = \frac{r_0}{1 + \gamma_0 K t}\left[\sin(\omega_c t), -\cos(\omega_c t)\right]. \tag{2.154}$$

The energy and trajectory are plotted with and without RR on Fig. 2.8 for an electron with initial quantum parameter $\chi_0 = 10^{-2}$.

2.5.4 *Charged Particle in a Plane-Wave Field with RR*

We consider the LL equation

$$a^\mu = f^{\mu\nu} u_\nu + \tau_0 \left[\partial_\eta f^{\mu\nu} u_\nu u^\eta - f^{\mu\nu} f_{\eta\nu} u^\eta + (f^{\nu\eta} u_\eta)(f_{\nu\alpha} u^\alpha) u^\mu\right].$$

The solution of this equation in a Plane-Wave field has been derived for the first time by [31]. We reproduce here the main steps (see Appendix A.8 for more details). Following the same steps as for the case without RR, we start by computing $d(ku)/d\tau$. This quantity ku is however here not constant and follows

$$-\frac{1}{(ku)^2}\frac{(ku)}{d\phi} = \frac{\tau_0}{\rho_0} f^{2\mu\nu} u_{0,\nu}, \tag{2.155}$$

which readily integrates into $ku(\phi) = \frac{\rho_0}{h(\phi,\phi_0)}$, where $\rho_0 = ku_0$ and

$$h(\phi, \phi_0) = 1 + \tau_0 \int_{\phi_0}^{\phi} f^{2\mu\nu}(\varphi, \phi_0) u_{0,\nu}. \tag{2.156}$$

Defining the "reduced" speed as $\tilde{u}^\mu = h(\phi, \phi_0) u^\mu$, we can show that it satisfies the following differential equation

$$\frac{d\tilde{u}^\mu}{d\phi} = \frac{1}{\rho_0}\left[h(\phi, \phi_0) f^{\mu\nu}(\phi) + \tau_0\rho_0 f'^{\mu\nu}(\phi)\right]\tilde{u}_\nu + h(\phi, \phi_0)\frac{\tau_0}{\rho_0} f^{2\mu\nu}(\phi) u_{0,\nu}, \tag{2.157}$$

where the prime denotes the derivation with respect to the phase ϕ. The solution to this equation can be obtained using Picard's method[22] and the orthogonal properties of the constant field tensors $\mathsf{f}_i^{\mu\nu}$ (see Appendix A.3.2). The exact solution is then given by

$$\boxed{\begin{aligned} h(\phi, \phi_0) u^\mu(\phi) = u_0^\mu &+ \frac{\mathcal{F}_{\mathrm{RR}}^{\mu\nu}(\phi, \phi_0) u_{0,\nu}}{ku_0} + \frac{\mathcal{F}_{\mathrm{RR}}^{2\mu\nu}(\phi, \phi_0) u_{0,\nu}}{2(ku_0)^2} \\ &+ \frac{1}{2(ku_0)}\left[h^2(\phi, \phi_0) - 1\right] k^\mu \end{aligned}} \tag{2.158}$$

with

$$\mathcal{F}_{\mathrm{RR}}^{\mu\nu}(\phi, \phi_0) \equiv \mathcal{I}_i(\phi, \phi_0)\mathcal{F}^{\mu\nu}(\phi, \phi_0), \tag{2.159}$$

$$\mathcal{I}_i(\phi, \phi_0) \equiv \frac{1}{[\psi_i(\phi) - \psi_i(\phi_0)]} I_i(\phi, \phi_0), \tag{2.160}$$

$$\begin{aligned} I(\phi) &\equiv \int_{\phi_0}^{\phi} d\varphi \left[h(\varphi)\psi_i'(\varphi) + \tau_0\rho_0\psi_i''(\varphi)\right] \\ &= h(\phi)\psi_i(\phi) - \psi_i(\phi_0) + \tau_0\rho_0\left[\psi_i'(\phi) - \psi_i'(\phi_0)\right] \\ &+ \tau_0\rho_0 \int_{\phi_0}^{\phi} d\varphi\, \psi_i(\varphi) \sum_{j=1,2} a_j^2 \psi_j'^2(\varphi). \end{aligned} \tag{2.161}$$

The trajectory of the electron is then deduced by

$$x^\mu(\phi) = x_0^\mu + \frac{1}{\rho_0}\int_{\phi_0}^{\phi} d\varphi\, h(\varphi) u^\mu(\varphi). \tag{2.162}$$

[22] Only two iterations are needed.

In the limit where $\tau_0 \to 0, h(\phi) = \mathcal{I}_i(\phi) = 1 \implies \mathcal{F}^{\mu\nu}_{\text{RR}} \to \mathcal{F}^{\mu\nu}$ and we recover the case without RR (2.79) as expected.

In the case of an initially ultra-relativistic electron counter-propagating with the plane-wave propagating in the direction $\hat{\mathbf{k}} = -\hat{\mathbf{z}}$, we have $\beta_0 \simeq 1 - 1/(2\gamma_0)$ and therefore $ku_0 = \omega_0 h_0 = \omega_0\gamma_0(1+\beta_0)$. We consider the two simple cases of a linearly and circularly polarized plane-wave.

- Linearly polarized plane-wave ($\delta = 1$)

$$\begin{aligned}
\gamma(\phi) &= \frac{1}{h(\phi)}\left[\gamma_0 + \frac{1}{2h_0}\left[\left(h^2(\phi)-1\right) + a_0^2 I_1(\phi)^2\right]\right],\\
u_x(\phi) &= -\frac{1}{h(\phi)} a_0 I_1(\phi),\\
u_y(\phi) &= 0,\\
u_z(\phi) &= \frac{1}{h(\phi)}\left[\gamma_0\beta_0 - \frac{1}{2h_0}\left[\left(h^2(\phi)-1\right) + a_0^2 I_1(\phi)^2\right]\right],\\
h(\phi) &= 1 - \frac{1}{2}\tau_0\rho_0 a_0^2 \sin 2\phi - \frac{1}{2}\tau_0\rho_0 a_0^2\phi,\\
I_1(\phi) &= \sin\phi - \frac{1}{2}\tau_0\rho_0 a_0^2\phi\sin\phi + \tau_0\rho_0\left[1 - \frac{3}{8}a_0^2\right]\cos\phi - \frac{1}{6}\tau_0\rho_0 a_0^2\cos 3\phi,\\
I_2(\phi) &= 0.
\end{aligned}$$

- Circularly polarized plane-wave ($\delta = 1/\sqrt{2}$)

$$\begin{aligned}
\gamma(\phi) &= \frac{1}{h(\phi)}\left[\gamma_0 + \frac{1}{2h_0}\left(h^2(\phi)-1\right) + \frac{a_0^2}{4h_0}\left(I_1(\phi)^2 + I_2(\phi)^2\right)\right],\\
u_x(\phi) &= -\frac{1}{h(\phi)}\frac{a_0}{\sqrt{2}} I_1(\phi),\\
u_y(\phi) &= -\frac{1}{h(\phi)}\frac{a_0}{\sqrt{2}} I_2(\phi),\\
u_z(\phi) &= \frac{1}{h(\phi)}\left[\gamma_0\beta_0 - \frac{1}{2h_0}\left(h^2(\phi)-1\right) - \frac{a_0^2}{4h_0}\left(I_1(\phi)^2 + I_2(\phi)^2\right)\right],\\
I_1(\phi) &= \frac{1}{\sqrt{2}}\sin\phi - \frac{1}{2\sqrt{2}}\tau_0\rho_0 a_0^2\phi\sin\phi + \frac{1}{\sqrt{2}}\tau_0\rho_0\left[1 - \frac{1}{2}a_0^2\right]\cos\phi,\\
I_2(\phi) &= -\frac{1}{\sqrt{2}}\cos\phi + \frac{1}{2\sqrt{2}}\tau_0\rho_0 a_0^2\phi\cos\phi + \frac{1}{\sqrt{2}}\tau_0\rho_0\sin\phi,
\end{aligned}$$

2.5.5 *Limit of Applicability of CED and Classical Radiation Dominated Regime (CRDR)*

All the results derived in this Chapter are valid in the realm of CED, that is to say when the electron quantum parameter is small compared to unity. As will be seen in the next Chapter, when the quantum parameter approaches unity, most of the electron's energy can be emitted in one single photon, making the classical picture of the electromagnetic field no longer valid. Moreover, in such a situation, it is impossible to model RR like a continuous friction force like was done in this Chapter. Although it is impossible, in the realm of CED, that the electron loses most of its energy in one single photon emission, it is possible to radiate most of its energy in one laser period in many photons, each having a small fraction of the electron's energy. This is the so-called Classical Radiation Dominated Regime (CRDR) [25, 31–33]. Using Eq. (2.62) to evaluate the classical radiated power, the Classical Radiation Dominated Regime (CRDR) is described by

$$R_c = \alpha \chi_0 a_0 \sim 1, \quad \chi_0 \ll 1. \tag{2.163}$$

References

1. Jackson JD (1999) Classical Electrodynamics, 3rd edn. Wiley, Hoboken
2. Landau LD, Lifshitz EM (1947) The Classical Theory of Fields. Butterworth-Heinemann, Oxford
3. Rohrlich F (1964) Classical Charged Particles. World Scientific Publishing Co. Inc.
4. Landau LD, Lifshitz EM (1960) Mechanics, vol 1. Pergamon Press, Oxford
5. Lagrange J-L (1811) Mécanique analytique. Courcier
6. Noether E (1918) Invariante variationsprobleme. Math Phys Klasse 1918:235–257
7. Maggiore M (2005) A Modern Introduction to Quantum Field Theory. Oxford University Press, Oxford
8. Soper DE (1976) Classical Field Theory. Dover Books on Physics. Dover Publications
9. Bañados M, Reyes I (2016) A short review on Noether's theorems, gauge symmetries and boundary terms. Int J Mod Phys D 25(10):1630021
10. Poincaré H (1906) Sur la dynamique de l'électron. Rendiconti del Circolo matematico di Palermo 21:129–176
11. Peskin ME, Schroeder DV (1995) An Introduction to Quantum Field Theory. Addison-Wesley Publishing Company, Boston
12. Schwartz M (2014) Quantum Field Theory and the Standard Model. Quantum Field Theory and the Standard Model. Cambridge University Press, Cambridge
13. Zee A (2016) Group Theory in a Nutshell for Physicists. Princeton University Press, Princeton
14. Maxwell JC (1865) A dynamical theory of the electromagnetic field. Philos Trans R Soc B 155:459–512
15. Belinfante FJ (1940) On the current and the density of the electric charge, the energy, the linear momentum and the angular momentum of arbitrary fields. Physica 7:449
16. Rosenfeld L (1940) Sur le tenseur d'impulsion-energie. Acad. Roy. Belg. Memoirs de classes de Science 18
17. Rougé A (2001) Introduction à la relativité. Ecole Polytechnique

18. Purcell E (2011) Electricity and Magnetism. Cambridge University Press, Cambridge
19. Coleman S (1961) Classical electron theory from a modern standpoint. RAND Memorandum RM-2820-PR
20. Sarachik ES, Schappert GT (1970) Classical theory of the scattering of intense laser radiation by free electrons. Phys Rev D 1:2738
21. Meyer JW (1971) Covariant classical motion of electron in a laser beam. Phys Rev D 3:621
22. Esarey E (1993) Nonlinear Thomson scattering of intense laser pulses from beams and plasmas. Phys Rev E 48:3003
23. Salamin Y (1998) Generation of Compton harmonics by scattering linearly polarized light of arbitrary intensity from free electrons of arbitrary initial velocity. J Phys A Math Gen 31(4):1319
24. Schwinger J (1949) On the classical radiation of accelerated electrons. Phys Rev 75:1912
25. Piazza D, Müller C, Hatsagortsyan K, Keitel CH (2012) Extremely high-intensity laser interactions with fundamental quantum systems. Rev Mod Phys 84:1177
26. Dirac PAM (1938) Classical theory of radiating electrons. Proc R Soc Lond A 167:148–169
27. Nakhleh CW (2013) The Lorentz-Dirac and Landau-Lifshitz equations from the perspective of modern renormalization theory. Am J Phys 81:180
28. Weinberg S (2005) The Quantum Theory of Fields. Cambridge University Press, Cambridge
29. Bhabha HJ (1946) On the expansibility of solutions in powers of the interaction constants. Phys Rev 70(9–10):759
30. Spohn H (2000) The critical manifold of the Lorentz-Dirac equation. EPL (Europhys Lett) 50(3):287
31. Di Piazza A (2008) Exact solution of the Landau-Lifshitz equation in a plane wave. arXiv
32. Shen CS (1970) Synchrotron emission at strong radiative damping. Phys Rev Lett 24(8):410
33. Koga J, Esirkepov TZ, Bulanov SV (2005) Nonlinear Thomson scattering in the strong radiation damping regime. Phys Plasmas 12:093106

Chapter 3
Quantum Electrodynamics

There are no real one-particle systems in nature, not even few-particle systems. The existence of virtual pairs and of pair fluctuations shows that the days of fixed particle numbers are over.

Victor Weisskopf

3.1 Outline

In the previous chapter, we described Classical Field Theory (CFT) and its application to the electromagnetic interaction: Classical ElectroDynamics (CED). This description was purely classical. However, we know that when the energy of the photon emitted by an accelerated electron is no longer negligible compared to the energy of the electron itself, the particle-like character of light starts to manifest itself and we need a quantum description to study this phenomenon. Quantum Field Theory (QFT) is the synthesis of quantum mechanics and CFT and it has been discussed in many textbooks (See, e.g. the seminal books by [1–3]). However, QFT and more precisely Quantum Electrodynamics (or *QED*, the QFT of the electromagnetic interaction) only starts recently to be of use in laser-plasma interaction. Our goal here is therefore to provide a self-consistent summary for researchers and students that are not familiar with this theory. In order to maintain the presentation both logical and compact, not all derivations presented here are straightforward: some results may require quite some algebra before being derived. I have however tried to state clearly all assumptions used in the derivations, and to provide physically intuitive pictures when possible in order to get a qualitative idea before going to the computational part. Moreover, the focus of this chapter is also Strong Field QED (SFQED), which differs from vacuum QED, in that the former takes place in a strong background

F. Niel, *Classical and Quantum Description of Plasma and Radiation in Strong Fields*, Springer Theses, https://doi.org/10.1007/978-3-030-73547-0_3

electromagnetic field (instead of vacuum). Strong field QED has (perhaps surprisingly) been discussed in very few books and one usually has to refer to articles or PhD thesis. We list here some references that can help the interested reader. The seminal results of SFQED were derived by [4]. The excellent textbook by [5] gives a consistent exposure of SFQED from first principle (In addition, it treats the case of unstable vacuum which we don't consider in this work). Finally, I would like to note that I have benefited a lot from the excellent thesis by [6–9].

This chapter is structured as follows. Section 3.2 introduces some general concepts of QFT, such as canonical quantization, Fock spaces and the scattering experiment that is central to the S-matrix. We then apply these results to QED and in particular, we discuss the Dirac Lagrangian, equation of motion and propagator in Sect. 3.3. We then turn to SFQED which is central to this thesis. In particular, the background electromagnetic field is described by a coherent state, which is the quantum state that is the closest to a classical one. The Fock states are generalized to *semi-coherent* states in order to take into account the presence of the background field (Sect. 3.4.1). Finally the Feynman rules in the Furry picture are enunciated and used to derive the different probabilities for nonlinear Compton scattering (Sect. 3.5) and nonlinear Breit-Wheeler pair production (Sect. 3.6).

3.2 Quantum Field Theory

3.2.1 Canonical Quantization of Classical Fields

In order to quantize a set of classical non-interacting bosonic $[\phi_i(x)]_{i=1..n}$ and fermionic $[\psi_j(x)]_{j=1..m}$ fields described by a free Lagrangian density $\mathcal{L}_0(\phi_i, \partial_\mu\phi_i, \psi_i, \partial_\mu\psi_i)$, we need to promote them to operators and to impose to them the equal time commutation (for bosonic fields) and anticommutation[1] (for fermionic fields) relations [1–3]

$$\left[\phi_i(x), \Pi_j(x')\right]_{x_0=x_0'} = i\,\delta_{ij}\,\delta^3(\mathbf{x}-\mathbf{x}'), \tag{3.1a}$$

$$\left\{\psi_i(x), \Xi_j(x')\right\}_{x_0=x_0'} = i\,\delta_{ij}\,\delta^3(\mathbf{x}-\mathbf{x}'), \tag{3.1b}$$

while

$$\left[\phi_i(x), \phi_j(x')\right] = 0, \qquad \left[\Pi_i(x), \Pi_j(x')\right] = 0, \tag{3.2a}$$

$$\left\{\psi_i(x), \psi_j(x')\right\} = 0, \qquad \left\{\Xi_i(x), \Xi_j(x')\right\} = 0, \tag{3.2b}$$

where Π_i (resp. Ξ_j) is the conjugate momentum associated to ϕ_i (resp. ψ_j) defined as

[1]As imposed by the spin-statistics theorem.

$$\Pi_i \equiv \frac{\partial \mathcal{L}_0}{\partial(\partial_0 \phi_i)}, \tag{3.3a}$$

$$\Xi_i \equiv \frac{\partial \mathcal{L}_0}{\partial(\partial_0 \psi_i)}. \tag{3.3b}$$

Expanding ϕ_i and ψ_j in Fourier components yields

$$\phi_i(x) = \int \frac{d^3k}{(2\pi)^3\, 2k^0} \left(a_{i,\mathbf{k}}\, \tilde{\phi}_{i,\mathbf{k}}(x) + a^\dagger_{i,\mathbf{k}}\, \tilde{\phi}^*_{i,\mathbf{k}}(x) \right), \tag{3.4a}$$

$$\psi_j(x) = \int \frac{d^3p}{(2\pi)^3\, 2p^0} \left(b_{j,\mathbf{p}}\, \tilde{\psi}_{j,\mathbf{p}}(x) + b^\dagger_{j,\mathbf{p}}\, \tilde{\psi}^*_{j,\mathbf{p}}(x) \right), \tag{3.4b}$$

where $\tilde{\phi}_{i,\mathbf{k}}(x)$ and $\tilde{\psi}_{i,\mathbf{p}}(x)$ are solutions of the momentum space bosonic and fermionic free equations of motion (2.2). The coefficients $a^\dagger_{\mathbf{k}}$ and $a_{\mathbf{k}}$ (resp. $b^\dagger_{\mathbf{p}}$ and $b_{\mathbf{p}}$) are interpreted as bosonic (resp. fermionic) creation and destruction operators which satisfy the commutation (resp. anticommutation) relations[2]

$$[a_{i,\mathbf{k}}, a^\dagger_{j,\mathbf{k}'}] = (2\pi)^3\, \delta_{ij}\, \delta^{(3)}(\mathbf{k} - \mathbf{k}'), \qquad [a_{i,\mathbf{k}}, a_{j,\mathbf{k}'}] = 0 \qquad [a^\dagger_{i,\mathbf{k}}, a^\dagger_{j,\mathbf{k}'}] = 0, \tag{3.5a}$$

$$\{b_{i,\mathbf{p}}, b^\dagger_{j,\mathbf{p}'}\} = (2\pi)^3\, \delta_{ij}\, \delta^{(3)}(\mathbf{p} - \mathbf{p}'), \qquad \{b_{i,\mathbf{p}}, b_{j,\mathbf{p}'}\} = 0 \qquad \{b^\dagger_{i,\mathbf{p}}, b^\dagger_{j,\mathbf{p}'}\} = 0. \tag{3.5b}$$

3.2.2 *Fock Space*

In QFT, particles are defined as excitations of quantum fields[3] and more precisely, the creation and annihilation operators defined in (3.4a) can be used to define a space of eigenstates of the free Hamiltonian.[4] For example, one-particle states would be defined as $|\mathbf{p}\rangle = a^\dagger_{\mathbf{p}}|0\rangle$. More generally, the state of a system of any given number $n_{\mathbf{k}}$ of bosons in the momentum mode $\mathbf{k}$ and $n_{\mathbf{p}}$ fermions in the momentum mode $\mathbf{p}$ is described by the so-called *Fock state* [10]

$$|\mathbf{k}_1..., \mathbf{p}_1, ...\rangle = \prod_{\mathbf{k},\mathbf{p}} \frac{(a^\dagger_{\mathbf{k}})^{n_{\mathbf{k}}}}{\sqrt{n_{\mathbf{k}}!}} \frac{(b^\dagger_{\mathbf{p}})^{n_{\mathbf{p}}}}{\sqrt{n_{\mathbf{p}}!}} |0\rangle\,, \tag{3.6}$$

---

[2] Note that this quantization procedure can be complicated by the vector nature of the field. We meet this difficulty in the quantization of the spin-1 particles.

[3] The consistency of quantum mechanics with special relativity indeed forces us to abandon the interpretation of the single-particle interpretation of the wave function (see Klein paradox and Schwinger effect in Chap. 1).

[4] For an interacting system, it is not possible to enumerate the number of particles in a given state since quantum fluctuations may temporarily create additional virtual particles. From a mathematical point of view, the equation of motion of an interacting field will be non-linear. The simple plane-wave expansion (3.4a) used for non-interacting fields where the coefficients are interpreted as creation and annihilation operators will therefore no longer be possible which renders the following construction impossible.

where $|0\rangle$ is the vacuum state of the free theory, defined as the state annihilated by all destruction operators $a_{\mathbf{k}} |0\rangle = b_{\mathbf{k}} |0\rangle = 0$.

3.2.3 The S-Matrix

So far, we have only considered non-interacting systems, which are of course of limited interest in practice. We will develop here the formalism used to describe interacting systems.

The Scattering Experiment

A key experiment in QFT is the scattering experiment in which we assume that the actual interaction region is confined in a small region in space and time [11]. The interaction term is therefore assumed to turn off asymptotically in the remote past and future and that, there, the theory is a free theory whose spectrum is made of eigenstates of the free Hamiltonian. We write ϕ the full interacting field and ϕ_{in} and ϕ_{out} the free fields in the far past and future. We then have[5]

$$\lim_{x_0 \to -\infty} \phi(x) = \phi_{\text{in}}(x), \tag{3.7a}$$

$$\lim_{x_0 \to +\infty} \phi(x) = \phi_{\text{out}}(x). \tag{3.7b}$$

Assuming that the Lagrangian density $\mathcal{L}$ of the interacting system is such that $\mathcal{L} \equiv \mathcal{L}_0 + \mathcal{L}_{\text{int}}$ where $\mathcal{L}_0$ is the free Lagrangian density and $\mathcal{L}_{\text{int}}$ the interaction term, then the corresponding Hamiltonian density reads $\mathcal{H} = \mathcal{H}_0 + \mathcal{H}_{\text{int}}$, where[6] $\mathcal{H}_{\text{int}} = -\mathcal{L}_{\text{int}}$. The evolution of the different fields is then determined, in the Heisenberg picture, by

$$\phi(t, \mathbf{x}) = e^{iH(t-t_0)}\, \phi(t_0, \mathbf{x})\, e^{-iH(t-t_0)}, \tag{3.8a}$$

$$\phi_{\text{in}}(t, \mathbf{x}) = e^{iH_0(t-t_0)}\, \phi_{\text{in}}(t_0, \mathbf{x})\, e^{-iH_0(t-t_0)}, \tag{3.8b}$$

$$\phi_{\text{out}}(t, \mathbf{x}) = e^{iH_0(t-t_0)}\, \phi_{\text{out}}(t_0, \mathbf{x})\, e^{-iH_0(t-t_0)}, \tag{3.8c}$$

where the Hamiltonian H is related to the Hamiltonian density $\mathcal{H}$ by $H = \int d^3x\, \mathcal{H}$. The physical quantities that we wish to compute are the transition amplitudes

$${}_{\text{out}}\langle \mathbf{q}_1, \mathbf{q}_2, ..., \mathbf{q}_i | \mathbf{p}_1, \mathbf{p}_2, ..., \mathbf{p}_f \rangle_{\text{int}}, \tag{3.9}$$

whose square modulus will enter in cross-sections [1]. The states $|\mathbf{p}_1, \mathbf{p}_2, ..., \mathbf{p}_f\rangle_{\text{in}}$ and $|\mathbf{q}_1, \mathbf{q}_2, ..., \mathbf{q}_i\rangle_{\text{out}}$ are asymptotic states that belong to the free in and out Fock

[5] We ignore here the issue of field renormalization for the sake of simplicity.

[6] Note that this relation holds only if the interaction Lagrangian does not contain any derivative of the fields.

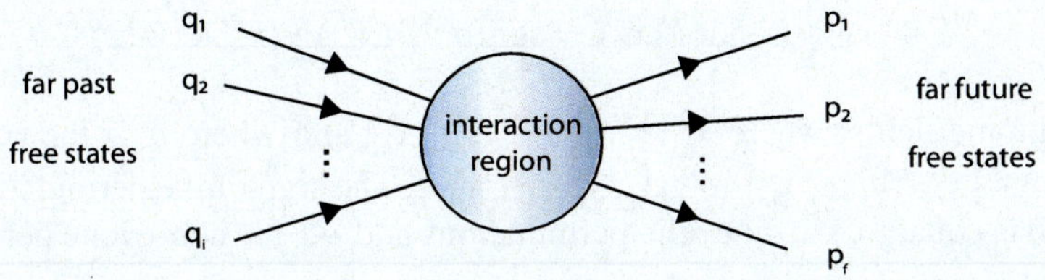

Fig. 3.1 Because of quantum fluctuations, an interacting system does not have, in QFT a given, fixed number of particle. In order to overcome this problem, we consider *scattering experiments* in which we only consider asymptotic free states. Given a free state in the far past, we want to compute the probability of each outcome in the far future. This is done using the *S matrix* (3.18)

spaces (3.6). In order to compute these transition amplitudes, we need to relate these free states to the full interacting state ϕ.

The Interaction Picture

In order to compute (3.9), we need to relate the interacting field ϕ to a field ϕ_I, called the *interaction picture* field, that evolves according to the free Hamiltonian H_0

$$\phi_I(t, \mathbf{x}) = e^{iH_0\tau}\, \phi_I(t_0, \mathbf{x})\, e^{-iH_0\tau}, \tag{3.10}$$

where $\tau = t - t_0$. We can therefore link this field to the full interacting field as

$$\phi(t, \mathbf{x}) = e^{iH\tau}\, \phi(t_0, \mathbf{x})\, e^{-iH\tau} = e^{iH\tau} e^{-iH_0\tau}\, \phi_I(t_0, \mathbf{x})\, e^{iH_0\tau} e^{-iH\tau}, \tag{3.11}$$

from which we define the time evolution operator

$$U(t, t_0) \equiv e^{iH_0\tau} e^{-iH\tau}. \tag{3.12}$$

It is possible to show that it satisfies the following differential equation

$$i\frac{d}{dt}U(t, t_0) = H_I(t)U(t, t_0), \tag{3.13}$$

where $H_I = e^{iH_0\tau} H_{\text{int}} e^{-iH_0\tau}$ is the interaction picture Hamiltonian. We thus deduce

$$U(t, t_0) = \mathcal{T} \exp\left[-i\int_{t_0}^{t} dt' H_I(t')\right], \tag{3.14}$$

where $\mathcal{T}$ is the so-called *time-ordering operator* defined as

$$\mathcal{T}\, \phi(x_1)\phi(x_2)...\phi(x_n) = \phi(x_{\sigma_1})\phi(x_{\sigma_2})...\phi(x_{\sigma_n}) \tag{3.15}$$

for bosonic fields and

$$\mathcal{T}\,\psi(x_1)\psi(x_2)...\psi(x_n) = \mathrm{sgn}(\sigma)\,\psi(x^0_{\sigma_1})\psi(x^0_{\sigma_2})...\psi(x^0_{\sigma_n}) \tag{3.16}$$

for fermionic fields, where $x^0_{\sigma_1} \leq x^0_{\sigma_2} \leq \ldots \leq x^0_{\sigma_n}$ and where σ is the permutation that brings $\{x^0_1, x^0_2, ..., x^0_n\}$ to $\{x^0_{\sigma_1}, x^0_{\sigma_2}, ..., x^0_{\sigma_n}\}$. The sign of the permutation $\mathrm{sgn}(\sigma)$ in (3.16) is equal to $+1$ for cyclic permutations and -1 for non-cyclic permutations.

The S-Matrix

Writing $|i\rangle_{\mathrm{in}}$ and $|f\rangle_{\mathrm{out}}$ a generic in and out state, they are therefore related by

$$|f\rangle_{\mathrm{out}} = U(-\infty, +\infty)\,|i\rangle_{\mathrm{in}}\,. \tag{3.17}$$

The operator $S \equiv U(-\infty, +\infty)$ is called the *S-matrix*. Using $\mathcal{L}_I = -\mathcal{H}_I$ the interaction picture Lagrangian, it reads

$$\boxed{S = \mathcal{T}\exp\left[i\int d^4x\,\mathcal{L}_I\right]} \tag{3.18}$$

However, no exact solution of S is known. We therefore need to rely on perturbation theory. We write

$$S = \sum_{n=0}^{+\infty} S^{(n)}, \tag{3.19}$$

where $S^{(n)} = \frac{i^n}{n!}\int d^4x_1 \int d^4x_2 ... \int d^4x_n\,\mathcal{T}\,[\mathcal{L}_I(x_1)\mathcal{L}_I(x_2)...\mathcal{L}_I(x_n)]$. Going back to the transition (3.9), we have

$$_{\mathrm{out}}\langle f|i\rangle_{\mathrm{int}} = {}_{\mathrm{out}}\langle f|S|i\rangle_{\mathrm{out}} = {}_{\mathrm{in}}\langle f|S|i\rangle_{\mathrm{in}}\,, \tag{3.20}$$

which in perturbation theory yields

$$S_{fi} = \sum_{n=0}^{+\infty} S^{(n)}_{fi}, \tag{3.21}$$

with $S^{(n)}_{fi} = {}_{\mathrm{in}}\langle f|S^{(n)}|i\rangle_{\mathrm{in}}$. This procedure leads to the well known Feynman diagrams [2, 12] that we will describe in more details in the case of strong field QED (SFQED).

3.3 Quantum Electrodynamics

QED is described by the following Lagrangian density [13]

$$\mathcal{L}_{\text{QED}} = \underbrace{\bar{\Psi}(i\partial\!\!\!/ - m)\Psi}_{\mathcal{L}_{\text{Dirac}}} \underbrace{-\frac{1}{4}F^{\mu\nu}F_{\mu\nu}}_{\mathcal{L}_{\text{field}}} \underbrace{-eA^{\mu}\bar{\Psi}\gamma^{\mu}\Psi}_{\mathcal{L}_{\text{int}}}. \tag{3.22}$$

In the same way as for CED, all the terms in this Lagrangian density will now be discussed one by one.

3.3.1 Lagrangian

Spin $\frac{1}{2}$ particles are described in QED by the so-called *Dirac Lagrangian* [14]

$$\mathcal{L}_{\text{Dirac}} = \bar{\Psi}(i\partial\!\!\!/ - m)\Psi, \tag{3.23}$$

where Ψ is the Dirac field, $\bar{\Psi} = \Psi^{\dagger}\gamma^{0}$ its adjoint and γ^{μ} the so-called *gamma matrices* (see Appendix B.1). Note that this Lagrangian is invariant by global $U(1)$ symmetry, that is to say it is invariant by

$$\Psi \to e^{ie\theta}\Psi, \tag{3.24}$$

where[7] θ an e are two constants.

Spin 1 particles, and in particular here, photons, are described by the Lagrangian of the electromagnetic field

$$\mathcal{L}_{\text{field}} = -\frac{1}{4}F^{\mu\nu}F_{\mu\nu}. \tag{3.25}$$

In the same way as for CED this Lagrangian is gauge invariant, i.e. invariant by

$$A^{\mu} \to A^{\mu} - \partial^{\mu}\theta. \tag{3.26}$$

The Lagrangian obtained by combining (3.23) and (3.25) is still invariant by global $U(1)$ symmetry, but under a local one. It is indeed transformed as

$$\mathcal{L}_{\text{Dirac}} + \mathcal{L}_{\text{field}} \to \mathcal{L}_{\text{Dirac}} + \mathcal{L}_{\text{field}} + ie\,\partial_{\mu}\theta\,\bar{\Psi}\gamma^{\mu}\Psi, \tag{3.27}$$

where θ is a function of x. In order to make this Lagrangian invariant under local $U(1)$ symmetry, we change all the partial derivatives ∂_{μ} by covariant derivatives D_{μ}. This yields

$$\mathcal{L} = \bar{\Psi}(iD\!\!\!\!/ - m)\Psi - \frac{1}{4}F^{\mu\nu}F_{\mu\nu}, \tag{3.28}$$

[7] We will see that q is associated to the electric charge.

where

$$D_\mu \Psi \equiv (\partial_\mu - ieA_\mu)\Psi. \tag{3.29}$$

We say that we have *gauged* the global $U(1)$ symmetry.[8] Expanding the covariant derivative in this Lagrangian, we get the Lagrangian of QED

$$\boxed{\mathcal{L}_{\text{QED}} = \bar{\Psi}(i\not{\partial} - m)\Psi - \frac{1}{4}F^{\mu\nu}F_{\mu\nu} - eA^\mu\bar{\Psi}\gamma^\mu\Psi} \tag{3.30}$$

Thus, the electromagnetic field is coupled to the Dirac field by the interacting term

$$\mathcal{L}_{\text{int}} = -eA^\mu\bar{\Psi}\gamma^\mu\Psi. \tag{3.31}$$

To the Dirac field, we therefore associate the following current

$$j^\mu_{\text{Dirac}} = \bar{\Psi}\gamma^\mu\Psi, \tag{3.32}$$

from which we deduce the conserved charge

$$Q_{\text{Dirac}} = \int d^3x\, \Psi^\dagger\Psi. \tag{3.33}$$

This current can also be derived from Noether's theorem applied to the global $U(1)$ symmetry.

3.3.2 Equations of Motion

Application of ELE to Eq. (3.23) yields the Dirac equation

$$\boxed{(i\not{\partial} - m)\Psi = 0} \tag{3.34}$$

and the corresponding one for its adjoint

$$\bar{\Psi}(i\overleftarrow{\not{\partial}} + m) = 0. \tag{3.35}$$

The solution of the free Dirac equation is given by

$$\Psi(x) = u_s(\mathbf{p})\, e^{-ipx}, \tag{3.36a}$$
$$\Psi(x) = v_s(\mathbf{p})\, e^{ipx}. \tag{3.36b}$$

[8]The obtained theory is called a *gauge theory* and A^μ a *gauge field*.

They are interpreted respectively as particles and anti-particles of energy $\epsilon_p = \sqrt{\mathbf{p}^2 + m^2}$ and spin $s = \pm\frac{1}{2}$. The components $u_s(\mathbf{p})$ and $v_s(\mathbf{p})$ are called *Dirac spinors* (see Appendix B.3 for more details.).

As mentioned Chap. 1, in QFT wave functions are no longer sufficient to describe the physical situation and we need to turn to quantum fields. To do so, we follow the procedure described in Sect. 3.2 and promote Ψ and $\bar{\Psi}$ to operators and impose to them the canonical anti-commutation relations

$$\left\{\Psi_i(x), \Psi_j^\dagger(x')\right\}_{x_0=x_0'} = \delta_{ij}\, \delta^{(3)}(\mathbf{x} - \mathbf{x}'), \tag{3.37}$$

since

$$\Pi_\Psi = \frac{\partial \mathcal{L}_{\text{Dirac}}}{\partial(\partial_0 \Psi)} = i\bar{\Psi}\gamma^0 = i\Psi^\dagger. \tag{3.38}$$

We obtain the quantized Dirac field

$$\Psi(x) = \int \frac{d^3k}{(2\pi)^3 2k^0} \sum_{s=\pm\frac{1}{2}} \left[c_{\mathbf{p},s} u_s(p) e^{-ipx} + d^\dagger_{\mathbf{p},s} v_s(p) e^{ipx}\right], \tag{3.39a}$$

$$\bar{\Psi}(x) = \int \frac{d^3k}{(2\pi)^3 2k^0} \sum_{s=\pm\frac{1}{2}} \left[c^\dagger_{\mathbf{p},s} \bar{u}_s(p) e^{ipx} + d_{\mathbf{p},s} \bar{v}_s(p) e^{-ipx}\right], \tag{3.39b}$$

where the creation and anhilation operators $\left(c_{\mathbf{p},s}, c^\dagger_{\mathbf{p}',s'}, d_{\mathbf{p},s}, d^\dagger_{\mathbf{p}',s'}\right)$ obey the following anti-commutation relations

$$\{c_{\mathbf{p},s}, c^\dagger_{\mathbf{p}',s'}\} = (2\pi)^3\, (2k_0)\, \delta^{(3)}(\mathbf{p} - \mathbf{p}')\, \delta_{ss'}, \tag{3.40a}$$

$$\{d_{\mathbf{p},s}, d^\dagger_{\mathbf{p}',s'}\} = (2\pi)^3\, (2k_0)\, \delta^{(3)}(\mathbf{p} - \mathbf{p}')\, \delta_{ss'}. \tag{3.40b}$$

Application of ELE to Eq. (3.25) yields Maxwell's equations exactly like in CED (see Chap. 2.). We then need to quantize the obtained field. However, we see that in this case the canonical quantization procedure cannot be straightforwardly applied.[9] In particular, we have

$$\Pi_{A^0} = \frac{\partial \mathcal{L}_{\text{field}}}{\partial \dot{A}^0} = 0. \tag{3.41}$$

Following the *covariant quantization* procedure, we add the so-called *gauge fixing term* $\frac{1}{2\xi}\left(\partial_\mu A^\mu\right)^2$ to Eq. (3.25) and require that $\partial_\mu A^\mu$ vanishes between physical states [15, 16]. Then the obtained quantized electromagnetic field reads

[9] There are many other subtleties in the quantization of gauge fields which we will not detail here. The interested reader is referred to [1–3, 13, 15, 16].

$$A^\mu(x) = \int \frac{d^3k}{(2\pi)^3 2\omega_k} \sum_{\sigma=1,2} \left[c_{\mathbf{k},\sigma}\, \epsilon^\mu_{\mathbf{k},\sigma}\, e^{-ikx} + c^\dagger_{\mathbf{k},\sigma}\, \epsilon^\mu_{\mathbf{k},\sigma}\, e^{ikx} \right], \tag{3.42}$$

where $\epsilon^\mu_{\mathbf{k},\sigma}$ are orthogonal polarization 4-vectors

$$\epsilon^\mu_{\mathbf{k},\sigma}\, \epsilon_{\mu,\mathbf{k}',\sigma'} = -\delta_{\sigma\sigma'}, \tag{3.43}$$

and the creation and annihilation operators follow the commutation relation

$$[c_{\mathbf{k},\sigma}, c^\dagger_{\mathbf{k},\sigma}] = -(2\pi)^3\, 2\omega_{\mathbf{k}}\, \delta^{(3)}(\mathbf{k} - \mathbf{k}')\, \eta_{\sigma\sigma'}. \tag{3.44}$$

3.4 Strong Field QED

The success of vacuum QED relies on the smallness of the fine-structure constant which makes the perturbative development of the S-matrix extremely efficient [1–3, 13]. However, in presence of a strong external electromagnetic field, this perturbative expansion breaks down and we need to develop non-perturbative methods [4, 5, 17–19]. Indeed, the Dirac propagator in presence of an external field $\mathcal{A}_{\text{ext}}$ satisfies the equation

$$\left(i\partial\!\!\!/ - e\mathcal{A}\!\!\!/_{\text{ext}} - m\right) S(x - x') = \delta^{(4)}(x - x'). \tag{3.45}$$

Performing a naive perturbative expansion in $\mathcal{A}_{\text{ext}}$ the so-called *dressed propagator* $S(x - x')$ (represented diagrammatically by the double horizontal line on Fig. 3.2) is expressed as an infinite sum of diagrams involving free electron propagators (simple horizontal lines on Fig. 3.2) and vertices (black dots on Fig. 3.2) representing the coupling with the external field $\mathcal{A}_{\text{ext}}$ (denoted by ⊗ on Fig. 3.2). For each coupling, we obtain one more vertex and free propagator.[10] Each vertex corresponds to $-ie\mathcal{A}\!\!\!/_{\text{ext}} \sim e\mathcal{A}_{\text{ext}}$ while each free propagator is $1/(p\!\!\!/ - m) \sim 1/m$. Hence, the expansion parameter is of the order $e\mathcal{A}_{\text{ext}}/m \equiv a_0 \gg 1$ which makes the perturbative expansion blow up for strong fields. The regime $a_0 \gg 1$ is therefore referred to as *non-perturbative*. We also talk about *multi-photon* or *nonlinear* regime. This is due to the fact that (going back temporarily to SI units)

$$a_0 = \frac{eE_{\text{ext}}}{mc\omega} = \frac{eE\lambda_c}{\hbar\omega}, \tag{3.46}$$

represents the number of photons (of energy $\hbar\omega$ where ω is the frequency of the external field, which we will refer to as "the laser field") absorbed over one Compton wavelength λ_c.

[10] A classical background field is incapable of forming loops.

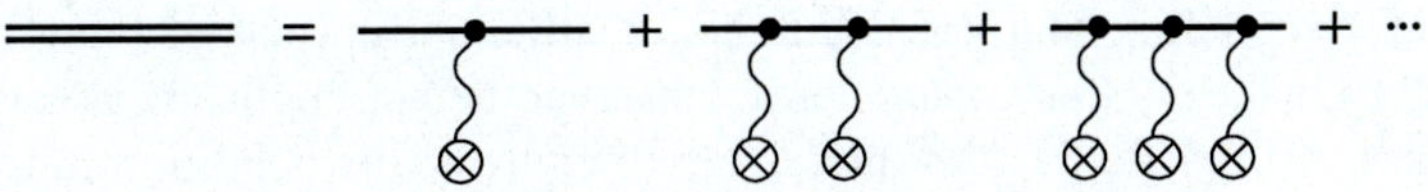

Fig. 3.2 Perturbative expansion of the dressed Dirac propagator (double horizontal line) with respect to the external field $\mathcal{A}_{\text{ext}}$ (denoted by $\otimes$). Each order adds one vertex (black) and one free Dirac propagator (simple black horizontal line) making the nth order being of magnitude $\sim a_0^n$

3.4.1 Coherent States

The usual formulation of QED in which the states belong to a Fock space as described in Sect. 3.3 is inadequate in SFQED. Indeed, the number of photons in a laser field is huge (see previous Sec.) and the usual perturbative treatment in this case would yield a tremendous amount of diagrams to be taken into account. Actually, for such a big amount of photons, the electromagnetic field will behave classically. Indeed, when the number of photons $N_{\mathbf{k}}$ in the state of momentum $\mathbf{k}$ is large, so will be the coefficients $c_{\mathbf{k},\sigma}$ and $c^{\dagger}_{\mathbf{k},\sigma}$ in Eq. (3.42). Therefore, neglecting the RHS of Eq. (3.44) which is of the order of 1 yields

$$c_{\mathbf{k},\sigma} c^{\dagger}_{\mathbf{k},\sigma} \simeq c^{\dagger}_{\mathbf{k},\sigma} c_{\mathbf{k},\sigma}\,, \tag{3.47}$$

and the creation and annihilation operators thus become the classical commuting coefficients $c_{\mathbf{k},\sigma}$ and $c^{*}_{\mathbf{k},\sigma}$. Intuitively, this is due to the fact that for strong enough fields, their amplitude is so large that the quantum fluctuations on top of them can safely be neglected. More quantitatively, let us evaluate the number of photons of the laser field. We consider an interval of time Δt during which the field E does not vary too much. According to the uncertainty relation, such a field will have frequencies up to $\omega \sim 1/\Delta t$. The density of modes with frequency ω is therefore given by

$$\frac{\omega^3}{c^3} \sim \frac{1}{(c\Delta t)^3}\,. \tag{3.48}$$

According to Eq. (2.39a), the energy density of the field is proportional to $\langle E \rangle^2$, from which we deduce the order of magnitude of number of photons as

$$N_{\mathbf{k}} \sim \frac{\langle E \rangle^2 c^3}{\hbar \omega^4}\,. \tag{3.49}$$

The condition of classicality $N_{\mathbf{k}} \gg 1$ is therefore equivalent to

$$\langle E \rangle \gg \frac{\sqrt{\hbar c}}{(c\,\Delta t)^2}\,. \tag{3.50}$$

We therefore conclude that variable fields, if sufficiently weak, can never be treated classically. On the contrary, static fields, whatever their strength, are always classical. In the case of laser fields, it is interesting to express this condition in terms of the laser intensity. We get [13]

$$I \gg 6 \times 10^5 \ \mathrm{W/cm^2} \left(\frac{\hbar\omega}{1 \ \mathrm{eV}} \right)^4 . \tag{3.51}$$

In the case of relativistic ($I > 10^{18}$ W/cm^2) optical lasers that we will consider in this thesis, the classical condition will thus always be satisfied.

The other assumption that must be verified is that the external field does not get depleted during the interaction, that is to say, the number of photons that are absorbed must be small compared to the total amount of photons in the laser field. A laser contains approximately 6×10^{18} photons of 1 eV per Joule of laser energy. As explained in Sect. 2.4.8, an electron in an external electromagnetic field emits harmonics up to $n_c \sim a_0^3$. Quantum mechanically, this means that the number of photons emitted is of the order of a_0^3. For a typical laser strength of $a_0 \sim 10^2$, we therefore have 10^6 photons absorbed per electron. For a typical number of 10^9 electrons, the total number of absorbed photons is around 10^{15} which leaves a typical laser beam with 100 J almost unaffected.[11]

Under these conditions, we will model the laser field by a *coherent state* [5, 22, 23]. Quantum mechanically, a coherent state of the photon field has the form

$$|A\rangle = \hat{D}\,|0\rangle , \tag{3.52}$$

where $\hat{D}$ is the so-called *displacement operator* and $|0\rangle$ the vacuum state of the photon Fock state. If the classical 4-potential associated with this displacement operator is

$$\mathcal{A}^\mu(x) = \int \frac{d^3k}{(2\pi)^3 2\omega_k} \sum_{\sigma=1,2} \left[C_{\mathbf{k},\sigma}\, \epsilon^\mu_{\mathbf{k},\sigma}\, e^{-ikx} + C^*_{\mathbf{k},\sigma}\, \epsilon^\mu_{\mathbf{k},\sigma}\, e^{ikx} \right], \tag{3.53}$$

where $C_{\mathbf{k},\sigma}$ are a c-numbers, then the displacement operator reads

$$\boxed{\hat{D} = \exp\left[\sum_{\sigma=1,2} \int \frac{d^3k}{(2\pi)^3 2k^0} \left(C_{\mathbf{k},\sigma} c^\dagger_{\mathbf{k},\sigma} - C^*_{\mathbf{k},\sigma} c_{\mathbf{k},\sigma} \right) \right]} \tag{3.54}$$

where $c^\dagger_{\mathbf{k},\sigma}$ and $c_{\mathbf{k},\sigma}$ are the creation and annihilation operators of the quantized photon field (3.42). Using the commutation relation (3.44), we can show that

$$\hat{D}^{-1} c_{\mathbf{k},\sigma} \hat{D} = c_{\mathbf{k},\sigma} + C_{\mathbf{k},\sigma}, \tag{3.55a}$$

$$\hat{D}^{-1} c^\dagger_{\mathbf{k},\sigma} \hat{D} = c^\dagger_{\mathbf{k},\sigma} + C^*_{\mathbf{k},\sigma}, \tag{3.55b}$$

[11] The reader interested about the treatment of depletion is referred to [20, 21].

from which we deduce that

$$\langle A|\mathcal{A}^{\mu}(x)|A\rangle = A^{\mu}(x). \tag{3.56}$$

It is in that sense that we say that the coherent state is "the most classical state" of the photon field. Finally, application of the displacement operator on the photon field operator $A^{\mu}(x)$ yields

$$\boxed{\hat{D}A^{\mu}(x)\hat{D}^{-1} = A^{\mu}(x) + \mathcal{A}^{\mu}(x)} \tag{3.57}$$

In order to describe scattering processes in a background field, we will thus have to generalize the Fock states defined in Sect. 3.3 in order to incorporate the coherent states that will describe the classical background field. These generalized states are written $|N; C\rangle$ obtained by acting the displacement $\hat{D}$ on a Fock state $|N\rangle$

$$|N; C\rangle = \hat{D}\,|N\rangle\,, \tag{3.58}$$

with

$$|N\rangle = |\{n_{\mathbf{k},\sigma}\}\rangle = \prod_{\mathbf{k},\sigma} \frac{(c^{\dagger}_{\mathbf{k},\sigma})^{n_{\mathbf{k},\sigma}}}{\sqrt{n_{\mathbf{k},\sigma}!}}\,|0\rangle\,. \tag{3.59}$$

These states are called *semi-coherent* and satisfy

$$\frac{\langle N; C|A^{\mu}(x)|N; C\rangle}{\langle N|N\rangle} = \mathcal{A}^{\mu}(x), \tag{3.60a}$$

$$\frac{\langle N; C|\hat{N}_{\mathbf{k},\sigma}|N; C\rangle}{\langle N|N\rangle} = |C_{\sigma}(\mathbf{k})|^2 + n_{\mathbf{k},\sigma}\,, \tag{3.60b}$$

where $\hat{N}_{\mathbf{k},\sigma} = c^{\dagger}_{\mathbf{k},\sigma}c_{\mathbf{k},\sigma}$ is the photon number operator. A semi-coherent state can therefore be pictured as N photons over a classical background $\mathcal{A}^{\mu}$. We will thus consider scattering processes of the form $|C; i\rangle \to |C; f\rangle$ (see Fig. 3.3) and so we will be led to compute matrix elements of the form

$${}_{\text{out}}\langle f; C|i; C\rangle_{\text{in}} = {}_{\text{in}}\langle f; C|S|i; C\rangle_{\text{in}} \equiv {}_{\text{in}}\langle f; C|S\left[\Psi, \bar{\Psi}, A\right]|i; C\rangle_{\text{in}}\,. \tag{3.61}$$

Using the definition of the semi-coherent state (3.58) and Eq. (3.57), we deduce that

$$\boxed{{}_{\text{in}}\langle f; C|S\left[\Psi, \bar{\Psi}, A\right]|i; C\rangle_{\text{in}} = {}_{\text{in}}\langle f|S\left[\Psi, \bar{\Psi}, A + \mathcal{A}\right]|i\rangle_{\text{in}}} \tag{3.62}$$

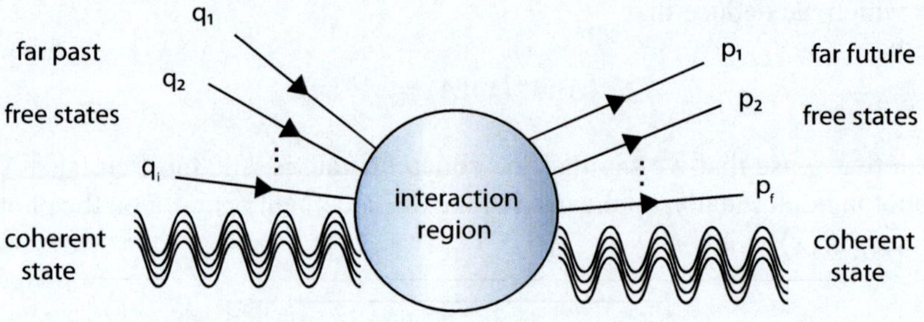

Fig. 3.3 In the presence of an external field, the scattering experiment described for vacuum QED in Fig. 3.1 is modified by the presence of the external field both in initial and final states (if depletion is neglected) modelled by a coherent state (3.52)

Equation (3.62) shows that computing S-matrix elements between coherent states, is equivalent to compute S-matrix elements between usual vacuum states with the shift $A^\mu \to A^\mu_{\text{rad}} + \mathcal{A}^\mu$ to the Lagrangian (3.30), which yields

$$\boxed{\mathcal{L}_{\text{SFQED}} = \bar{\Psi}(i\not{\partial} - e\not{\mathcal{A}} - m)\Psi - \frac{1}{4}F^{\mu\nu}F_{\mu\nu} - e\bar{\Psi}\not{A}_{\text{rad}}\Psi} \tag{3.63}$$

We will use this fact extensively in the rest of this thesis in order to compute processes in strong background fields.

3.4.2 *Furry Picture*

Following the procedure described in Sect. 3.2.3, we split the Lagrangian (3.63) into a free and an interacting part like

$$\mathcal{L}_{\text{SFQED}} = \mathcal{L}_0 + \mathcal{L}_{\text{int}}, \tag{3.64}$$

with

$$\mathcal{L}_0 = \bar{\Psi}\left(i\not{\partial} - e\not{\mathcal{A}} - m\right)\Psi - \frac{1}{4}F^{\mu\nu}F_{\mu\nu}, \tag{3.65a}$$

$$\mathcal{L}_{\text{int}} = -e\bar{\Psi}\not{A}_{\text{rad}}\Psi. \tag{3.65b}$$

The interaction picture with such a splitting of the Lagrangian is called *Furry picture* [24]. The equation of motion for Ψ corresponding to the free Lagrangian $\mathcal{L}_0$ is then

$$\boxed{\left(i\not{\partial} - e\not{\mathcal{A}} - m\right)\Psi = 0} \tag{3.66}$$

3.4.3 Volkov States

In order to apply the S-matrix expansion described in Sect. 3.2.3 with the splitting (3.64), we need to have a solution of (3.66). This is provided by the Volkov states [25] (see Appendix B.4 for more details)

$$\Psi^V_{p,r}(x) = \sqrt{\frac{1}{2\epsilon_p V}}\, E_p(x) u_r(\mathbf{p}), \tag{3.67}$$

where $u_r(\mathbf{p})$ is the free Dirac spinor defined in Sect. 3.3.2, while

$$E_p(x) = \left[1 + \frac{e\not{k}\not{\mathcal{A}}(\phi)}{2(kp)}\right] e^{iS_p(x)}, \tag{3.68}$$

is the so-called *Ritus matrix*. The phase $S_p(x)$ is the classical Hamilton-Jacobi function for a charged particle in a potential $\mathcal{A}$ and is given by

$$S_p(x) = -px - \int_{\phi_0}^{\phi} d\varphi \left[\frac{e(p\mathcal{A}(\varphi))}{(kp)} - \frac{e^2\mathcal{A}(\varphi)^2}{2(kp)}\right]. \tag{3.69}$$

The Volkov states are therefore quasi-classical states. Finally, the constant $\sqrt{\frac{1}{2\epsilon_p V}}$ is chosen so that the states are normalized to one particle per volume V

$$\int_V \Psi^{V\dagger}_{p,r} \Psi^V_{p,r} = 1. \tag{3.70}$$

Let us analyze in more detail each term of this solution. First of all, we notice that for $\mathcal{A}(\varphi) \to 0$, $\Psi^V_{p,r}(x) \to \Psi_{p,r}(x)$ as we would expect. Multiplying Eq. (3.66) on the left by $\not{p} - e\not{\mathcal{A}} + m$, yields (see Appendix B.4)

$$\left[(p - e\mathcal{A})^2 - m^2 - \frac{i}{2} e F_{\mu\nu}\sigma^{\mu\nu}\right]\psi = 0, \tag{3.71}$$

where $\frac{1}{2}\sigma^{\mu\nu}$ represents the spin of the electron. Therefore, taking $\sigma^{\mu\nu} = 0$ yields the spinless Gordon-Volkov equation

$$\left[(p - e\mathcal{A})^2 - m^2\right]\psi = 0, \tag{3.72}$$

whose solution reads

$$\Psi^G_{p,r}(x) = \sqrt{\frac{1}{2\epsilon_p V}}\, e^{iS_p(x)} u_r(\mathbf{p}), \tag{3.73}$$

where $S_p(x)$ is given by (3.69). We therefore see that the term $\frac{e\not{k}\mathcal{A}(\phi)}{2(kp)}$ represents the change in the spin of the electron due to the external field. Moreover, it is easy to see that the spin of the electron returns to its initial state after leaving the field.[12] Computing the associated Volkov current yields

$$j_\mu^V = \bar{\Psi}^V \gamma_\mu \Psi^V \sim p^\mu - eA^\mu + \left(\frac{epA}{kp} - \frac{e^2 A^2}{2kp}\right) k^\mu = p_{\rm cl}^\mu(\phi), \tag{3.74}$$

which is exactly the 4-momentum of a classical particle in a plane-wave. Before entering the plane-wave, the electron wave-function is a plane-wave field $\Psi_{p,r}(x)$. Then, when entering the plane-wave, the centroid of the wave function follows the classical trajectory (3.74) while its phase is shifted like [6] $px \rightarrow px + \int_{\phi_0}^{\phi} d\varphi \left[\frac{e(p\mathcal{A}(\varphi))}{(kp)} - \frac{e^2\mathcal{A}(\varphi)^2}{2(kp)}\right]$.

Finally, following the procedure described in Sect. 3.2, the quantized Dirac-Volkov field is given by

$$\Psi(x) = \int \frac{d^3k}{(2\pi)^3 2k^0} \sum_{s=\pm\frac{1}{2}} \left[c_{\mathbf{p},s}\, E_p(x) u_s(p) + d^\dagger_{\mathbf{p},s}\, E_{-p}(x) v_s(p)\right]. \tag{3.75}$$

3.4.4 *Position Space SFQED Feynman Rules*

In the same way as vacuum QED, the perturbative expansion of the S matrix can be represented diagrammatically by Feynman diagrams [5, 12]. The rules to build these Feynman diagrams in SFQED are formally obtained from vacuum QED by replacing free spinors $u_{p,r}$ by the Volkov dressed spinors $E_p(x)u_{p,r}$, while the fermion free propagators are replaced by the Volkov-Dirac propagator. On the contrary, the photon wave function and propagator are left unchanged by the background field. These rules[13] are summarized on Table 3.1

3.5 Non-linear Compton Scattering

We apply SFQED Feynman rules 3.1 to the Compton scattering process described by the Feynman diagram 3.4. The S-matrix element for this process is given by

$$S_{fi} = -ie \int d^4x\, \overline{\psi}_{p',s'}\, \not{A}^*_{k',\lambda}\, \psi_{p,s}, \tag{3.76}$$

[12] In addition to the classical fact that the momentum of the electron also returns to its initial value after leaving the pulse (See Chap. 2).

[13] These rules apply to stable vacua, i.e. in the presence of fields with $\zeta_1 = \zeta_2 = 0$ [Eq. (2.22)]. For Feynman rules in unstable vacuum, the reader is referred to [5].

Table 3.1 Strong field QED Feynman rules for the calculation of S-matrix elements in the presence of a strong background electromagnetic field. The double line in the fermion states and propagators stress the fact that the interaction with the background field is taken into account non perturbativly

Vertex	$-ie\gamma^{\mu}$	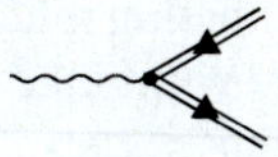
Photon propagator	$-iG^{\mu\nu}(x-x')$	
Dirac propagator	$iS^{V}(x-x')$	
Incoming fermion	$E_p(x)\, u_{p,\sigma}$	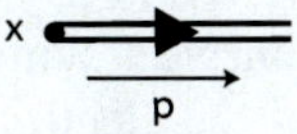
Outgoing fermion	$\overline{u}_{p,\sigma}\, \overline{E}_p(x)$	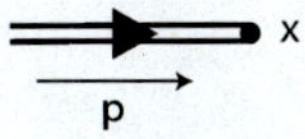
Incoming anti-fermion	$\overline{v}_{p,\sigma}\, \overline{E}_{-p}(x)$	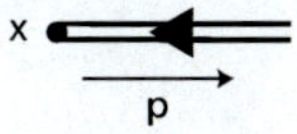
Outgoing anti-fermion	$E_p(x)\, v_{p,\sigma}$	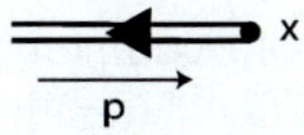
Incoming photon	$\epsilon^{\mu}\, e^{-ikx}$	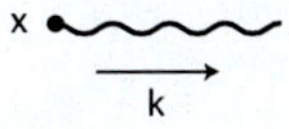
Outgoing photon	$\epsilon^{*\mu}\, e^{ikx}$	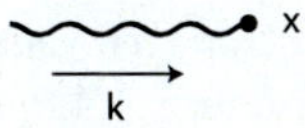

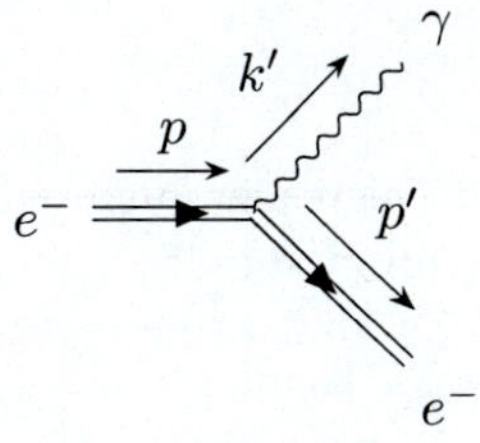

Fig. 3.4 Feynman diagram for the nonlinear Compton scattering $e^- + n\gamma_0 \to e^- + \gamma$ in which an initial electron with momentum p in a background laser field yields a photon with momentum k and a final electron with momentum p' in the same background laser field. This Feynman diagram is associated to the S-matrix element (3.76)

in which we insert Eq. (3.67) and the gamma quantum described as a single mode $\{\mathbf{k}_\gamma, \lambda_\gamma\}$ of a quantized radiation field A^{μ}

$$A^{\mu}_{\gamma} = \sqrt{\frac{1}{2\,V\omega_\gamma}}\, e^{-ik_\gamma x}\epsilon^{\mu}_{\gamma}. \tag{3.77}$$

Moreover, we make the following assumptions: (i) the electron interacts with a slowly varying external field compared to the formation time of the radiated photon [this is the so-called local constant field approximation (LCFA)]. The local constant field approximation relies on the possibility of neglecting space-time variations of the external field within the region of formation of the considered process (here a high-energy photon, see also Ref. [26]). It is possible to show that the photon formation length is inversely proportional to $a_0 = e\,|A^\mu|/(mc^2)$, which makes this condition valid for $a_0 \gg 1$. The LCFA has been the focus of many recent papers. The interested reader is referred to [27–29].

In addition, we assume that : (ii) the external field is undercritical, i.e. both Lorentz invariants of the electromagnetic fields (2.22) are small with respect to the corresponding invariants of the critical field of QED $E_s = \alpha\, E_{cr} \simeq 1.3 \times 10^{18}$ V/m

$$\zeta_1 \ll E_s^2, \tag{3.78a}$$

$$\zeta_2 \ll E_s^2, \tag{3.78b}$$

and (iii) small compared to the square of the electron quantum parameter χ and normalized field strength a_0

$$\max(\zeta_1, \zeta_2) \ll \chi^2, a_0^2. \tag{3.79}$$

Under condition (ii), vacuum is stable and one can use the usual Volkov states (3.67) and Furry picture 3.4.2 to compute the desired SFQED cross sections. Moreover, under condition (iii) together with condition (ii), all the emission probabilities (that we generically denote W) will depend only on χ and a_0

$$W(a_0, \chi, \{\zeta_i\}) = W(a_0, \chi, 0) + O(\{\zeta_i\}). \tag{3.80}$$

The reader interested in cases where vacuum is unstable under particle creation,[14] is referred to [5].

Under these assumptions, Eq. (3.76) yields after lengthy calculation (See Appendix 3.5 for more detail) the Lorentz invariant differential probability of photon emission [4, 6, 7]

$$\boxed{\frac{d^2 N_{\text{nCs}}}{d\tau\, d\chi_\gamma} = \frac{2}{3}\frac{\alpha^2}{\tau_e}\frac{G(\chi, \chi_\gamma)}{\chi_\gamma}} \tag{3.81a}$$

$$\boxed{G(\chi, \chi_\gamma) = \frac{\sqrt{3}}{2\pi}\frac{\chi_\gamma}{\chi}\left[\int_\nu^{+\infty} dy\, \mathrm{K}_{5/3}(y) + \frac{3}{2}\chi_\gamma\, \nu\, \mathrm{K}_{2/3}(\nu)\right]} \tag{3.81b}$$

where $G(\chi, \chi_\gamma)$ is the so-called quantum emissivity (see Fig. 3.5b), and $\nu = 2\,\chi_\gamma/[3\chi\,(\chi - \chi_\gamma)]$.

---

[14] These are electron-positron in QED.

In Eq. 3.81a, the differential probability of photon emission is written in a manifestly Lorentz invariant way using the proper time τ and the photon quantum parameter χ_γ. However, it can be convenient in practice to write it as a function of the laboratory time t and normalized energy γ_γ

$$\left.\frac{d^2 N_{\mathrm{nCs}}}{dt d\gamma_\gamma}\right|_\chi (\gamma_\gamma, \gamma) = \frac{2}{3}\frac{\alpha^2}{\tau_e}\frac{\widetilde{G}(\chi, \xi)}{\gamma\gamma_\gamma}, \tag{3.82}$$

where:

$$\widetilde{G}(\chi, \xi) = \frac{\sqrt{3}}{2\pi}\xi\left[\int_\nu^{+\infty} dy\, \mathrm{K}_{5/3}(y) + \frac{3}{2}\chi_\gamma \nu\, \mathrm{K}_{2/3}(\nu)\right], \tag{3.83}$$

with $\xi = \chi_\gamma/\chi = \gamma_\gamma/\gamma$ and $\nu = 2\xi/[3\chi(1-\xi)]$.

Multiplying Eq. (3.82) by γmc^2 yields the radiated spectrum (Fig. 3.6)

$$\frac{d^2\mathcal{E}}{d\tau d\gamma_\gamma} = P_0\,\alpha^2\, G(\chi, \chi_\gamma), \tag{3.84}$$

which integrating over all photon energies gives the total quantum radiated power

$$P_{\mathrm{rad}} = \int_0^{+\infty} d\gamma_\gamma \frac{1}{\gamma}\frac{d^2\mathcal{E}}{d\tau d\gamma_\gamma} = P_0\,\alpha^2\chi^2\, g(\chi), \tag{3.85}$$

where (see Fig. 3.5a)

$$g(\chi) = \int_0^{+\infty} d\chi_\gamma \frac{G(\chi, \chi_\gamma)}{\chi^3} = \frac{9\sqrt{3}}{8\pi}\int_0^{+\infty} d\nu \left[\frac{2\nu^2\, \mathrm{K}_{5/3}(\nu)}{(2+3\nu\chi)^2} + \frac{4\nu\,(3\nu\chi)^2}{(2+3\nu\chi)^4}\mathrm{K}_{2/3}(\nu)\right]. \tag{3.86}$$

The quantum emissivity G is represented as a function of χ and $\xi = \chi_\gamma/\chi$ in Fig 3.5b. We see that it presents a cut-off at $\xi = 1$ (electrons cannot emit photons with an energy greater than their own energy), which is not the case of the classical emissivity (2.125) (Fig. 3.5c) for which electrons can radiate frequencies corresponding to an energy greater than the electron's energy. The total quantum radiated power (3.85) is attenuated by a factor $g(\chi)$ [given by Eq. (3.86) and represented in Fig. 3.5a] compared to the classical one (2.62). To better see this effect, let us represent the radiation spectrum as a function of ξ for different fixed χ. For $\chi = 10^{-2}$ (Fig. 3.9a), the difference between the classical and quantum radiation spectra is very small and both spectra are exponentially attenuated below $\xi = 1$. For $\chi = 10^{-1}$ (Fig. 3.9b), we notice a small difference in the total radiated power and in the peak of the two spectra. There is barely any radiation above $\xi = 1$ in the classical spectrum. Finally, for $\chi = 1$ (Fig. 3.9c), there is a big difference in the total radiated powers and in the

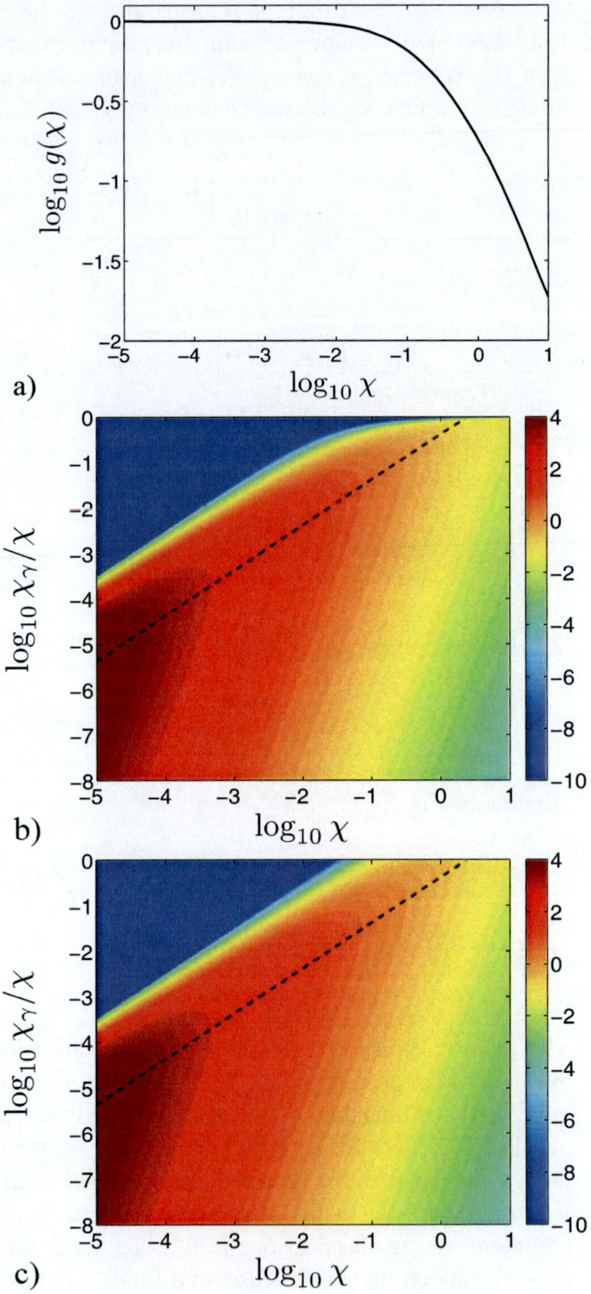

Fig. 3.5 (a) Dependence of $g(\chi)$ on the electron quantum parameter χ leading to a reduction of the emitted power due to quantum effects. (b) Quantum emissivity $G(\chi/\chi_\gamma)/\chi^2$ and (c) its classical limit as a function of χ and $\chi_\gamma/\chi = \gamma_\gamma/\gamma$. Dashed lines in panel (b) and (c) show $\chi_\gamma \simeq 0.435\,\chi^2$ for which the classical limit of $G(\chi, \chi_\gamma)$ is maximum [30]

maximum of the spectra. Moreover, the classical predicts (incorrectly) non negligible radiation above $\xi = 1$.

In the quantum case, it is possible to evaluate the maximum of the radiation as $\xi_{\max} \sim \frac{1}{2}\chi$. For χ above unity, since we cannot radiate above $\xi = 1$, $\xi_{\max}$ of course saturates to 1:

$$\xi_{\max} \sim \begin{cases} \frac{1}{2}\chi\,, & \chi \lesssim 1 \\ 1\,, & \chi > 1 \end{cases} \tag{3.87}$$

As we can see from this simple formula, for small values of χ, electrons radiate photons with an energy much smaller than the energy of the electron itself. In these conditions, the electron loses energy in a succession of small jumps that can correctly be modelled by a classical friction force (Fig. 2.8) such as the LL force (2.140) introduced in the previous chapter 2. On the contrary, for χ approaching unity or above, electrons emit photons with energies close to their own energy. The time evolution of the electron's energy will therefore be composed of a succession of large jumps which cannot be modelled by a classical force. More precisely, the proportion of emitted photons having an energy larger than 10% of the electron's energy is represented in Fig. 3.7.

From the time-energy differential probability of emission (3.82) we can deduce the emission rate as

$$\frac{dN}{dt}(\gamma, \chi) = \frac{2}{3}\frac{\alpha^2}{\tau_e}\int_0^{+\infty} d\gamma_\gamma \frac{\widetilde{G}(\chi, \xi)}{\gamma\gamma_\gamma} = \frac{2}{3}\frac{\alpha^2}{\tau_e}\frac{W_{\mathrm{nCs}}(\chi)}{\gamma}, \tag{3.88}$$

where W_{nCs} admits the following asymptotic expressions

$$W_{\mathrm{nCs}}(\chi) \simeq \begin{cases} \frac{5\sqrt{3}A}{4}\chi\,, & \chi \ll 1 \\ 2.19\,A\chi^{2/3}. & \chi \gg 1 \end{cases}$$

3.6 Non-linear Breit-Wheeler Pair Production

We apply SFQED Feynman rules 3.1 to the Breit-Wheeler process described by the Feynman diagram 3.8. The S-matrix element for this process is given by

$$S_{fi} = -ie \int d^4x\, \overline{\psi}_{p,s}\, \not{A}_{k,\lambda}\, \psi_{-p',-s'}, \tag{3.89}$$

in which we insert Eq. (3.67) and (3.77). After lengthy calculation [4, 8], we obtain the differential probability of pair production

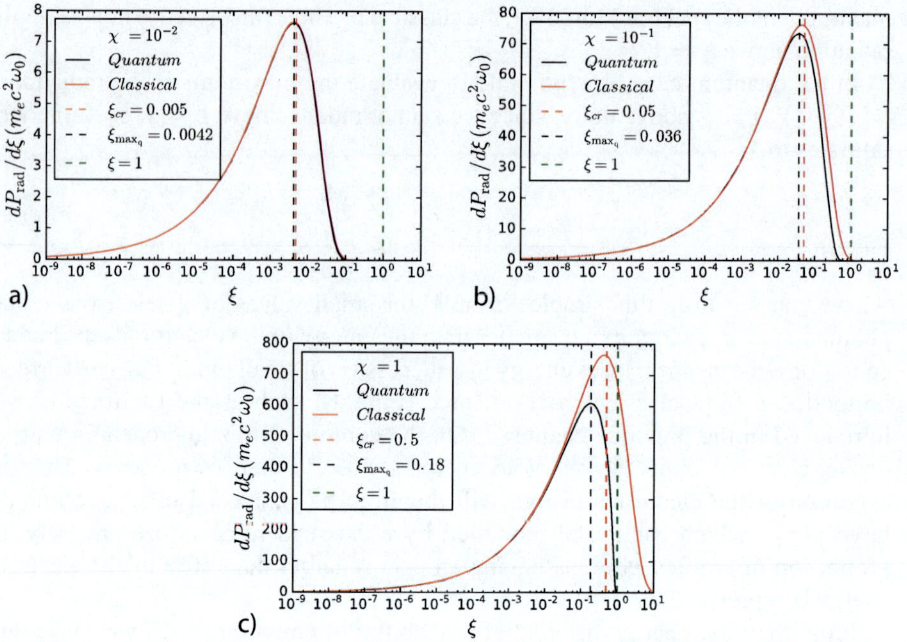

Fig. 3.6 Comparaison of the radiation spectra calculated in the classical and quantum frameworks for $\chi = 10^{-2}$ (a), $\chi = 10^{-1}$ (b) and $\chi = 1$ (c). The red dashed line represents the critical frequency (maximum of the classical spectrum) and the black line the maximum of the quantum spectrum. The green dashed line represents the limit $\xi = 1$

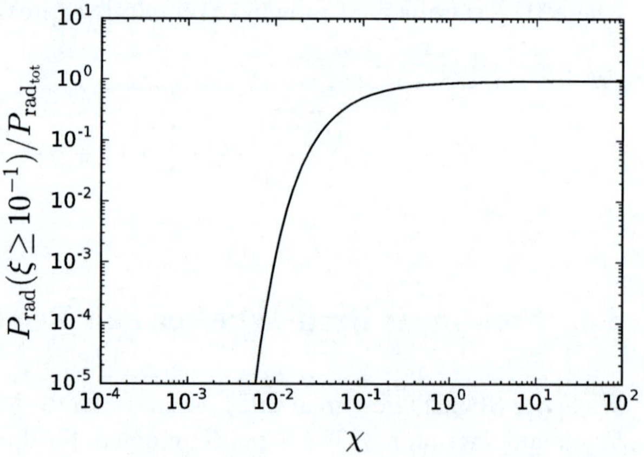

Fig. 3.7 Ratio of the radiated power coming from high-energy photons (for which $\xi \geq 10^{-1}$). For small values of χ most of the energy is emitted in low energy photons (the total power emitted in photons with more than 10% of the energy of the electron goes to 0) while, as χ approaches unity (and above) this quantity tends to 1, i.e. all the energy is emitted in photons with an energy larger than 10% of the electron's energy

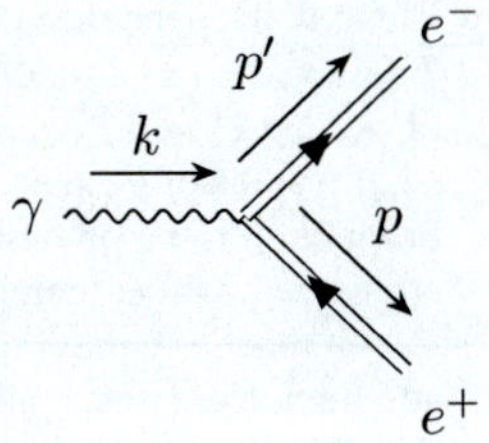

Fig. 3.8 Feynman diagram for the nonlinear Breit-Wheeler pair production $\gamma + n\gamma_0 \to e^- + e^+$ in which an initial high energy photon with momentum k in a background laser field yields an electron-positron pair with momentum p and p' in the same background laser field. This Feynman diagram is associated to the S-matrix element (3.89)

$$\boxed{\begin{aligned} \frac{d^2 N_{\mathrm{nBW}}}{dt d\chi} &= \frac{2}{3}\frac{\alpha^2}{\tau_e}\frac{T_{\mathrm{nBW}}(\chi_\gamma, \chi)}{\chi_\gamma \gamma_\gamma} && (3.90a) \\ T_{\mathrm{nBW}}(\chi_\gamma, \xi') &= \frac{\sqrt{3}}{2\pi}\left[-\int_{\nu'}^{+\infty} dy\, \mathrm{K}_{5/3}(y) + \frac{3}{2}\chi_\gamma \nu' \,\mathrm{K}_{2/3}(\nu')\right] && (3.90b) \end{aligned}}$$

where $\nu' = \frac{2}{3}\frac{\chi_\gamma}{(\chi_\gamma - \chi)\chi}$. Similarly to the case of nonlinear Compton scattering (previous section), we can rewrite Eq. (3.90a) in terms of laboratory quantities as

$$\left.\frac{d^2 N_{\mathrm{nBW}}}{dt d\gamma}\right|_{\chi_\gamma} (\gamma, \gamma_\gamma) = \frac{2}{3}\frac{\alpha^2}{\tau_e}\frac{\tilde{T}_{\mathrm{nBW}}(\chi_\gamma, \xi')}{\gamma \gamma_\gamma}, \tag{3.91}$$

where

$$\tilde{T}_{\mathrm{nBW}}(\chi_\gamma, \xi') = \frac{\sqrt{3}}{2\pi}\xi'\left[-\int_{\nu'}^{+\infty} dy\, \mathrm{K}_{5/3}(y) + \frac{3}{2}\chi_\gamma \nu' \,\mathrm{K}_{2/3}(\nu')\right]. \tag{3.92}$$

Finally, the production rate is given by

$$\frac{dN_{\mathrm{nBW}}}{dt} = A\frac{W_{\mathrm{nBW}}(\chi_\gamma)}{\gamma_\gamma}, \tag{3.93}$$

with

$$W_{\mathrm{nBW}}(\chi_\gamma) = \frac{\sqrt{3}}{2\pi}\int_0^1 d\xi'\left[-\int_{\nu'}^{+\infty} dy\, \mathrm{K}_{5/3}(y) + \frac{3}{2}\chi_\gamma \nu' \,\mathrm{K}_{2/3}(\nu')\right]. \tag{3.94}$$

The function (3.94) has the following asymptotic forms (See Appendix C.3)

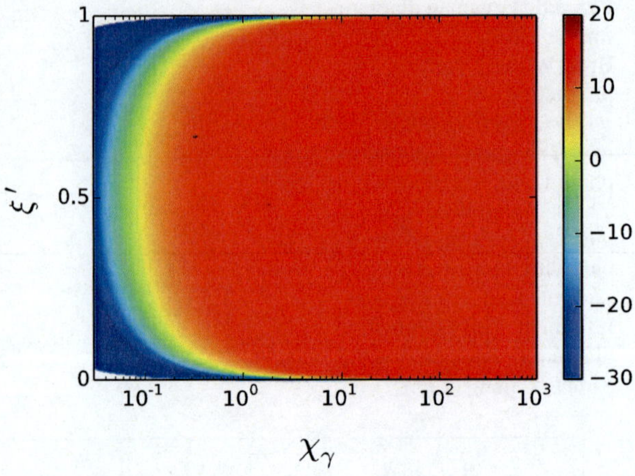

Fig. 3.9 Dependence of $\tilde{T}_{\mathrm{nBW}}(\chi_\gamma, \xi')$ as a function of χ_γ and $\xi' = \chi/\chi_\gamma$. For small values of χ_γ pair production is exponentially attenuated, while, increasing χ_γ, 'it reaches a maximum and then decreases again slowly

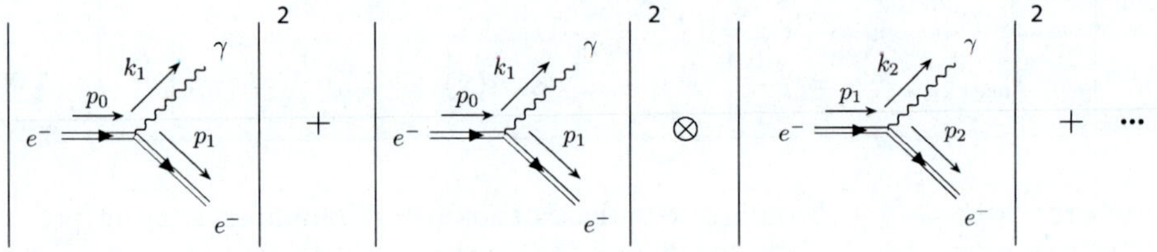

Fig. 3.10 In the moderately quantum regime [36] ($a_0 \gg 1$ and $\chi \lessapprox 1$), RR stems from multiple incoherent photon emission. At each photon emission, the incoming electron momentum is changed by the previous photon emission. The dots represent all the higher order terms representing 3, 4 etc. successive incoherent photon emissions

$$W_{\mathrm{nBW}}(\chi_\gamma, \gamma_\gamma) \simeq \begin{cases} 0.345A\,\frac{\chi_\gamma}{\gamma_\gamma}\exp\left[-\frac{8}{3\chi_\gamma}\right], & \chi_\gamma \ll 1 \\ 0.57A\,\frac{\chi_\gamma^{2/3}}{\gamma_\gamma}, & \chi_\gamma \gg 1 \end{cases}$$

The function $\tilde{T}_{\mathrm{nBW}}$ is represented in Fig. 3.9 as a function of χ_γ and ξ'. For small values of χ_γ, the total pair production rate goes to 0 exponentially (no pair creation in the classical regime), while, increasing χ_γ, it reaches a maximum and then decreases again slowly.

3.7 Quantum Radiation-Reaction

In CED, the problem of radiation reaction consists in solving consistently both the equation of motion for the electron and Maxwell's equation for the external and radiated fields (2.127). By analogy, the equivalent problem in QED would amount to completely determine both the S-matrix in the Furry picture (3.18) and the final state

$|f\rangle$ for an initial state composed of an electron alone $|i\rangle = |e^-\rangle$. The first-order term of the expansion of the S-matrix corresponds to nonlinear Compton scattering 3.4 while the higher-order terms correspond either to radiative corrections [31, 32] or to higher-order coherent or incoherent processes, that is to say processes involving more than one basic quantum process (such as the photon emission by an electron/positron or the electron/positron pair creation by a high-energy photon). Higher-order coherent processes correspond to processes occurring in the same formation length [33–35] while incoherent higher-order processes are processes composed of several basic quantum processes but all occurring in different formation regions. In the regime where $\chi \lesssim 1$ (and still in the LCFA approximation $a_0 \gg 1$), the coherent higher-order processes are negligible because they scale with α. Moreover, since $\chi \lesssim 1$, the subsequent emitted photons have $\chi_\gamma \lesssim 1$ and are unable to produce pairs. We therefore define the *non-linear moderately quantum regime* [19, 36] as corresponding to $\chi \lesssim 1$ and $a_0 \gg 1$. In this regime, radiation reaction in the QED framework reduces to the overall electron energy and momentum loss due to the emission of many photons consecutively, and incoherently (See Fig. 3.10).

References

1. Peskin ME, Schroeder DV (1995) An introduction to quantum field theory. Addison-Wesley Publishing Company, Boston
2. Weinberg S (2005) The quantum theory of fields. Cambridge University Press, Cambridge
3. Schwartz M (2014) Quantum field theory and the standard model. Cambridge University Press, Cambridge
4. Ritus VI (1979) Quantum effects of the interaction of elementary particles with an intense electromagnetic field. Ph.D. thesis
5. Fradkin ES, Gitman DM, Shvartsman SM (1991) Quantum electrodynamics with unstable vacuum. Springer, Heidelberg
6. Seipt D (2012) Strong-field QED processes in short laser pulses. Ph.D. thesis, Institut fÂšur Theoretische Physik Fachrichtung Physik FakultÂšat Mathematik und Naturwissenschaften der Technischen UniversitÂšat Dresden
7. Mackenroth KF (2012) Quantum radiation in ultra-intense laser pulses. Ph.D. thesis, Combined Faculties for the Natural Sciences and Mathematics of the Ruperto-Carola University of Heidelberg, Germany
8. Meuren S (2015) Nonlinear quantum electrodynamic and electroweak processes in strong laser fields. Ph.D. thesis, Combined Faculties of the Natural Sciences and Mathematics of the Ruperto-Carola-University of Heidelberg, Germany
9. Torgrimsson G (2016) Pair production, vacuum birefringence and radiation reaction in strong field QED. Ph.D. thesis, Department of Physics, Chalmers University of Technology
10. Fock V (1932) Konfigurationsraum und zweite quantelung. Z Phys 75:622–647
11. Heisenberg W (1943) Die beobachtbaren grossen in der theorie der elementarteilche. Zeitschrift für Physik 120(7)
12. Feynman RP (1949) Space-time approach to quantum electrodynamics. Phys Rev 76(6):769
13. Landau L, Lifshitz EM (2012) Quantum electrodynamcis. Ellipses
14. Dirac PAM (1928) The quantum theory of the electron. Proc Roy Soc Lond A 117(778):610–624
15. Gupta SN (1950) Theory of longitudinal photons in quantum electrodynamics. Proc Phys Soc Sect A 63(7):681

16. Bleuler K (1950) Eine neue methode zur behandlung der longitudinalen und skalaren photonen. Helv Phys Acta 23(5):567–586
17. Mitter H (1975) Quantum electrodynamics in laser fields. In: Acta physica Austriaca supplementum XIV, p 397
18. Baier VN, Katkov VM, Strakhovenko VM (1998) Electromagnetic processes at high energies in oriented single crystals
19. Di Piazza A, Müller C, Hatsagortsyan K, Keitel CH (2012) Extremely high-intensity laser interactions with fundamental quantum systems. Rev Mod Phys 84:1177
20. Ilderton A, Seipt D (2018) Backreaction on background fields: a coherent state approach. Phys Rev D 97:016007
21. Heinzl T, Ilderton A, Seipt D (2018) Mode truncations and scattering in strong fields. Phys Rev D 98:016002
22. Glauber RJ (1963) Coherent and incoherent states of the radiation field. Phys Rev 131(6):2766
23. Harvey C, Heinzl T, Ilderton A (2009) Signatures of high-intensity Compton scattering. Phys Rev A 79:063407
24. Furry WH (1951) On bound states and scattering in positron theory. Phys Rev 81:115
25. Volkov DM (1935) über eine Klasse von Lösungen der Diracschen Gleichung. Z Phys 94:250
26. Baier V, Katkov V (2005) Concept of formation length in radiation theory. Phys Rep 409(5):261–359
27. Di Piazza A, Tamburini M, Meuren S, Keitel CH (2018) Implementing nonlinear Compton scattering beyond the local-constant-field approximation. Phys Rev A 98:012134
28. Ilderton A, King B, Seipt D (2018) High-intensity QED beyond the locally constant field approximation. arXiv arXiv:1808.10339v1
29. Di Piazza A, Tamburini M, Meuren S, Keitel CH (2019) Improved local-constant-field approximation for strong-field QED codes. arXiv arXiv:1811.05834v2
30. Niel F, Riconda C, Amiranoff F, Duclous R, Grech M (2018) From quantum to classical modeling of radiation reaction: a focus on stochasticity effects. Phys Rev E 97:043209
31. Meuren S, Di Piazza A (2011) Quantum electron self-interaction in a strong laser field. Phys Rev Lett 107:260401
32. Meuren S, Keitel CH, Di Piazza A (2013) Polarization operator for plane-wave background fields. Phys Rev D 88:013007
33. Seipt D, Kämpfer B (2012) Two-photon Compton process in pulsed intense laser fields. Phys Rev D 85:101701
34. Mackenroth F, Di Piazza A (2013) Nonlinear double Compton scattering in the ultrarelativistic quantum regime. Phys Rev Lett 110:070402
35. King B (2015) Double Compton scattering in a constant crossed field. Phys Rev A 91:033415
36. Di Piazza A, Hatsagortsyan KZ, Keitel CH (2010) Quantum radiation reaction effects in multiphoton Compton scattering. Phys Rev Lett 105:220403

Chapter 4
Kinetic Description

This reminds him of the way blood plasma carries around red and white corpuscles and germs. So he proposed to call our "uniform discharge" a "plasma". Of course we all agreed.

Mott-Smith (1971), referring to Irving Langmuir coining the word plasma in 1927

4.1 Outline

In the previous chapters, we described Classical and Quantum Field Theory (CFT/QFT) as well as their application to the electromagnetic interaction, namely Classical and Quantum ElectroDynamics (CED/QED). These descriptions concerned only single particles. However, in real situations, the interaction of a laser will take place either with an electron beam or with a plasma, both being formed by many particles. The kinetic theory describes ensemble of particles statistically through the use of distribution functions (rather than by following individually the position and momentum of each particles at each time). It has been the focus of many textbooks (see, e.g. the books by [1–3]). Numerically, such systems are simulated by Particle-In-Cell (PIC) codes, which reproduce the behavior of the assembly of charged particles which constitutes the plasma, by consistently solving Maxwell's equations together with the equations of motion for the charges. They have been discussed in textbooks such as [4, 5].

However, the recent progress in the peak laser intensity and the emergence of ultra-intense laser-plasma interaction has made it mandatory to take into account radiation-reaction as well as SF-QED processes in addition to classical plasma physics both in the theoretical framework as well as in numerical schemes. Our goal here is therefore to show how to introduce RR as well as QED effects both in the kinetic theory

F. Niel, *Classical and Quantum Description of Plasma and Radiation in Strong Fields*, Springer Theses, https://doi.org/10.1007/978-3-030-73547-0_4

and in the numerical schemes used to simulate such plasmas. In order to maintain the presentation both logical and compact, not all derivations presented here are straightforward: some results may require quite some algebra before being derived. For a more thorough exposure of the numerical treatment of SF-QED processes, the reader is referred to the PhD thesis of [6] or the review article by [7].

This chapter is structured as follows. Section 4.2 introduces the general Klimontovich equation from which we derive the Vlasov equation following a mean-field procedure. In Sect. 4.3 we then show how the Vlasov equation fails to incorporate the radiation by ultra-relativistic particles and how to modify it to include such radiation as well as its counter-action under the form of both the classical and quantum RR. Finally, in Sect. 4.4, we discuss how to modify the classical PIC loop in order to simulate plasmas at ultra-high intensities where SF-QED are no longer negligible.

4.2 Kinetic Description

4.2.1 The Klimontovich Equation

Clearly, the complete description of a plasma (or any set of particles e.g. an electron beam) would involve the knowledge of the positions $\mathbf{x}_i(t)$ and momenta $\mathbf{p}_i(t)$ of each particle i at each time t. Such a knowledge can be put for each species α into the form of a density of particles

$$f_{M_\alpha}(\mathbf{x}, \mathbf{p}, t) = \sum_i \delta^{(3)}[\mathbf{x} - \mathbf{x}_i(t)]\, \delta^{(3)}[\mathbf{p} - \mathbf{p}_i(t)], \tag{4.1}$$

where $\mathbf{x}$ and $\mathbf{p}$ are the Eulerian coordinates of the phase space and $\mathbf{x}_i$ and $\mathbf{p}_i$ the Lagrangian coordinates of the ith particle and where the subscript M stands for *microscopic* [1–3]. Such a function then completely specifies the microscopic state of the system of particles.[1] To this density, we can associate a charge density and a current

$$\rho_M(\mathbf{x}, t) = \sum_\alpha q_\alpha \int d^3v\, f_{M_\alpha}(\mathbf{x}, \mathbf{p}, t), \tag{4.2a}$$

$$\mathbf{J}_M(\mathbf{x}, t) = \sum_\alpha q_\alpha \int d^3v\, \mathbf{v}\, f_{M_\alpha}(\mathbf{x}, \mathbf{p}, t), \tag{4.2b}$$

where q_α is the charge of the species α. The electric and magnetic fields being described by Maxwell's equations, we then have [8]

---

[1]This function is very spiky and is either equal to infinity (when taken on a particle trajectory) or to 0 (where there is no particle).

$$\nabla \cdot \mathbf{E}_M = \frac{\rho_M}{\epsilon_0}, \tag{4.3a}$$

$$\nabla \times \mathbf{B}_M = \frac{1}{c^2}\,\partial_t \mathbf{E}_M + \mu_0\,\mathbf{J}_M, \tag{4.3b}$$

$$\nabla \times \mathbf{E}_M = -\partial_t \mathbf{B}_M, \tag{4.3c}$$

$$\nabla \cdot \mathbf{B}_M = 0, \tag{4.3d}$$

with

$$\rho_M = \sum_\alpha q_\alpha \int d^3p\, f_{M_\alpha}(\mathbf{x}, \mathbf{p}, t), \tag{4.4a}$$

$$\mathbf{J}_M = \sum_\alpha q_\alpha \int d^3p\, \mathbf{v}\, f_{M_\alpha}(\mathbf{x}, \mathbf{p}, t), \tag{4.4b}$$

where the subscript M emphasizes the fact that $\mathbf{E}_M$ and $\mathbf{B}_M$ are microscopic fields, while the dynamic of each particle is describes by Newton-Lorentz equations

$$\frac{d\boldsymbol{x}_i}{dt} = \mathbf{v}_i, \tag{4.5a}$$

$$\frac{d\boldsymbol{v}_i}{dt} = \frac{q_i}{m_i}\left(\mathbf{E}_M + \mathbf{v}_i \times \mathbf{B}_M\right). \tag{4.5b}$$

Deriving (4.1) with respect to time, we get [using Eqs. (4.5a)] the so-called *Klimontovich-Dupree equation* [1]

$$\boxed{\frac{\partial f_{M_\alpha}}{\partial t} + \mathbf{v} \cdot \frac{\partial f_{M_\alpha}}{\partial \mathbf{x}} + \frac{q_\alpha}{m_\alpha}\left(\mathbf{E}_M + \mathbf{v} \times \mathbf{B}_M\right) \cdot \frac{\partial f_{M_\alpha}}{\partial \mathbf{p}} = 0} \tag{4.6}$$

Although exact, this description is not very useful. Indeed, the Klimontovich equation basically just states the conservation of the number of particles in phase-space

$$\frac{df_{M_\alpha}}{dt} = 0. \tag{4.7}$$

Moreover, this equation provides far more information than what we can treat in practice. For example, 1 μm^3 of a hydrogen plasma at $10^{-3}\, n_c$ contains more than 10^6 particles. We will thus been instead interested in averaged quantities.

4.2.2 *The Vlasov Equation*

Let us decompose the different quantities into their average (smooth) part and a fluctuation (spiky) part

$$f_{M_\alpha}(\mathbf{x}, \mathbf{p}, t) = f_\alpha(\mathbf{x}, \mathbf{p}, t) + \delta f_\alpha(\mathbf{x}, \mathbf{p}, t), \tag{4.8a}$$
$$\mathbf{E}_M(\mathbf{x}, t) = \mathbf{E}(\mathbf{x}, t) + \delta\mathbf{E}(\mathbf{x}, t), \tag{4.8b}$$
$$\mathbf{B}_M(\mathbf{x}, t) = \mathbf{B}(\mathbf{x}, t) + \delta\mathbf{B}(\mathbf{x}, t), \tag{4.8c}$$

with

$$f_\alpha(\mathbf{x}, \mathbf{p}, t) \equiv \langle f_{M_\alpha}(\mathbf{x}, \mathbf{p}, t)\rangle, \tag{4.9a}$$
$$\mathbf{E}(\mathbf{x}, t) \equiv \langle \mathbf{E}_M(\mathbf{x}, t)\rangle, \tag{4.9b}$$
$$\mathbf{B}(\mathbf{x}, t) \equiv \langle \mathbf{B}_M(\mathbf{x}, t)\rangle, \tag{4.9c}$$

and where $\langle . \rangle$ represents the ensemble average over a set of realisations of the plasma in identical prescribed macroscopic conditions. Thus taking the average of the Klimontovich equation (4.6) and using the definitions (4.9) yields

$$\frac{\partial f_\alpha}{\partial t} + \mathbf{v} \cdot \frac{\partial f_\alpha}{\partial \mathbf{x}} + \frac{q_\alpha}{m_\alpha}(\mathbf{E} + \mathbf{v} \times \mathbf{B}) \cdot \frac{\partial f_\alpha}{\partial \mathbf{p}} = -\frac{q_\alpha}{m_\alpha}\left\langle(\delta\mathbf{E} + \mathbf{v} \times \delta\mathbf{B}) \cdot \frac{\partial f_{M_\alpha}}{\partial \mathbf{p}}\right\rangle. \tag{4.10}$$

In this equation, the left hand side describes the collective behaviour of the plasma while the right hand side describes the discrete-particle nature of the plasma which thus contains all the collisional behaviour.

Collisional effects stem from the Coulomb force associated to individual charges while kinetic effects are described by the kinetic energy. A plasma will thus be qualified of *collisionless* if

$$\frac{3}{2}k_B T_e \gg \frac{e^2}{4\pi\epsilon_0} n_e^{1/3} \iff \Lambda = n_e \lambda_D^3 \gg 1, \tag{4.11}$$

where T_e and n_e are the electron temperature and density (resp.), Λ is called the plasma parameter and $\lambda_D = \sqrt{\frac{\epsilon_0 k_B T_e}{e^2 n_e}}$ the Debye length. Throughout this work, we will consider ultra-relativistic electron beams and plasmas, for which the condition (4.11) will be verified. Neglecting the rhs, we thus get

$$\boxed{\frac{\partial f_\alpha}{\partial t} + \mathbf{v} \cdot \frac{\partial f_\alpha}{\partial \mathbf{x}} + \frac{q_\alpha}{m_\alpha}(\mathbf{E} + \mathbf{v} \times \mathbf{B}) \cdot \frac{\partial f_\alpha}{\partial \mathbf{p}} = 0} \tag{4.12}$$

which is the so-called *Vlasov equation* [1]. The fields satisfy the ensemble averaged Maxwell's equations

$$\nabla \cdot \mathbf{E} = \frac{\rho}{\epsilon_0}, \tag{4.13a}$$

$$\nabla \times \mathbf{B} = \frac{1}{c^2}\,\partial_t \mathbf{E} + \mu_0\,\mathbf{J}, \tag{4.13b}$$

$$\nabla \times \mathbf{E} = -\partial_t \mathbf{B}, \tag{4.13c}$$

$$\nabla \cdot \mathbf{B} = 0, \tag{4.13d}$$

where

$$\rho_M = \sum_\alpha q_\alpha \int d^3p\, f_\alpha(\mathbf{x}, \mathbf{p}, t), \tag{4.14a}$$

$$\mathbf{J}_M = \sum_\alpha q_\alpha \int d^3p\, \mathbf{v}\, f_\alpha(\mathbf{x}, \mathbf{p}, t), \tag{4.14b}$$

are the averaged charge and current. The resulting system of equations Eqs. (4.12)–(4.13d) is the so-called *Vlasov-Maxwell system*. In 4-vector notation, the Vlasov equation rewrites

$$\boxed{p^\mu \frac{\partial f_\alpha}{\partial x^\mu} + q_\alpha F^{\mu\nu} p_\nu \frac{\partial f_\alpha}{\partial p^\mu} = 0} \tag{4.15}$$

4.3 Kinetic Description of Radiation-Reaction

4.3.1 Motivation

The Vlasov equation is an equation on the average density of particles α in an elementary volume of phase-space $\Delta^3 x \Delta^3 p$. For this equation to make sense, the fluctuations $\delta f_\alpha(\mathbf{x}, \mathbf{p}, t)$ must therefore be small compared to the average $f_\alpha(\mathbf{x}, \mathbf{p}, t) \equiv \langle f_{M_\alpha}(\mathbf{x}, \mathbf{p}, t)\rangle$. This is achieved if the number of particles in $\Delta^3 x$ is big enough. More quantitatively, we need $\Delta x \gg d_\alpha$, where d_α is the mean inter-particle distance that we evaluate as $d_\alpha = n_\alpha^{-1/3}$ with n_α the density of the species α. We consider for definiteness the interaction between an optical laser at $\lambda_0 \sim 1\ \mu\text{m}$ and a plasma at $n_e \sim 100\, n_c$ where

$$n_c = \frac{m_e \epsilon_0 \omega_0^2}{e^2} \simeq 1.115 \times 10^9 \left(\frac{1\ \mu\text{m}}{\lambda}\right)\ \mu\text{m}^{-3}, \tag{4.16}$$

is the critical density. The mean inter-particle distance is then of the order of $2.10^{-4}\ \mu\text{m}$. In terms of frequency $\omega_e = 2\pi c/d_e$, this corresponds to an energies of about $\hbar\omega_e \simeq 10\,\text{keV}$. Considering a typical ultra-relativistic electron with $\gamma_e \simeq 2000$ in an external field corresponding to a quantum parameter $\chi_e \simeq 10^{-2}$, the typical radiated energy is $\hbar\omega_c = \frac{1}{2}\epsilon_0\chi_e \simeq 10$ MeV [see Eq. (3.87)], which is 3 order of magnitude above the limit of validity of the Vlasov equation. In this context, we then need to modify it in order to take into account explicitly this radiation.

4.3.2 Vlasov Equation with Classical RR

As shown in Chap. 2, RR is described classically for a single particle by the LL equation

$$\frac{d\mathbf{p}}{dt} = \mathbf{F}_{\text{Lorentz}} + \mathbf{F}_{\text{LL}}. \tag{4.17}$$

Put in matrix form, we get

$$\frac{d}{dt}\begin{pmatrix}\mathbf{x}\\ \mathbf{p}\end{pmatrix} = \begin{pmatrix}\frac{\mathbf{p}}{p_0}\\ \mathbf{F}_{\text{Lorentz}} + \mathbf{F}_{\text{LL}}\end{pmatrix}. \tag{4.18}$$

Using the equivalence between deterministic differential equations and Liouville equations [9], we deduce the equation of evolution for the electron distribution function by

$$\frac{\partial f_e}{\partial t} + \frac{\mathbf{p}}{m}\cdot\frac{\partial f_e}{\partial \mathbf{x}} + \frac{\partial\left[(\mathbf{F}_{\text{Lorentz}} + \mathbf{F}_{\text{LL}})\, f_e\right]}{\partial \mathbf{p}} = 0, \tag{4.19}$$

where we interpreted the electron distribution function as the probability density function of the electron momentum and where d/dt is the Vlasov operator (4.12).

Contrarily to the "traditional" Vlasov equation where $\nabla_{\mathbf{p}}\cdot\mathbf{F}_{\text{Lorentz}} = 0$, we have for the RR force $\nabla_{\mathbf{p}}\cdot\mathbf{F}_{RR} \neq 0$. This means that, when included, RR tends to change the volume of phase-space that is occupied by the system.[2] More precisely, we can show that [10]

$$\begin{aligned}\nabla_{\mathbf{p}}\cdot\mathbf{F}_{RR} \sim -\Bigg\{&\left[\nabla_{\mathbf{x}}\cdot\mathbf{E} - \mathbf{v}\cdot\left(\nabla_{\mathbf{x}}\times\mathbf{B} - \frac{\partial\mathbf{E}}{\partial t}\right)\right] + 2\left[\frac{\mathbf{E}^2+\mathbf{B}^2}{\gamma_e}\right]\\ &+ 4\gamma_e\left[(\mathbf{v}\times\mathbf{E})^2 + (\mathbf{v}\times\mathbf{B})^2 - 2\mathbf{v}\cdot(\mathbf{E}\times\mathbf{B})\right]\Bigg\} \geq 0.\end{aligned} \tag{4.20}$$

The phase-space is thus contracted due to RR in the classical regime. We could have concluded the same thing by looking at the form of the RR force (2.143). The dominant term is in $-\gamma^2$, which means that the more energetic particles loose more energy. The net effect is then to narrow the distribution function of the electron bunch. The "improved" Vlasov equation including classical RR then reads

[2] Contrarily to the case without RR where the volume of phase-space is conserved.

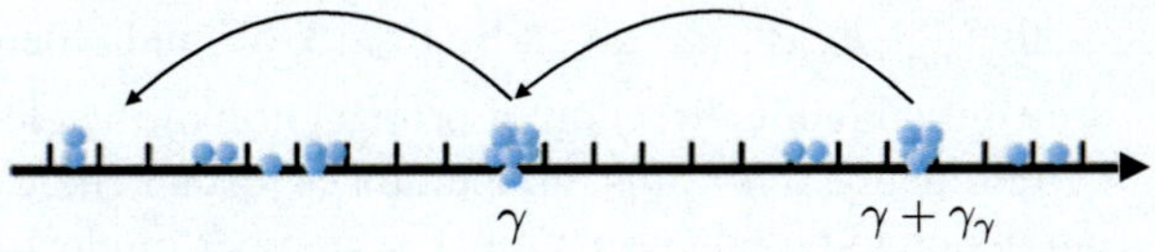

Fig. 4.1 Schematic representation of the variation in the electron energy distribution due to photon emission by nonlinear Compton scattering. The number of electrons at an energy γ is increased by electrons having an energy $\gamma + \gamma_\gamma$ and emitting a photon with an energy γ_γ [first term of the RHS of Eq. (4.23)], while it is decreased by electrons at this energy γ emitting a photon with any energy γ_γ [second term of the RHS of Eq. (4.23)] (see Fig. 4.1)

$$\boxed{\frac{\partial f_e}{\partial t} + \mathbf{v} \cdot \frac{\partial f_e}{\partial \mathbf{x}} + (\mathbf{F}_{\text{Lorentz}} + \mathbf{F}_{\text{LL}}) \cdot \frac{\partial f_e}{\partial \mathbf{p}} = -f_e \frac{\partial \mathbf{F}_{\text{LL}}}{\partial \mathbf{p}}} \tag{4.21}$$

4.3.3 Vlasov Equation with Quantum RR

As stated in Chap. 3, when the quantum parameter of the electron approaches unity, the radiation can no longer be modelled as a continuous electromagnetic field but its discrete nature has to be taken into account. In order to model photon emission, we use the discrete jump part of the differential Chapman-Kolmogorov equation [9, 11] with the differential probability given by SFQED (3.81) and described by the Feynman diagram 3.4. We get

$$\boxed{\frac{\partial f_e}{\partial t} + \mathbf{v} \cdot \frac{\partial f_e}{\partial \mathbf{x}} + \mathbf{F}_{\text{Lorentz}} \cdot \frac{\partial f_e}{\partial \mathbf{p}} = \mathcal{C}_{\text{nCs}}[f_e]} \tag{4.22}$$

where the nonlinear Compton scattering (nCs) collisional operator is given by

$$\boxed{\begin{aligned} \mathcal{C}_{\text{nCs}}[f_e] \equiv & \int_0^{+\infty} d\gamma_\gamma \frac{d^2 N_{\text{nCs}}}{dt d\gamma_\gamma}(\chi, \gamma + \gamma_\gamma, \gamma_\gamma)\, f_e(t, \mathbf{x}, \gamma + \gamma_\gamma, \mathbf{\Omega}) \\ & - f_e(t, \mathbf{x}, \gamma, \mathbf{\Omega}) \int_0^{+\infty} d\gamma_\gamma \frac{d^2 N_{\text{nCs}}}{dt d\gamma_\gamma}(\chi, \gamma, \gamma_\gamma), \end{aligned}} \tag{4.23}$$

with $\frac{d^2 N_{\mathrm{nCs}}}{dtd\gamma_\gamma}(\chi, \gamma, \gamma_\gamma)$ given by Eq. (3.90) and where we assumed that the photons are emitted in the direction of propagation of the electron.[3]

Intuitively, this collision operator is just an energy balance on the electron population due to photon emission. The number of electrons at an energy γ is increased by electrons having an energy $\gamma + \gamma_\gamma$ and emitting a photon with an energy γ_γ [first term of the RHS of Eq. (4.23)], while it is decreased by electrons at this energy γ emitting a photon with any energy γ_γ [second term of the RHS of Eq. (4.23)] (see Fig. 4.1).

4.3.4 Vlasov Equation with Pair-Production

When the quantum parameter of the emitted photons is of the order or above unity, they can decompose into an electron-positron pair by stimulated (or nonlinear) Breit-Wheeler pair production according to (3.90). Pair production can be taken into account in the same way as nonlinear Compton scattering, using a collision operator to follow the electron, positron and photon distributions in the energy space. However, the situation is more complicated here since the emitted photons can now decompose into electron/positron pairs, which in turn radiate new photons etc. This then yields a coupled system between electrons, positrons and photons. Writing f_e, f_p and f_γ the electron, positron and photon distribution functions (resp.), we get

$$\frac{\partial f_e}{\partial t} + \mathbf{v} \cdot \frac{\partial f_e}{\partial \mathbf{x}} + \mathbf{F}_{\mathrm{Lorentz}} \cdot \frac{\partial f_e}{\partial \mathbf{p}} = C_{\mathrm{nCs}}[f_e] + C^e_{\mathrm{nBW}}[f_\gamma] \tag{4.24a}$$

$$\frac{\partial f_p}{\partial t} + \mathbf{v} \cdot \frac{\partial f_p}{\partial \mathbf{x}} + \mathbf{F}_{\mathrm{Lorentz}} \cdot \frac{\partial f_p}{\partial \mathbf{p}} = C_{\mathrm{nCs}}[f_p] + C^e_{\mathrm{nBW}}[f_\gamma] \tag{4.24b}$$

$$\frac{\partial f_\gamma}{\partial t} + \mathbf{c} \cdot \frac{\partial f_\gamma}{\partial \mathbf{x}} = C_\gamma[f_e] + C_\gamma[f_p] - C^\gamma_{\mathrm{nBW}}[f_\gamma] \tag{4.24c}$$

with

$$C^e_{\mathrm{nBW}}[f_\gamma] \equiv \int_0^{+\infty} d\gamma_\gamma \, \frac{d^2 N_{\mathrm{nBW}}}{dtd\gamma}(\chi_\gamma, \gamma_\gamma, \gamma) f_\gamma(t, \mathbf{x}, \gamma_\gamma, \mathbf{\Omega}) \tag{4.25a}$$

$$C^\gamma_{\mathrm{nBW}}[f_\gamma] \equiv \int_1^{+\infty} d\gamma \, \frac{d^2 N_{\mathrm{nBW}}}{dtd\gamma}(\chi_\gamma, \gamma_\gamma, \gamma) f_\gamma(t, \mathbf{x}, \gamma_\gamma, \mathbf{\Omega}) \tag{4.25b}$$

$$C_\gamma[f_e] \equiv \int_1^{+\infty} d\gamma \, \frac{d^2 N_{\mathrm{nCs}}}{dtd\gamma_\gamma}(\chi, \gamma + \gamma_\gamma, \gamma_\gamma) \, f_e(t, \mathbf{x}, \gamma + \gamma_\gamma, \mathbf{\Omega}) \tag{4.25c}$$

---

[3]This is a good approximation for ultra-relativistic electrons as explained in Chap. 2.

where we assumed that electron-positron pairs are emitted in the direction of propagation of the photon.

4.4 Extended PIC Method for SFQED

As previously explained, the usual Vlasov equation is not sufficient to describe physics in the ultra-relativistic regime and one needs to improve it to include QED processes. These processes thus need to be solved in the PIC loop. We therefore add what we call the *QED modules* to the classical PIC loop. For a review of the implementation of the QED modules in the usual PIC loop, the reader is referred to [7]. We just remind here the two main schemes to simulate classical and quantum RR.

4.4.1 Classical RR Pusher

The first and simplest pusher is the deterministic radiation reaction pusher that allows one to describe the radiating electron dynamics in the framework of classical electrodynamics. Its implementation closely follows that proposed by Tamburini *et al.* [12] with the difference that it relies on the equations of motions (5.10a) and (5.10b).

This pusher (and all pushers discussed here) are based on the *leap-frog* technique and assume that forces and momenta are known at integer (n) and half-integer ($n - \frac{1}{2}$) time-steps, respectively. Following Ref. [12], we first compute the effect of the Lorentz and radiation reaction force separately. Starting from momentum $\mathbf{p}^{(n-\frac{1}{2})}$ and considering the Lorentz force $\mathbf{f}_L^{(n)}$, the first step is performed using the standard Boris pusher [13] giving:

$$\mathbf{p}_L = \mathbf{p}^{(n-\frac{1}{2})} + \mathbf{f}_L^{(n)}\,\Delta t, \tag{4.26}$$

with Δt the time-step. In a second step, we compute the effect of the radiation reaction force:

$$\mathbf{p}_R = \mathbf{p}^{(n-\frac{1}{2})} + \mathbf{f}_{\rm rad}^{(n)}\,\Delta t, \tag{4.27}$$

where $\mathbf{f}_{\rm rad}^{(n)} = P_{\rm cl}\,\mathbf{u}/(c\mathbf{u}^2)$ [Eqs. (5.10b) and (5.11)] is computed using the particles properties at time step ($n - \frac{1}{2}$) and current value of the fields to estimate χ. Finally, the momentum at time-step ($n + \frac{1}{2}$) is computed as:

$$\mathbf{p}^{(n+\frac{1}{2})} = \mathbf{p}_L + \mathbf{p}_R - \mathbf{p}^{(n-\frac{1}{2})}. \tag{4.28}$$

This pusher has been validated (not shown) against analytical solutions for the cases of a radiating electron in a constant, homogeneous magnetic field [14] and a plane-wave [15].

4.4.2 Monte-Carlo Module

Our implementation closely follows that presented in Refs. [16–18]. To treat the discontinuous process of high-energy photon emission each electron is first assigned a final optical depth τ_f sampled from $\tau_f = -\ln(r)$ with $0 < r \leq 1$ a uniform random number. At the same time, a *current* optical depth τ_c is assigned to each electron, which is initialized at 0 and updated in time (possibly using a sub-cycling with respect to the main, Lorentz, loop at Δt) following

$$\frac{d\tau_c}{dt} = \int_0^{\chi} \frac{d^2 N_{\rm nCs}}{d\chi_\gamma dt} d\chi_\gamma. \tag{4.29}$$

When τ_c reaches the final optical depth τ_f, the electron emits a photon. The emitted photon quantum parameter is computed inverting the cumulative distribution function

$$\mathrm{CDF}(\chi_\gamma) = \frac{\int_0^{\chi_\gamma} G(\chi, \chi'_\gamma)/\chi'_\gamma \, d\chi'_\gamma}{\int_0^{\chi} G(\chi, \chi'_\gamma)/\chi'_\gamma \, d\chi'_\gamma}, \tag{4.30}$$

with χ the electron quantum parameter at the time of emission. From Eq. (5.19), this uniquely defines the energy of the emitted photon $\varepsilon_\gamma = mc^2 \gamma \chi_\gamma / \chi$ (with γ the energy of the radiating electron), and the electron momentum right after emission $\mathbf{p}^+$ is then updated considering forward emission $\mathbf{\Omega} = \mathbf{p}/|\mathbf{p}|$:

$$\mathbf{p}^+ = \mathbf{p} - \frac{\varepsilon_\gamma}{c} \mathbf{\Omega}. \tag{4.31}$$

Note that this implementation, which conserves momentum, does not exactly conserve energy. The error made on the energy is however small for ultra-relativistic electrons ($\gamma \ll 1$) [6]. Finally, we note that, in between emission events, the electron dynamics governed by the Lorentz force is updated as in the previous pushers using the Boris approach.

The Monte-Carlo has been validated (not shown here) against direct resolution of the partial-integro differential equation (4.23) for the cases of a radiating electron in a constant, homogeneous magnetic field as well as in a plane electromagnetic wave.

References

1. Krall NA (1973) Principles of Plasma Physics. McGraw-Hill, New York
2. Landau LD, Lifshitz EM (1979) Physical kinetics. Robert Maxwell, M.C
3. Nicholson DR (1983) Introduction to plasma theory. Wiley, Hoboken
4. Tajima T (1989) Computational plasma physics: with applications to fusion and astrophysics. CRC Press, Boca Raton
5. Birdsall CK, Langdon A (1991) Plasma physics via computer simulation. CRC Press, Boca Raton
6. Lobet M (2015) Radiative and QED effects in ultra-relativistic laser-matter interaction. Ph.D. thesis, Université de Bordeaux
7. Gonoskov A, Bastrakov S, Efimenko E, Ilderton A, Marklund M, Meyerov I, Muraviev A, Sergeev A, Surmin I, Wallin E (2015) Extended particle-in-cell schemes for physics in ultrastrong laser fields: Review and developments. Phys Rev E 92:023305
8. Jackson JD (1999) Classical electrodynamics, 3rd edn. Wiley, Hoboken
9. Gardiner CW (1989) Handbook of stochastic methods. Springer, Heidelberg
10. Tamburini M (2011) Radiation reaction effects in superintense laser-plasma interaction. Ph.D. thesis, University of Pisa
11. Kloeden PE, Platen E (1991) Numerical solution of stochastic differential equations. Springer, Heidelberg
12. Tamburini M, Pegoraro F, Piazza AD, Keitel CH, Macchi A (2010) Radiation reaction effects on radiation pressure acceleration. New J Phys 12(12):123005
13. Boris JP (1970) Plasma physics laboratory
14. Elkina NV, Fedotov AM, Herzing C, Ruhl H (2014) Accurate numerical simulation of radiation reaction effects in strong electromagnetic fields. arXiv
15. Di Piazza (2008) Exact solution of the Landau-Lifshitz equation in a plane wave. arXiv
16. Duclous R, Kirk JG, Bell AR (2011) Monte Carlo calculations of pair production in high-intensity laser-plasma interactions. Plasma Phys Control Fusion 53(1):015009
17. Arber TD, Bennett K, Brady CS, Lawrence-Douglas A, Ramsay MG, Sircombe NJ, Gillies P, Evans RG, Schmitz H, Bell AR, Ridgers CP (2015) Contemporary particle-in-cell approach to laser-plasma modelling. Plasma Phys Control Fusion 57:113001
18. Lobet M, d'Humières E, Grech M, Ruyer C, Davoine X, Gremillet L (2016) Modeling of radiative and quantum electrodynamics effects in PIC simulations of ultra-relativistic laser-plasma interaction. J Phys: Conf Ser 688(1):012058

Part II
From Quantum to Classical Radiation Reaction

Chapter 5
Effect of RR on the Electron Distribution Function

5.1 Outline

Over the last years, high-energy photon emission by ultra-relativistic particles and its back-reaction on the particle dynamics, also known as radiation reaction, has received a large interest from the strong-field physics, laser-plasma interaction and astrophysics communities.

The interest of the laser-plasma community is driven by the development of multi-petawatt laser facilities such as ELI [1] or APOLLON [2]. Within the next decade, these laser systems will deliver light pulses with a peak power up to 10 PW and durations in the femtosecond regime, thus allowing to reach on-target intensities close to 10^{23} W/cm^2. This opens a novel regime of relativistic laser-matter interaction ruled by both collective and quantum electrodynamics (QED) effects [3]. Among the latter, high-energy photon emission and electron-positron pair production have attracted a lot of attention [4–12]. Some of these QED processes have been observed in recent laser-plasma experiments, some involving pair production in the Coulomb field of highly-charged ions [13, 14], and most recently in link to the problem of radiation reaction [15, 16]. This line of study is at the center of various proposals for experiments on forthcoming multi-petawatt laser facilities.

Radiation reaction has also been shown to be of importance in various scenarios relating to relativistic astrophysics. Kinetic plasma simulations have demonstrated that it can alter the physical nature of radiation-dominated relativistic current sheets at ultra-high magnetization [17]. Its importance was also demonstrated for the interpretation and modeling of gamma-ray flares in the Crab-Nebulae [18] and of pulsars [19].

These developments have motivated various theoretical works devoted to the treatment of radiation reaction in both classical electrodynamics [20–26], and QED [27–32] (see also Ref. [3] for a review). Radiation reaction, treated either using a radiation friction force in the framework of classical electrodynamics or a Monte-Carlo procedure to account for the quantum process of high-energy photon emission, has also recently been implemented in various kinetic simulation codes, in particular Particle-In-Cell (PIC) codes [4–6, 33–35]. These numerical tools have been used to tackle

F. Niel, *Classical and Quantum Description of Plasma and Radiation in Strong Fields*, Springer Theses, https://doi.org/10.1007/978-3-030-73547-0_5

various problems, from laser-plasma interaction under extreme light conditions to relativistic astrophysics.

QED effects are negligible (so-called classical regime) when the energy of the emitted photons remains small with respect to that of the emitting electron, and radiation reaction can then be treated as a continuous friction force acting on the particles, as proposed e.g. by Landau and Lifshitz [36]. In the quantum regime, photons with energies of the order of the energy of the emitting electrons can be produced [3]. The on-set of QED effects has two important consequences [37]: first, the instantaneous power radiated away by an electron is reduced with respect to the "classical" prediction, second, the discrete and stochastic nature of photon emission impacts the electron dynamics (so-called *straggling*) which cannot be treated using the continuous friction force[1] and thus motivated the development of Monte-Carlo procedures [33, 35].

Considering an ultra-relativistic electron beam interacting with a counter-propagating high-intensity laser pulse, Neitz and Di Piazza have demonstrated that, even in the limit $\chi \lll 1$, the stochastic nature of photon emission cannot always be neglected [39]. Using a Fokker-Planck approach, the authors show that the stochastic nature of high-energy photon emission can lead to an energy spreading (i.e. effective heating) of the electron beam while a purely classical treatment using the radiation friction force would predict only a cooling of the electron beam [25, 40, 41]. Vranic and collaborators have further considered this scenario in Ref. [42] to study the competition between this effective heating and classical cooling of the electron beam distribution.

The present study focuses on the effects of this stochastic nature of high-energy photon emission on radiation reaction. In contrast with previous works [39, 42], we extend the study to $\chi \lesssim 1$ and arbitrary configurations, i.e. we do not restrict ourselves to the study of an electron beam with a counter-propagating light pulse, and demonstrate the existence of an intermediate regime, henceforth referred to as the intermediate quantum regime. To do so, we rely on a statistical approach of radiation reaction, starting from a linear Boltzmann description of photon emission and its back-reaction (from which the Monte-Carlo procedure derives), then studying in detail its Fokker-Planck limit. This procedure and a systematic comparison with the linear Boltzmann description allow to highlight different effects related to the quantum nature of photon emission, among which are the stochastic energy spreading and quantum quenching of radiation losses. The appropriate model that needs to be used in different physical situations and the relevant measurable quantities are discussed.

In particular, beyond the situation of a plane-wave counterpropagating with a narrow electron bunch as considered by [39, 42], we develop in this Chapter a formalism that can be applied to a wide variety of situations of interest for the strong-field physics, laser-plasma interaction and astrophysics communities.

This Chapter is structured as follows. In Sect. 5.2, we summarize the classical treatment of the radiation emission and its back-reaction on the electron dynamics and

[1]Another purely quantum effect is the so-called spin force (see, e.g., Ref. [38]) though this is not considered in this work.

show that, in the case of ultra-relativistic electrons, the momentum and energy evolution equations take a simple and intuitive form. This form of the radiation friction force has the advantage to conserve the on-shell condition while being straightforward to implement numerically. High-energy photon emission and its back-reaction as inferred from the quantum approach is then summarized in Sect. 5.2.3, which introduces the key-quantities that appear in the statistical descriptions we develop in the following Sections. In Sect. 5.3, starting from a kinetic master equation, we derive a Fokker-Planck (FP) equation where quantum effects appear both as a correction on the friction force (drift term) and in a diffusion term, the latter accounting for the stochastic nature inherent to the quantum emission process. Interestingly, the leading term of the Landau-Lifshitz equation with a quantum correction naturally appears from the FP expansion. The domain of validity of the FP expansion is then studied in detail, and for arbitrary conditions of interaction. In Sect. 5.4, the equations of evolution for the successive moments of the electron distribution function are discussed considering the classical, FP and linear Boltzmann descriptions. These equations allow for some analytical predictions on the average energy and energy dispersion when considering, but not limited to, an electron beam interacting with a high-intensity laser field. It also sheds light on other processes such as the quantum quenching of radiation losses observed in recent numerical simulations [43]. Finally, conclusions are given in Sect. 5.6.

5.2 Dynamics of a Radiating Electron in Classical and Quantum Electrodynamics

5.2.1 *Dynamics of a Classical Radiating Electron*

As explained in Chap. 2, the dynamic of a radiating electron is described in CED by the LL equation

$$\frac{dp^\mu}{d\tau} = -\frac{e}{mc}\,F^{\mu\nu}\,p_\nu - \frac{2}{3}\,\tau_e\left[\frac{e}{m^2\,c}\,\partial_\eta F^{\mu\nu}\,p_\nu\,p^\eta + \frac{e^2}{m^2\,c^2}\,F^{\mu\nu}\,F_{\eta\nu}\,p^\eta \right.$$
$$\left. - \frac{e^2}{m^4\,c^4}\,(F^{\nu\eta}\,p_\eta)\,(F_{\nu\alpha}\,p^\alpha)\,p^\mu\right], \tag{5.1}$$

where c is the speed of light in vacuum, τ the proper time, $\tau_e = r_e/c$ the time for light to travel across the classical radius of the electron $r_e = e^2/(4\pi\epsilon_0\,mc^2)$, ϵ_0 the permittivity of vacuum and $p^\mu = (\gamma mc, \mathbf{p})$ the electron four-momentum with $\gamma = \sqrt{1+\mathbf{p}^2/(mc)^2}$ the electron Lorentz factor [SI units will be used throughout this part]. Moreover, we consider a single electron with charge $-e$ and mass m. In 3-vector notation, Eq. (5.1) gives the equations of evolution of the energy and momentum of the electron, respectively

$$\begin{aligned}
mc^2\frac{d\gamma}{dt} &= -ec\,\boldsymbol{\beta}\cdot\mathbf{E} - \frac{2}{3}ec\tau_e\gamma\,\dot{\mathbf{E}}\cdot\boldsymbol{\beta} \\
&+ \frac{2}{3}\frac{ec}{E_{cr}}\mathbf{E}\cdot(\mathbf{E}+\boldsymbol{\beta}\times\mathbf{H}) \\
&- \frac{2}{3}\frac{ec}{E_{cr}}\gamma^2\left[(\mathbf{E}+\boldsymbol{\beta}\times\mathbf{H})^2-(\boldsymbol{\beta}\cdot\mathbf{E})^2\right],
\end{aligned} \tag{5.2a}$$

$$\begin{aligned}
\frac{d\mathbf{p}}{dt} &= -e\,(\mathbf{E}+\boldsymbol{\beta}\times\mathbf{H}) - \frac{2}{3}\,e\,\tau_e\,\gamma\,\left(\dot{\mathbf{E}}+\boldsymbol{\beta}\times\dot{\mathbf{H}}\right) \\
&+ \frac{2}{3}\frac{e}{E_{cr}}\left[(\boldsymbol{\beta}\cdot\mathbf{E})\,\mathbf{E}-\mathbf{H}\times(\mathbf{E}+\mathbf{u}\times\mathbf{H})\right] \\
&- \frac{2}{3}\frac{e}{E_{cr}}\gamma^2\left[(\mathbf{E}+\boldsymbol{\beta}\times\mathbf{H})^2-(\boldsymbol{\beta}\cdot\mathbf{E})^2\right]\boldsymbol{\beta},
\end{aligned} \tag{5.2b}$$

where $\boldsymbol{\beta} = \mathbf{p}/(\gamma mc)$ is the normalized electron velocity, $E_{cr} = 4\pi\epsilon_0 m^2c^4/e^3 \simeq 1.8\times10^{20}$ V/m the critical field of CED, $\mathbf{E}$ and $\mathbf{H}$ are the electric and magnetic fields, respectively, and dotted fields are (totally) differentiated with respect to time t. We see that the last term of the RR force is γ^2 bigger than the other terms. Following this observation, the LL force is often approximated by its last term as

$$mc^2\frac{d\gamma}{dt} = -ec\,\boldsymbol{\beta}\cdot\mathbf{E} - P_{\rm cl}, \tag{5.3a}$$

$$\frac{d\mathbf{p}}{dt} = -e\,(\mathbf{E}+\boldsymbol{\beta}\times\mathbf{H}) - P_{\rm cl}\,\boldsymbol{\beta}, \tag{5.3b}$$

where $P_{\rm cl} = \frac{2}{3}\frac{e}{E_{cr}}\gamma^2\left[(\mathbf{E}+\boldsymbol{\beta}\times\mathbf{H})^2-(\boldsymbol{\beta}\cdot\mathbf{E})^2\right]$. Here, we shall be more careful and analyze the consequences of such an approximation. As a guide line, the physical quantities that we want to conserve are : the on-shell condition, and the momentum radiated by the electron.

Let us first analyze the on-shell condition. In the following, we will write g^μ the LL force

$$\begin{aligned}
g^\mu \equiv -\frac{2}{3}\tau_e\Bigg[&\frac{e}{m^2c}\,\partial_\eta F^{\mu\nu}\,p_\nu\,p^\eta + \frac{e^2}{m^2c^2}\,F^{\mu\nu}\,F_{\eta\nu}\,p^\eta \\
&- \frac{e^2}{m^4c^4}\,(F^{\nu\eta}\,p_\eta)\,(F_{\nu\alpha}\,p^\alpha)\,p^\mu\Bigg],
\end{aligned} \tag{5.4}$$

and g_i^μ its ith term of g^μ. The first term g_1^μ, also known as the Schott term, stands as a four-force (i.e. it is perpendicular to the four-momentum)

$$g_1^\mu\,p_\mu = -\frac{2}{3}\tau_e\frac{e}{m_e^2c}\partial_\eta F^{\mu\nu}\,p_\nu\,p^\eta\,p_\mu = 0, \tag{5.5}$$

by antisymmetry of $F^{\mu\nu}$. Moreover, the last two terms of the bracket have to be kept together to form a four-force f^μ such that $f^\mu\,p_\mu = 0$. Indeed

$$(g_2^\mu + g_3^\mu)\, p_\mu = -\frac{2}{3}\tau_e\left[\frac{e^2}{m_e^2c^2}F^{\mu\nu}\, F_{\eta\nu}\, p^\eta\, p_\mu - \frac{e^2}{m_e^4c^4}(F^{\mu\nu}\, p^\eta)(F_{\nu\alpha}\, p^\alpha)\, p^\mu\, p_\mu\right] = 0, \quad (5.6)$$

using the on-shell condition. So while the first term can safely be neglected, we need to keep the last two terms for the orthogonality condition to be verified.

However, if we now add the constraint that the 4-momentum emitted by the electron $\Delta p^\mu = \int g^\mu d\tau$ must be the same in Eq. (5.3b) as in (5.2b), we are led to compute each term $\int g_i^\mu d\tau$ separately. We obtain

$$\begin{aligned}\int (g^\mu)_1 d\tau &= -\frac{2}{3}\tau_e\frac{e}{m_e^2c}\int \partial_\eta F^{\mu\nu} p_\nu p^\eta d\tau = -\frac{2}{3}\tau_e\frac{e}{m_e^2c}\int \frac{dF^{\mu\nu}}{d\tau}p_\nu d\tau \\ &= -\frac{2}{3}\tau_e\frac{e}{m_e^2c}\left[[F^{\mu\nu}p_\nu] - \int\frac{dp_\nu}{d\tau}F^{\mu\nu}d\tau\right] \qquad (5.7a)\\ &= \frac{2}{3}\tau_e\frac{e^2}{m_e^2c^2}\int F^{\mu\nu}F_{\eta\nu}p^\eta d\tau,\end{aligned}$$

$$\int (g^\mu)_2 d\tau = -\frac{2}{3}\tau_e\frac{e^2}{m_e^2c^2}\int F^{\mu\nu}F_{\eta\nu}p^\eta d\tau, \qquad (5.7b)$$

$$\int (g^\mu)_3 d\tau = \frac{2}{3}\tau_e\frac{e^2}{m_e^4c^4}\int (F^{\nu\eta}p_\eta)(F^{\nu\alpha}p^\alpha)p^\mu d\tau. \qquad (5.7c)$$

Therefore, the first two terms cancel each other while the last term corresponds to the total four-momentum radiated away by the particle

$$\Delta p^\mu = \frac{P_0}{c^2}\int\left|\frac{F^{\nu\eta}}{E_{cr}}\frac{p_\eta}{mc}\right| dx^\mu, \qquad (5.8)$$

with $P_0 = 2\,mc^2/(3\,\tau_e)$ and x^μ the electron four-position. Therefore, it is not possible, in general, to consider each term of the LL force separately. However, it is possible to rearrange parts of g_i^μ so that the constraints (on-shell condition and emitted 4-momentum) are satisfied.

5.2.2 *Radiation Friction Force Acting on an Ultra-Relativistic Electron*

To consider the action of the force on an ultra-relativistic electron, we develop the radiation reaction force $\mathbf{f}_{\mathrm{rad}}$ [three last terms in Eq. (5.4)] as a longitudinal force (acting in the direction of the electron velocity) and a transverse force (acting in the normal direction to the electron velocity). One obtains $\mathbf{f}_{\mathrm{rad}} = \mathbf{f}_{\mathrm{rad}}^{\parallel} + \mathbf{f}_{\mathrm{rad}}^{\perp}$ with

$$\begin{aligned}\mathbf{f}_{\rm rad}^{\parallel} &= -\frac{2}{3}\, e\tau_e\gamma(\dot{\mathbf{E}}\cdot\boldsymbol{\beta})\,\boldsymbol{\beta}/\beta^2 \\ &+ \frac{2}{3}\frac{e}{E_{cr}}\,\mathbf{E}\cdot(\mathbf{E}+\boldsymbol{\beta}\times\mathbf{H})\,\boldsymbol{\beta}/\beta^2 \\ &- \frac{2}{3}\frac{e}{E_{cr}}\,\gamma^2\left[(\mathbf{E}+\boldsymbol{\beta}\times\mathbf{H})^2-(\boldsymbol{\beta}\cdot\mathbf{E})^2\right]\boldsymbol{\beta}/\beta^2,\end{aligned} \tag{5.9a}$$

$$\begin{aligned}\mathbf{f}_{\rm rad}^{\perp} &= -\frac{2}{3}\, e\tau_e\gamma\left[(\dot{\mathbf{E}})_\perp+\boldsymbol{\beta}\times\dot{\mathbf{H}}\right] \\ &+ \frac{2}{3}\frac{e}{E_{cr}}\left[(\boldsymbol{\beta}\cdot\mathbf{E})\,\mathbf{E}_\perp+(\boldsymbol{\beta}\cdot\mathbf{H})\,\mathbf{H}_\perp+(\mathbf{E}\times\mathbf{H})_\perp\right],\end{aligned} \tag{5.9b}$$

where $\parallel$ ($\perp$) denotes the vector component parallel (perpendicular) to the electron velocity $\boldsymbol{\beta}$. Let us stress at this point that the last term in Eq. (5.9a) contains not only the contribution of the last term of the LL radiation reaction force Eq. (5.4), but also part of its second term.

For an ultra-relativistic electron ($\gamma \gg 1$), the last terms in Eqs. (5.4) and (5.9a) give the important contributions, and all other terms together with the perpendicular component of the radiation reaction force [Eq. (5.9b)] can be neglected.[2] One then obtains the equations of evolution for the ultra-relativistic electron energy and momentum

$$mc^2\frac{d\gamma}{dt} = -ec\,\boldsymbol{\beta}\cdot\mathbf{E} - P_{\rm cl}, \tag{5.10a}$$

$$\frac{d\mathbf{p}}{dt} = -e\,(\mathbf{E}+\boldsymbol{\beta}\times\mathbf{H}) - P_{\rm cl}\,\boldsymbol{\beta}/(c\beta^2), \tag{5.10b}$$

where $P_{\rm cl}$ denotes the classical instantaneous power radiated away by the electron

$$P_{\rm cl} = P_0\,\eta^2, \tag{5.11}$$

for which we have introduced

$$\eta = \frac{\gamma}{E_{cr}}\sqrt{(\mathbf{E}+\boldsymbol{\beta}\times\mathbf{H})^2-(\boldsymbol{\beta}\cdot\mathbf{E})^2}. \tag{5.12}$$

In covariant notation, η clearly appears as a Lorentz invariant

$$\eta = \left|\frac{F^{\mu\nu}}{E_{cr}}\frac{p_\nu}{mc}\right|. \tag{5.13}$$

In contrast with Eq. (5.3b), the radiation reaction force, last term in Eq. (5.10b), takes the form of a friction force $\mathbf{f}_{\rm rad} = \nu\boldsymbol{\beta}$ [with a nonlinear friction coefficient ν] that also has the property of conserving the on-shell condition $p^\mu p_\mu = m^2c^2$. This

[2]This is coherent with the well known result discussed in Chap. 2: the radiation of an ultra-relativistic electron is directed in a cone of angle $\sim 1/\gamma$ in the direction of its velocity.

can be seen by taking the scalar product of Eq. (5.10b) by $c\boldsymbol{\beta}$, which turns out to be consistent with the energy conservation Eq. (5.10a). This would not have been the case had we retained only the leading ($\propto \gamma^2$) terms in Eq. (5.1). Our formulation hence contrasts with that proposed, e.g. by Tamburini *et al.* [4], where the authors choose to retain both last two terms in Eq. (5.1) to preserve the on-shell condition. Beyond its simple and intuitive form, the radiation force given by Eq. (5.10b) is also straightforward to implement in numerical tools. It also allows to avoid the computation of two components in the radiation reaction force which have large ($\propto \gamma^2$) differences in their magnitude and that may lead to accumulation of round-off errors (see, e.g., Ref [44] for an accurate treatment of this problem).

5.2.3 Dynamics of a Quantum Radiating Electron

Let us now recall some notations as well as the results derived in Chap. 3 that will be useful in this chapter. We make here the same assumptions as made in Sect. 3.5 (we consider a stable vacuum, in the LCFA and in the nonlinear moderately quantum regime). Under these assumptions, the (Lorentz invariant) production rate of high-energy photons emitted by the electron can be written as [45]:

$$\frac{d^2N}{d\tau d\chi_\gamma} = \frac{2}{3}\frac{\alpha^2}{\tau_e}\frac{G(\chi,\chi_\gamma)}{\chi_\gamma}, \tag{5.14}$$

where

$$G(\chi,\chi_\gamma) = \frac{\sqrt{3}}{2\pi}\frac{\chi_\gamma}{\chi}\left[\int_\nu^{+\infty} \mathrm{K}_{5/3}(y)\,dy + \frac{3}{2}\chi_\gamma\,\nu\,\mathrm{K}_{2/3}(\nu)\right], \tag{5.15}$$

is the so-called quantum emissivity, $\nu = 2\,\chi_\gamma/[3\chi\,(\chi-\chi_\gamma)]$ and where

$$\chi = \left|\frac{F^{\mu\nu}}{E_s}\frac{p_\nu}{mc}\right| = \frac{\eta}{\alpha}, \tag{5.16}$$

is the Lorentz invariant quantum parameter for the electron. We will also need

$$a_0 = \frac{e\,|A^\mu|}{mc^2} \gg 1, \tag{5.17}$$

where A^μ is the four-potential corresponding to the electromagnetic field tensor $F^{\mu\nu} = \partial^\mu A^\nu - \partial^\nu A^\mu$. The production rate Eq. (5.14) only depends on the electron quantum parameter χ and on the (Lorentz invariant) quantum parameter for the emitted photon:

$$\chi_\gamma = \left| \frac{F^{\mu\nu}}{E_s} \frac{\hbar k_\nu}{mc} \right|, \tag{5.18}$$

where $k^\nu = (\hbar\omega/c, \hbar\mathbf{k})$ is the four-momentum of the emitted photon. Considering an ultra-relativistic electron, the photon quantum parameter χ_γ can be expressed in terms of the electron quantum parameter χ and the electron and photon energies as

$$\chi_\gamma = \frac{\gamma_\gamma}{\gamma} \chi, \tag{5.19}$$

where $mc^2\gamma_\gamma$ is the photon energy. Another Lorentz invariant can be derived from Eq. (5.14)

$$\frac{d^2\mathcal{E}}{d\tau d\gamma_\gamma} = P_0\, \alpha^2\, G(\chi, \chi_\gamma), \tag{5.20}$$

which denotes the emitted power distribution in terms of the photon normalized energy. The instantaneous power radiated away by the electron is another Lorentz invariant. It is obtained by integrating Eq. (5.20) over all photon energies giving

$$P_{\text{rad}} = \int_0^{+\infty} d\gamma_\gamma\, \frac{1}{\gamma} \frac{d^2\mathcal{E}}{d\tau d\gamma_\gamma} = P_0\, \alpha^2 \chi^2\, g(\chi), \tag{5.21}$$

where

$$g(\chi) = \int_0^{+\infty} d\chi_\gamma\, \frac{G(\chi, \chi_\gamma)}{\chi^3} = \frac{9\sqrt{3}}{8\pi} \int_0^{+\infty} d\nu \left[\frac{2\nu^2\, \mathrm{K}_{5/3}(\nu)}{(2 + 3\nu\chi)^2} + \frac{4\nu\, (3\nu\chi)^2}{(2 + 3\nu\chi)^4}\, \mathrm{K}_{2/3}(\nu) \right]. \tag{5.22}$$

Figure 3.5(a) shows $g(\chi)$ for χ ranging from 10^{-5} to 10.

Let us now stress that Eq. (5.21) is nothing but the classical instantaneous power radiated away by the electron [Eq. (5.11)] multiplied by $g(\chi)$ (we recall that $\chi = \eta/\alpha$):

$$P_{\text{rad}} = P_{\text{cl}}\, g(\chi). \tag{5.23}$$

The classical limit is recovered when the emitted photon energies remain much smaller than the emitting electron energy, i.e. by taking the limit $\chi_\gamma \ll \chi \ll 1$ [correspondingly $\nu \sim 2\chi_\gamma/(3\chi^2)$]. In this limit, $g(\chi) \sim 1$, and Eqs. (5.20) and (5.21) reduce to their classical forms Eqs. (2.125) and (5.11), respectively.

Therefore, to take into account the difference between the classical and quantum radiated spectrum in our classical equation of motion [Eqs. (5.10a) and (5.10b)], we can replace phenomenologically P_{cl} by its quantum expression P_{rad}, $g(\chi)$ giving a so-called quantum correction (see, e.g., [46, 47]) If this approach is here mainly heuristic, we will see in Sect. 5.3 that it has a rigorous interpretation, the statistical average of the quantum description providing the quantum correction naturally.

Finally, to highlight QED effects on the emitted radiation properties, we have plotted $G(\chi, \chi_\gamma)/\chi^2$ and its classical limit in Fig. 3.5(b) and 3.5(c), respectively. As can be seen, quantum effects mainly tend to decrease the photon emission rate at high-energies. In particular, emission of photons with an energy larger than the emitting photon energy (i.e. for $\chi_\gamma > \chi$) is prevented. As a result, the overall emitted power is reduced [see also Fig. 3.5(a)].

The process of photon emission can be described by three different (but related) random variables: (i) the electron energy itself, (ii) the number N_t of photon emission events in a time interval $[0, t]$, and (iii) the time T_n of the nth emission event. The last two variables are of course equivalent since $T_n \geq t \iff N_t \leq n$, both denoting that there are at least n emissions in the time interval $[0, t]$. It is possible to show that N_t follows a Poisson process of parameter

$$\tau(\chi, \gamma, t) = \int_0^t W(\chi, \gamma) dt' \tag{5.24}$$

which is usually referred to as the *optical depth*, and where:

$$W(\chi, \gamma) = \frac{2}{3} \frac{\alpha^2}{\tau_e \gamma} \int_0^{+\infty} d\xi \, \widetilde{G}(\chi, \xi)/\xi \tag{5.25}$$

is the instantaneous rate of photon emission. Hence, the probability for the electron to emit n photons during a time interval t is given by

$$P[N(\gamma, t) = n] = e^{-\tau(\chi,\gamma,t)} \frac{\tau(\chi, \gamma, t)^n}{n!}, \tag{5.26}$$

while the cumulative probability of the random variable T_n is given by

$$P[T_{n\geq 1} < t] = 1 - e^{-\tau(\chi,\gamma,t)}. \tag{5.27}$$

A discrete stochastic formulation of these discontinuous jumps can be rigorously deduced [48], leading to a Monte-Carlo description (see Sect. 4.4.2 and Refs. [33, 35] for more details).

5.3 From Quantum to Classical Radiation Reaction for Ultra-relativistic Electrons

In this Section, we propose a statistical description of high-energy photon emission and its back-reaction, starting from the quantum point of view, and getting toward the classical regime.

5.3.1 *Kinetic Point of View: The Linear Boltzmann Equation*

We assume here that all electrons are considered to emit high-energy radiation in an incoherent way, that is the radiation emission by an electron is not influenced by neighbor electrons. This is justified whenever the high-energy photon emission has a wavelength much shorter than the typical distance between two electrons $\propto n_e^{-1/3}$, with n_e the characteristic density of the electron population [49].

The equation of evolution for the electron distribution function $f_e(t, \boldsymbol{r}, \gamma, \boldsymbol{\Omega})$ accounting for the effect of high energy photon emission and the corresponding photon distribution function $f_\gamma(t, \boldsymbol{r}, \gamma, \boldsymbol{\Omega})$ can be written in the form:

$$\frac{d}{dt} f_e = \int_0^{+\infty} d\gamma_\gamma \, w_\chi(\gamma + \gamma_\gamma, \gamma_\gamma) \, f_e(t, \mathbf{x}, \gamma + \gamma_\gamma, \Omega)$$

$$- f_e(t, \mathbf{x}, \gamma, \Omega) \int_0^{+\infty} d\gamma_\gamma w_\chi(\gamma, \gamma_\gamma), \tag{5.28a}$$

$$\frac{d}{dt} f_\gamma = \int_1^{+\infty} d\gamma \, w_\chi(\gamma + \gamma_\gamma, \gamma_\gamma) \, f_e(t, \mathbf{x}, \gamma + \gamma_\gamma, \Omega), \tag{5.28b}$$

where it has been assumed that radiation emission (and its back-reaction) is dominated by the contribution of ultra-relativistic electrons (for which $\mathbf{p} \simeq mc\gamma\Omega$), and that such ultra-relativistic electrons emit radiation in the direction Ω of their velocity, and the total time derivatives in Eqs. (5.28a) and (5.28b) will be detailed in Sect. 5.3.2. The system of Eqs. (5.28a) and (5.28b) stands as a Master equation. It describes a discontinuous jump process with $w_\chi(\gamma, \gamma_\gamma)$ giving the rate of jump from a state of electron energy $mc^2\gamma$ to the state of energy $mc^2(\gamma - \gamma_\gamma)$, via the emission of a photon of energy $mc^2\gamma_\gamma$.

Equation (5.28a) is a linear Boltzmann equation, and its right-hand-side (rhs), henceforth denoted $C[f_e]$, acts as a *collision* operator (see also Refs. [23, 39, 50, 51]). It accounts for the effect of high-energy photon emission on the dynamics of an electron radiating in the electromagnetic fields **E** and **H**, that is for radiation reaction. It depends on $w_\chi(\gamma, \gamma_\gamma)$ which denotes the rate of emission of a photon with energy $mc^2\gamma_\gamma$ by an electron with energy $mc^2\gamma$ and quantum parameter χ. Note that the dependency on χ implicitly states that the emission rate is computed locally in space and time, i.e. taking the local value of the electromagnetic field at time t and position

x, for a given electron momentum direction $\mathbf{\Omega}$. Under the assumptions previously introduced (Sect. 5.2.3), this emission rate reads:

$$w_\chi(\gamma, \gamma_\gamma) = \left.\frac{d^2N}{dtd\gamma_\gamma}\right|_\chi (\gamma_\gamma, \gamma) = \frac{2}{3}\frac{\alpha^2}{\tau_e}\frac{\widetilde{G}(\chi, \gamma_\gamma/\gamma)}{\gamma\gamma_\gamma}, \tag{5.29}$$

where:

$$\widetilde{G}(\chi, \xi) = \frac{\sqrt{3}}{2\pi}\xi\left[\int_\nu^{+\infty} K_{5/3}(y)dy + \frac{\xi^2}{1-\xi}K_{2/3}(\nu)\right], \tag{5.30}$$

with $\xi = \chi_\gamma/\chi = \gamma_\gamma/\gamma$ and $\nu = 2\xi/[3\chi(1-\xi)]$. It is complemented by Eq. (5.28b) that describes the temporal evolution of the photon distribution function. In this work, photons are simply created and then propagate freely. The rhs of Eq. (5.28b) thus stands as a source term and will be denoted $\mathcal{S}[f_e]$ in the rest of this work.

In Sect. 5.4, we will show that Eq. (5.28a) conserves the total number of electrons, while Eq. (5.28b) predicts a total number of photons increasing with time as more and more photons are radiated away. It will also be demonstrated that the total energy lost by electrons due to radiation emission is indeed transferred to high-energy photons, that is, the system of Eqs. (5.28a) and (5.28b) does conserve the total energy in the system. While it allows to fully model high-energy photon emission and its back-reaction as depicted by the linear Boltzmann Eq. (5.28a) and Eq. (5.28b), the Monte-Carlo procedure (Sect. 4.4.2) has some limitations. Indeed, in regimes of intermediate χ parameters, numerous discrete events of small energy content may occur, giving rise to computational cost overhead. These events may however have a non-negligible cumulative effect. As will be shown in what follows, this case is precisely the operating regime of the Fokker-Planck approximation (a by-product of the master equation). In the following, we show that a Fokker-Planck approach can be used to treat many discrete events at once.

5.3.2 *Toward the Classical Limit: The Fokker-Planck Approach*

Let us now focus on the linear Boltzmann Eq. (5.28a) which we rewrite in the form (see Appendix C.1 for more details):

$$\begin{aligned}\partial_t f_e + \nabla\cdot[c\beta\mathbf{\Omega} f_e] - \frac{1}{mc^2}\partial_\gamma[ec\beta(\mathbf{\Omega}\cdot\mathbf{E})f_e] \\ - \frac{e}{p}\nabla_\Omega\cdot[(\mathbb{1} - \mathbf{\Omega}\otimes\mathbf{\Omega})\cdot(\mathbf{E}+\beta\mathbf{\Omega}\times\mathbf{H})f_e] = \mathcal{C}[f_e],\end{aligned} \tag{5.31}$$

where $\beta = \sqrt{\gamma^2 - 1}/\gamma$, $p = mc\sqrt{\gamma^2 - 1}$, $\nabla_{\mathbf{\Omega}}$ denotes the derivative with respect to $\mathbf{\Omega}$, $\mathbb{1}$ the rank 2 unit tensor and $\otimes$ stands for the dyadic product. While the left hand side of Eq. (5.31) is the standard Vlasov operator written for the energy-direction distribution function $f_e(t, \mathbf{r}, \gamma, \mathbf{\Omega})$ (see e.g. Ref. [52]), $C[f_e]$ is the *collision* operator given by the rhs side of Eq. (5.28a).

Rewriting the integrand in the first integral of $C[f_e]$ as a Taylor series in γ_γ/γ, the *collision* operator can be formally casted in the form of the Kramers-Moyal expansion:

$$\mathscr{C}[f_e] = \sum_{n=1}^{\infty} \frac{1}{n!} \partial_\gamma^n \left[A_n(\chi, \gamma) f_e \right], \tag{5.32}$$

with $A_n(\chi, \gamma) = \int d\gamma_\gamma \, \gamma_\gamma^n \, w_\chi(\gamma, \gamma_\gamma)$ the nth moment associated to the kernel $w_\chi(\gamma, \gamma_\gamma)$. For the particular kernel given by Eq. (5.29), we get for the associated moments:

$$A_n(\chi, \gamma) = \frac{2}{3} \frac{\alpha^2}{\tau_e} \gamma^{n-1} a_n(\chi) \tag{5.33}$$

with:

$$a_n(\chi) = \int_0^{+\infty} d\xi \, \xi^{n-1} \, \tilde{G}(\chi, \xi). \tag{5.34}$$

Plugging the expression for $\tilde{G}(\chi, \xi)$ (5.30) in Eq. (5.34) yields after integration by parts

$$a_n(\chi) = \frac{\sqrt{3}}{2\pi} 3^{n+1} \chi^{n+1} \int_0^{+\infty} d\nu \left[\frac{9\chi^2 \nu^{n+2}}{(2 + 3\nu\chi)^{n+3}} K_{2/3}(\nu) \right. \tag{5.35}$$

$$\left. + \frac{\nu^{n+1}}{(n+1)(2 + 3\nu\chi)^{n+1}} K_{5/3}(\nu) \right], \tag{5.36}$$

which in the limit $\chi \ll 1$ reduces to:

$$a_n(\chi) \sim \frac{\sqrt{3}}{4\pi(n+1)} 3^{n+1} \, \Gamma\left(\frac{n}{2} + \frac{1}{6}\right) \Gamma\left(\frac{n}{2} + \frac{11}{6}\right) \chi^{n+1},$$

with Γ the Gamma function (See Appendix C.2 for more details). In Fig. C.1 we plot $a_0(\chi)$ to $a_4(\chi)$ as well as their asymptotic expansions as given by (5.37) in dashed lines (Fig. 5.1).

Note that the first moment depends on γ only through χ and reads $A_1(\chi, \gamma) = \frac{2}{3} \frac{\alpha^2}{\tau_e} \chi^2 g(\chi)$ [where $\chi^2 g(\chi) = a_1(\chi)$, and $g(\chi)$ is the quantum correction given by Eq. (5.22)] (See Appendix C.2 for more details about the functions $a_n(\chi)$). In general, using expansion (5.32) for the operator in the linear Boltzmann Eq. (5.31)

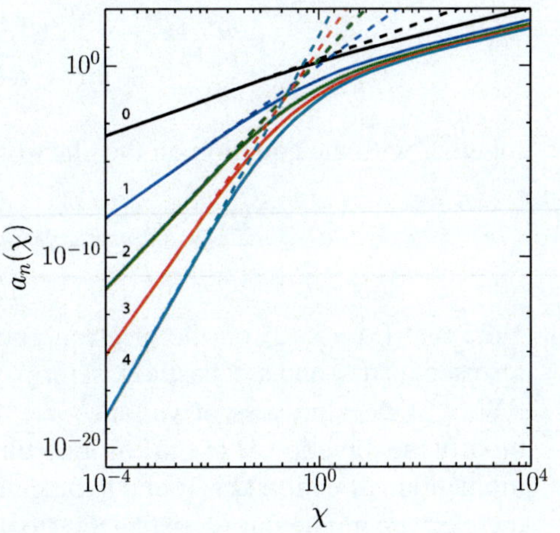

Fig. 5.1 Plot of $a_n(\chi)$ for $n = 0...4$ and their asymptotic expansion in dashed lines, for $\chi = 10^{-4}$ to 10^4.

would require to solve an infinite order partial differential equation. Therefore, it is common to truncate Eq. (5.32). This truncation cannot however be done properly by a finite and larger than two number of terms [53].

The truncation using the first two terms in Eq. (5.32) is actually justified in the limit $\gamma_\gamma \ll \gamma$. This is ensured for $\chi \ll 1$, i.e. in the classical regime of radiation emission, and the resulting truncation corresponds to a Fokker-Planck expansion. In this limit, the *collision* operator reduces to:

$$\mathcal{C}_{\mathrm{FP}}\,[f_e] = \partial_\gamma[S(\chi)f_e] + \frac{1}{2}\partial_\gamma^2\big[R(\chi,\gamma)f_e\big], \tag{5.37}$$

the first term being referred to as the *drift* term, and the second one as the *diffusion* term, and where we have introduced:

$$S(\chi) = A_1(\chi) = \int_0^{+\infty} d\gamma_\gamma\, \gamma_\gamma\, w_\chi(\gamma,\gamma_\gamma) = \frac{2}{3}\frac{\alpha^2}{\tau_e}\,\chi^2\, g(\chi), \tag{5.38a}$$

$$R(\chi,\gamma) = A_2(\chi) = \int_0^{+\infty} d\gamma_\gamma\, \gamma_\gamma^2\, w_\chi(\gamma,\gamma_\gamma) = \frac{2}{3}\frac{\alpha^2}{\tau_e}\,\gamma\, h(\chi), \tag{5.38b}$$

where $h(\chi) = a_2(\chi)$ reads:

$$h(\chi) = \frac{9\sqrt{3}}{4\pi}\int_0^{+\infty} d\nu\left[\frac{2\chi^3\nu^3}{(2+3\nu\chi)^3}\mathrm{K}_{5/3}(\nu) + \frac{54\chi^5\nu^4}{(2+3\nu\chi)^5}\mathrm{K}_{2/3}(\nu)\right]. \tag{5.39}$$

Equation (5.31) rewritten using the operator Eq. (5.37) is a Fokker-Planck equation. Mathematically, it is equivalent to the Ito stochastic differential equation for the random process $\gamma(t)$ [54]:

$$mc^2 d\gamma = -ec\,(\boldsymbol{\beta}\cdot\mathbf{E})\,dt - mc^2 S(\chi)dt + mc^2\sqrt{R(\chi,\gamma)}\,dW, \tag{5.40}$$

together with the equation on the electron momentum direction $\boldsymbol{\Omega}$:

$$\frac{d\boldsymbol{\Omega}}{dt} = -\frac{e}{p}(\mathbb{1} - \boldsymbol{\Omega}\otimes\boldsymbol{\Omega})\cdot(\mathbf{E} + \beta\boldsymbol{\Omega}\times\mathbf{H}). \tag{5.41}$$

Note that Eq. (5.40) on the electron energy contains both deterministic (first two terms in its rhs) and stochastic (last term) increments, the latter being modeled using dW, a Wiener process of variance dt. As high-energy photon emission does not modify the direction $\boldsymbol{\Omega}$ of the emitting ultra-relativistic electron, the equation on the momentum direction $\boldsymbol{\Omega}$ is found to be given by the Lorentz force only. It follows that the electron momentum satisfies the stochastic differential equation:

$$d\mathbf{p} = -e(\mathbf{E} + \boldsymbol{\beta}\times\mathbf{H})dt - mc^2\,S(\chi)\,\boldsymbol{\beta}/(c\boldsymbol{\beta}^2)dt + mc^2\sqrt{R(\chi,\gamma)}\,dW\,\boldsymbol{\beta}/(c\boldsymbol{\beta}^2). \tag{5.42}$$

Derived from the framework of quantum electrodynamics in the limit $\gamma_\gamma \ll \gamma$, Eqs. (5.40) and (5.42) are the generalization of the purely deterministic equations of motion Eqs. (5.10a) and (5.10b) derived in the framework of classical electrodynamics (CED). The first terms in the rhs of Eqs. (5.40) and (5.42) correspond to the effect of the Lorentz force. The second terms follow from the *drift* term of the Fokker-Planck operator [Eq. (5.37)] and account for the deterministic effect of radiation reaction on the electron dynamics. It is important to point out that, from Eq. (5.38a), we get

$$P_{\rm rad} = P_{\rm cl}\,g(\chi) = mc^2\,S(\chi). \tag{5.43}$$

The deterministic terms thus turn out to be *the leading terms of the LL equation including the quantum correction* introduced phenomenologically in Sect. 5.2.3, and here rigorously derived from the quantum framework. Note that this procedure fully and rigorously justifies the use of the quantum corrected LL friction force in Particle-In-Cell codes.

Finally, the last terms in Eqs. (5.40) and (5.42), which follow from the *diffusion* term of the Fokker-Planck operator [Eq. (5.37)], account for the stochastic nature of high-energy photon emission and its back-reaction on the electron dynamics. It is a purely quantum effect, which is not present in the framework of CED. As a result, Eqs. (5.40) and (5.42) extend the validity of Eqs. (5.10a) and (5.10b) from the classical regime of radiation reaction ($\chi \ll 1$) to the intermediate quantum regime ($\chi \lesssim 1$) by accounting for both the deterministic radiation friction force, and the stochastic nature of radiation emission. The domain of validity of this Fokker-Planck description and the extension of validity to the weakly quantum regime $\chi \lesssim 1$ will now be discussed in more details.

Before doing so, however, we want to briefly discuss how our findings fit in with respect to previous works. In contrast with the work of Elkina *et al.* [50] (in which the Master equation approach is applied to the cascades in circularly polarized laser fields), and that by Neitz and Di Piazza [39] (in which the authors provide a Fokker-Planck based analytical description of an electron beam colliding with an ultra-relativistic light pulse), our approach is more general. No assumption is here done on the electron and field configuration, which led us to treat the full Vlasov operator in Eq. (5.31) and allowed us to derive Eq. (5.42), valid for arbitrary geometries. As we will show in the next Sections, this will allow us to bring new useful insights and predictions, and opens new opportunities in the numerical treatment of radiation reaction in arbitrary geometries.

5.3.3 Domain of Validity of the Fokker-Planck and Quantum-Corrected Landau-Lifshitz Descriptions

Let us now study the domain of validity of the previously derived Fokker-Planck and quantum-corrected Landau-Lifshitz descriptions. We use as a starting point the Kramers-Moyal expansion [Eq. (5.32)] for the *collision* operator. If the high-order moments of the kernel in Eq. (5.32) do not give a proper description of the *collision* operator unless all accounted for, computing them still allows us to infer the limit of validity of the Fokker-Planck and corrected Landau-Lifshitz descriptions, and of Eqs. (5.40) and (5.42), in particular.

Here we derive an estimate of the relative importance of the successive terms in the Kramers-Moyal expansion (5.32), by computing:

$$B_n^{n+1} = \frac{n!}{(n+1)!} \frac{\partial_\gamma^{n+1}[A_{n+1} f_e]}{\partial_\gamma^n [A_n f_e]} \sim \frac{A_{n+1}}{\gamma (n+1) A_n}. \tag{5.44}$$

For the particular kernel Eq. (5.29), we can use Eq. (C.2), and the ratio of the $(n+1)^{th}$ to the nth contribution is found to depend on the χ parameter only:

$$B_n^{n+1} = \frac{1}{n+1} \frac{a_{n+1}(\chi)}{a_n(\chi)}. \tag{5.45}$$

In the limit $\chi \ll 1$, which ensures $\gamma_\gamma \ll \gamma$, the previous ratio reduces to $B_n^{n+1} \to b_n \chi$, where (introducing the Gamma function Γ)

$$b_n = \frac{3}{n+2} \frac{\Gamma(\frac{n}{2}+\frac{2}{3})\Gamma(\frac{n}{2}+\frac{7}{3})}{\Gamma(\frac{n}{2}+\frac{1}{6})\Gamma(\frac{n}{2}+\frac{11}{6})}, \tag{5.46}$$

slowly increases with n from $b_1 \simeq 1.07$, up to its asymptotic value $b_\infty = 3/2$ for $n \gg 1$. This ordering $\propto \chi^n$ confirms that the Fokker-Planck expansion (5.37) and

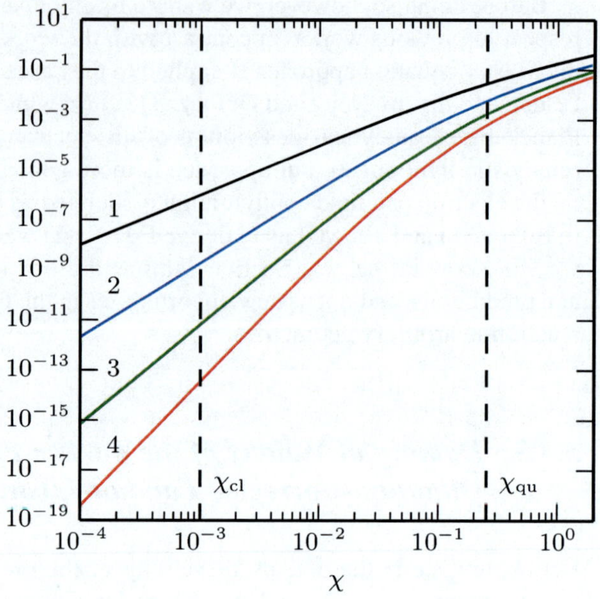

Fig. 5.2 Dependence with χ of the functions $a_n(\chi)/n!$ for $n = 1$ to 4, in black, blue, green and red (respectively). The left vertical line at $\chi = \chi_{cl} = 1 \times 10^{-3}$ indicates the threshold of the classical to the intermediate regime and the right vertical line at $\chi = \chi_{qu} = 2.5 \times 10^{-1}$ the limit of the intermediate to the quantum regime (see explicit definitions in the text). [55]

therefore Eqs. (5.40) and (5.42) are valid for small χ. As $\chi \to 1$, all terms in the Kramers-Moyal expansion (5.32) become of the same order, no truncation can be made in the *collision* operator, and the full linear Boltzmann Eq. (5.31) needs to be considered. It is however well known that this limit also corresponds to the onset of various other QED processes such as electron-positron pair production, and the present approach is not satisfying anymore.

To be more quantitative on the domain of applicability of the Fokker-Planck and quantum-corrected Landau-Lifshitz descriptions, a first general criterion can be obtained by comparing the relative importance of the various terms in the Kramers-Moyal expansion as given by the approximated expression Eq. (5.45). For this reason, we have plotted in Fig. 5.2 the functions $a_n(\chi)/n!$ for $n = 1$ to 4. The classical regime of radiation reaction can now be defined as the region $\chi < \chi_{cl} \simeq 1 \times 10^{-3}$ for which the second term in the Kramers-Moyal expansion (diffusion term) is at least three order of magnitude below the first (drift) term. The intermediate quantum regime is defined as the region $\chi_{cl} < \chi < \chi_{qu} \simeq 2.5 \times 10^{-1}$ where the diffusion term contribution is not negligible but the higher-order terms in the Kramers-Moyal expansion are. Finally, the quantum regime is the region $\chi > \chi_{qu}$ for which the third term in the Kramers-Moyal expansion becomes larger than a tenth of the diffusion term.

This criterion is based on a statistical description of the system and is consistent with the one proposed in the literature, which is solely based on the argument that the recoil of a single emitting photon can be neglected. This first criterion, that depends only on the quantum parameter χ, gives a first approximation for the validity of the deterministic (quantum corrected LL) or FP description when considering an arbitrary electron distribution function. In the next Chap. 6, we will derive a more general criterion for the domain of validity of the different models that accounts for the general energy distribution of an arbitrary distribution function by analyzing into more details the temporal evolution of its successive moments.

5.4 Local Temporal Evolution of Integrated Quantities

In this Section, we discuss the local temporal evolution of the successive moments of the electron distribution function as inferred from the different descriptions discussed in Sect. 5.3. These are defined as

$$\overline{q}_\alpha(t, \mathbf{x}) = \int d^2\Omega\, d\gamma\, q\, \hat{f}_e, \tag{5.47}$$

which denotes for a given quantity q its local average over the normalized electron distribution function $\hat{f}_e = f_e/n_e$ taken at a time t and position $\mathbf{x}$, with

$$n_e(t, \mathbf{x}) \equiv \int d^2\Omega\, d\gamma\, f_e, \tag{5.48}$$

the electron density at this time and position and where the index α indicates the method used to compute the distribution function [α = cLL, FP, MC for the deterministic (quantum corrected Landau-Lifshitz friction force), Fokker-Planck and linear Boltzmann (Monte-Carlo) approach, respectively]. The deterministic description here corresponds to Eq. (5.31) accounting for the drift term only in the *collision* operator that is equivalent to the LL formalism including the quantum correction $g(\chi)$ in the radiated power, hereafter indicated as "corrected Landau-Lifshitz" (cLL). The Fokker-Planck description corresponds to Eq. (5.31) with the collision operator given by Eq. (5.37), and the linear Boltzmann description corresponds to Eq. (5.31) with the full collision operator.

5.4.1 Local Energy Moments of the Collision Operators

Let us start by discussing briefly the first moments, in energy, of the *collision* operators. Computing the order 0 and 1 moments for all three descriptions leads to:

$$\int d^2\Omega \int_1^{+\infty} d\gamma\, C_\alpha[\hat{f}_e] = 0, \tag{5.49a}$$

$$\int d^2\Omega \int_1^{+\infty} d\gamma\, \gamma\, C_\alpha[\hat{f}_e] = \overline{S(\chi)}_\alpha. \tag{5.49b}$$

Concerning the second order moment of the *collision* operators, only the linear Boltzmann and Fokker-Planck methods give a similar form:

$$\int d^2\Omega\, d\gamma\, (\gamma-\overline{\gamma}_\alpha)^2 \mathcal{C}_\alpha[\hat{f}_e] = -2\overline{(\gamma-\overline{\gamma}_\alpha)S(\chi)}_\alpha + \overline{R(\chi,\gamma)}_\alpha \tag{5.50}$$

while the deterministic description simply leads to:

$$\int d^2\Omega\, d\gamma\, (\gamma-\overline{\gamma}_{\mathrm{cLL}})^2 \mathcal{C}_{\mathrm{cLL}}[\hat{f}_e] = -2\overline{(\gamma-\overline{\gamma}_{\mathrm{cLL}})S(\chi)}_{\mathrm{cLL}}. \tag{5.51}$$

The third order moment is the first for which all three descriptions lead to formally different equations of evolution:

$$\int d^2\Omega\, d\gamma\, (\gamma-\overline{\gamma}_\alpha)^3 \mathcal{C}_\alpha[\hat{f}_e] \tag{5.52}$$

$$= \begin{cases} -3\overline{(\gamma-\overline{\gamma}_\alpha)^2\, S(\chi)}_\alpha & \alpha = \mathrm{cLL} \\ -3\overline{(\gamma-\overline{\gamma}_\alpha)^2\, S(\chi)}_\alpha + 3\overline{(\gamma-\overline{\gamma}_\alpha)R(\chi,\gamma)}_\alpha & \alpha = \mathrm{FP} \\ -3\overline{(\gamma-\overline{\gamma}_\alpha)^2\, S(\chi)}_\alpha + 3\overline{(\gamma-\overline{\gamma}_\alpha)R(\chi,\gamma)}_\alpha \\ -\overline{A_3(\chi,\gamma)}_\alpha & \alpha = \mathrm{MC} \end{cases}$$

Finally, the nth order moment of the *collision* operator for any order n reads:

$$\int d^2\Omega\, d\gamma\, (\gamma-\overline{\gamma}_\alpha)^n \mathcal{C}_\alpha[\hat{f}_e] \tag{5.53}$$

$$= \begin{cases} -n\overline{(\gamma-\overline{\gamma}_\alpha)^{n-1}\, S(\chi)}_\alpha & \alpha = \mathrm{cLL} \\ -n\overline{(\gamma-\overline{\gamma}_\alpha)^{n-1}\, S(\chi)}_\alpha \\ +\frac{n}{2}(n-1)\overline{(\gamma-\overline{\gamma}_\alpha)^{n-2}\, R(\chi,\gamma)}_\alpha & \alpha = \mathrm{FP} \\ \sum_{k=1}^{n}(-1)^k\binom{n}{k}\overline{(\gamma-\overline{\gamma}_\alpha)^{n-k}\, A_k(\chi,\gamma)}_\alpha \\ & \alpha = \mathrm{MC} \end{cases}$$

and we can verify that for $n \geq 3$, the three description give different results.

Note that in the complete (MC) description, the evolution of the energy moment of order n involves the first n^{th} moments associated to the kernel of the Kramers-Moyal expansion, so that only the equation for the first two moments are formally the same for the FP and MC description.

5.4.2 Equations of Continuity for the Electrons and Photons

Let us first compute the moment of order 0 of Eq. (5.28a) according to the definition (5.47), we get

$$\partial_t n_e(t, \mathbf{r}) + \nabla_{\mathbf{r}} \cdot [\mathbf{v} \, n_e(t, \mathbf{r})] = 0. \tag{5.54}$$

This is the usual continuity equation, which is consistent with the fact that the number of electrons is conserved. Doing the same for photons, Eq. (5.28b) yields

$$\partial_t n_\gamma(t, \mathbf{r}) + \mathbf{c} \cdot \nabla_{\mathbf{r}} n_\gamma(t, \mathbf{r}) = \overline{W(\chi, \gamma)}_\alpha(t, \mathbf{r}), \tag{5.55}$$

where $W(\chi, \gamma) \equiv A_0(\chi, \gamma)$ as defined in Eq. (C.2) and (5.34). This equation shows that the number of photons at $\mathbf{r}$ in t increases with a rate equal to the mean over all the momenta of the electrons at $\mathbf{r}$ in t of the emission rate for a single electron as given in Sect. 5.2.3.

5.4.3 Equation of Evolution of the Mean Momentum

Computing the first order moment in momentum of Eq. (5.28a) with the definition (5.47) yields

$$\partial_t \overline{\mathbf{p}} + \nabla_{\mathbf{r}} \cdot \overline{\mathbf{pp}} = -e \left(\mathbf{E} + \overline{\mathbf{v}} \times \mathbf{B}\right) - mc \left[\overline{S(\chi)\, \boldsymbol{\Omega}}\right](t, \mathbf{r}).$$

which shows that the variation of the average momentum in a closed volume is due to the flux of momentum entering this closed volume, to the Lorentz force acting on the average velocity and to the average of the quantum corrected radiated power.

5.5 Global Temporal Evolution of Average Quantities

In this Section, we discuss the global temporal evolution of the successive moments of the electron distribution function, that is to say, we don't look at local quantities like in the previous Section, but we now integrate over all space. These are defined as

$$\langle \gamma \rangle_\alpha(t) = \frac{\int d^3x \, d^2\Omega \, d\gamma \; \gamma f_e}{\int d^3x \, d^2\Omega \, d\gamma \; f_e}, \tag{5.56}$$

and for $n > 1$,

$$\mu_n|_\alpha(t) = \frac{\int d^3x\, d^2\Omega\, d\gamma\, (\gamma - \langle\gamma\rangle_\alpha(t))^n f_e}{\int d^3x\, d^2\Omega\, d\gamma\, f_e}, \tag{5.57}$$

where the index α indicates the method used to compute the distribution function [$\alpha =$ cLL, FP, MC for the deterministic (quantum corrected Landau-Lifshitz friction force), Fokker-Planck and linear Boltzmann (Monte-Carlo) approach, respectively]. The deterministic description here corresponds to Eq. (5.31) accounting for the drift term only in the *collision* operator. The Fokker-Planck description corresponds to Eq. (5.31) with the collision operator given by Eq. (5.37), and the linear Boltzmann description corresponds to Eq. (5.31) with the full collision operator.

In this Section, we first give the equations of evolution for the successive moments of the electron distribution function and then discuss their implications on both the different (cLL, FP, MC) descriptions, and the physics they describe.

5.5.1 Conservation of the Number of Electrons and Total Energy, and Photon Production Rate

As shown in Sect. 5.4.1, the integral over all possible electron energies of the *collision* operators for all three description is identically zero. For all three descriptions, the photon emission process does not impact the electron density distribution in the $(\mathbf{x}, \boldsymbol{\Omega})$ phase-space and all three descriptions conserve the total number of electrons $N_e = \int d^3x\, d^2\Omega\, d\gamma f_e$, as expected.

The situation is obviously different when looking at the total number of photons. Indeed, integrating the source term, rhs of Eq. (5.28b), over all possible photon energies and directions provides us with the local photon production rate:

$$W_{\mathrm{loc}}(t, \mathbf{x}) = \int d^2\Omega\, d\gamma_\gamma\, S[f_e] = n_e(t, \mathbf{x}) \big\langle W(\chi, \gamma)\big\rangle_\alpha, \tag{5.58}$$

where $n_e(t, \mathbf{x})$ is the electron density, and $W(\chi, \gamma) \equiv A_0(\chi, \gamma)$ is the rate of photon emission [Eq. (5.25)] introduced in Sect. 5.2.3. As a result, the total number of photons $N_\gamma(t) = \int d^3x\, d^2\Omega_\gamma\, d\gamma_\gamma\, f_\gamma$ increases with time as:

$$\frac{d}{dt} N_\gamma = \int d^3x\, W_{\mathrm{loc}}(t, \mathbf{x}), \tag{5.59}$$

the rhs of the previous equation denoting the total photon production rate.

Finally, the equations of evolution for the total energy $U_{e,\gamma}(t) = mc^2 \int d^3x\, d^2\Omega\, d\gamma\, \gamma f_{e,\gamma}$ of the electron and photon population reads:

$$\frac{d}{dt}U_e = \int d^3x\, \mathbf{J}_e \cdot \mathbf{E} - P_{\rm rad}^{\rm tot}(t), \tag{5.60a}$$

$$\frac{d}{dt}U_\gamma = P_{\rm rad}^{\rm tot}(t), \tag{5.60b}$$

where:

$$P_{\rm rad}^{\rm tot}(t) = mc^2 \int d^3x\, n_e(t,\mathbf{x})\, \langle S(\chi)\rangle_\alpha, \tag{5.61}$$

is the total power radiated away by the electrons, and $\mathbf{J}_e(t,\mathbf{x}) = -e\, n_e(t,\mathbf{x})\mathbf{V}_e(t,\mathbf{x})$ is the electron current density [with $\mathbf{V}_e(t,\mathbf{x}) = c\,\langle u\,\boldsymbol{\Omega}\rangle_\alpha$ the electron mean velocity and where $u = v/c$ is the individual electron speed normalized to the speed of light c]. This confirms that all three descriptions conserve the total energy in the system, the only source of energy being the work of the external electromagnetic field.

5.5.2 Equations of Evolution

Let us first discuss the temporal evolution of the electron average energy. Dividing Eq. (5.60a) by the total (constant) number of electrons $N_e = \int d^3x\, n_e(t,\mathbf{x})$ (see Sect. 5.5.1), one finds that *all three descriptions lead to the same equation of evolution for the average electron energy*:

$$mc^2 \frac{d\langle\gamma\rangle_\alpha}{dt} = -ec\,\langle \boldsymbol{\beta}\cdot\mathbf{E}\rangle_\alpha - mc^2\,\langle S(\chi)\rangle_\alpha, \tag{5.62}$$

where $\langle q\rangle_\alpha(t) = \int d^3x\, d^2\Omega\, d\gamma\, q f_e / \int d^3x\, d^2\Omega\, d\gamma\, f_e$ stands as the total (including spatial) average of q over the distribution function. The first term in the rhs of Eq. (5.62) stands for the average work rate of the external field, and the second term denotes the power radiated away averaged over the whole distribution function.

It is important to stress that the distribution functions considering the different descriptions are not necessarily the same. Therefore, Eq. (5.62) does not, in general, predicts that all three approaches will give similar results on the average electron energy. We will later quantify this difference in Sect. 5.5.4.

Let us now turn to the derivation of the equation of evolution for the variance in energy. To do so, we focus on the radiation reaction effect and will thus neglect the effect of the external field on the energy dispersion which cannot be treated for arbitrary configurations. This is justified in particular for the case of an electron population evolving in a constant magnetic field or interacting with a plane wave. In those cases indeed, the additional Vlasov term is either zero or negligible and the numerical simulations presented in Sect. 6.4 will show very good agreement with the present analysis.

The equations of evolution of the energy variance are obtained by multiplying the Master equations for the electron distribution by $(\gamma - \langle\gamma\rangle)^2$, and integrating over γ, $\boldsymbol{\Omega}$ and space. In contrast with the previous case (mean energy), only the linear Boltzmann and Fokker-Planck descriptions formally give the same equation for the time evolution of $\sigma_\gamma^2 = \big\langle(\gamma - \langle\gamma\rangle)^2\big\rangle_\alpha$ [here α = FP, MC]:

$$\left.\frac{d\sigma_\gamma^2}{dt}\right|_\alpha = \langle R(\chi,\gamma)\rangle_\alpha - 2\big\langle(\gamma - \langle\gamma\rangle_\alpha)S(\chi)\big\rangle_\alpha, \tag{5.63}$$

while the deterministic description gives:

$$\left.\frac{d\sigma_\gamma^2}{dt}\right|_{\mathrm{cLL}} = -2\big\langle(\gamma - \langle\gamma\rangle_{\mathrm{cLL}})S(\chi)\big\rangle_{\mathrm{cLL}}. \tag{5.64}$$

Present in all three descriptions, the term $-2\,\langle(\gamma - \langle\gamma\rangle)S(\chi)\rangle_\alpha$ is, in most cases, negative since high-energy photon emission and its back-reaction is dominated by electrons at the highest energies. It will therefore lead to a decrease of σ_γ, i.e. to a cooling of the electron population.

In contrast, the term $\langle R(\chi,\gamma)\rangle$, which pertains to the stochastic nature of high-energy photon emission in the QED framework, is a purely quantum term, and as such is absent from the deterministic description. This quantum term is always positive and leads to a spreading of the energy distribution, i.e. to an effective heating of the electron population.

In the following, we will further discuss the relative importance of the deterministic (cooling) and quantum (energy spreading/heating) terms and their impact on the electron population.

Using Eq. (5.52) for the third moment of the *collision* operators, an equation of evolution for $\mu_3(t) = \langle(\gamma - \langle\gamma\rangle_\alpha)^3\rangle_\alpha$ can be derived. As the three descriptions lead to a different form for the third moment of the *collision* operator, they will also lead to different equations of evolution for μ_3 (note that the contribution from the Vlasov operator[3] is not considered, and one focuses on the radiation reaction contribution only).

The equation of evolution in the deterministic description is given by

$$\left.\frac{d\mu_3}{dt}\right|_{\mathrm{cLL}} = 3\,\langle S(\chi)\rangle\big\langle(\gamma - \langle\gamma\rangle)^2\big\rangle_{\mathrm{cLL}} - 3\big\langle(\gamma - \langle\gamma\rangle_{\mathrm{cLL}})^2\,S(\chi)\big\rangle_{\mathrm{cLL}}, \tag{5.65}$$

in the Fokker-Planck description by

[3]Rigorously, in Eqs. (5.63) and (5.64), an additional contribution from the Vlasov operator should be accounted for so that the lhs of these equations generally reads: $\frac{d\sigma_\gamma^2}{dt} + 2e/(mc)\,(\langle\gamma u\boldsymbol{\Omega}\cdot\mathbf{E}\rangle_\alpha - \langle\gamma\rangle_\alpha\langle u\boldsymbol{\Omega}\cdot\mathbf{E}\rangle_\alpha)$. Similarly, for any order n moments, an additional contribution from the Vlasov operator would appear in the left hand side term of the equation of evolution, so that it should read $d_t\mu_n|_\alpha + n\,e/(mc)\big\langle(\gamma - \langle\gamma\rangle_\alpha)^{n-1}\,u(\boldsymbol{\Omega}\cdot\mathbf{E})\big\rangle_\alpha$.

$$\left.\frac{d\mu_3}{dt}\right|_{\rm FP} = 3\,\langle S(\chi)\rangle\,\big\langle(\gamma-\langle\gamma\rangle)^2\big\rangle_{\rm FP} - 3\big\langle(\gamma-\langle\gamma\rangle_{\rm FP})^2\,S(\chi)\big\rangle_{\rm FP} + 3\big\langle(\gamma-\langle\gamma\rangle_{\rm FP})R(\chi,\gamma)\big\rangle_{\rm FP}, \tag{5.66}$$

and in the linear Boltzmann (MC) description by

$$\left.\frac{d\mu_3}{dt}\right|_{\rm MC} = 3\,\langle S(\chi)\rangle\,\big\langle(\gamma-\langle\gamma\rangle)^2\big\rangle_{\rm MC} - 3\big\langle(\gamma-\langle\gamma\rangle_{\rm MC})^2\,S(\chi)\big\rangle_{\rm MC} + 3\big\langle(\gamma-\langle\gamma\rangle_{\rm MC})R(\chi,\gamma)\big\rangle_{\rm MC} - \langle A_3(\chi,\gamma)\rangle_{\rm MC}. \tag{5.67}$$

As will be further discussed in Sect. 5.5.6, this third order moment relates to the skewness (asymmetry) of the electron distribution function. In particular, we will show that it provides a link to the so-called *quenching* of radiation losses, as introduced and discussed in Ref. [43].

Finally, for any order n, we get for the nth moment $\mu_n|_\alpha(t) = \langle(\gamma-\langle\gamma\rangle_\alpha)^n\rangle_\alpha$:

$$\left.\frac{d\mu_n}{dt}\right|_{\rm cLL} = n\,\langle S(\chi)\rangle_{\rm cLL}\,\mu_{n-1}|_{\rm cLL} - n\big\langle(\gamma-\langle\gamma\rangle_{\rm MC})^{n-1}S(\chi)\big\rangle_{\rm cLL}, \tag{5.68a}$$

$$\left.\frac{d\mu_n}{dt}\right|_{\rm FP} = n\,\langle S(\chi)\rangle_{\rm FP}\,\mu_{n-1}|_{\rm FP} - n\big\langle(\gamma-\langle\gamma\rangle_{\rm FP})^{n-1}S(\chi)\big\rangle_{\rm FP} + \frac{n(n-1)}{2}\big\langle(\gamma-\langle\gamma\rangle_{\rm FP})^{n-2}R(\chi,\gamma)\big\rangle_{\rm FP}, \tag{5.68b}$$

$$\left.\frac{d\mu_n}{dt}\right|_{\rm MC} = n\,\langle S(\chi)\rangle_{\rm MC}\,\mu_{n-1}|_{\rm MC} + \sum_{k=0}^{n-1}(-1)^{n-k}\binom{n}{k}\big\langle(\gamma-\langle\gamma\rangle_{\rm MC})^k A_{n-k}(\chi,\gamma)\big\rangle_{\rm MC}. \tag{5.68c}$$

Before investigating into more details the evolution of the first two moments, let us discuss more closely these equation of evolution for an arbitrary order n.

At any order n, the purely deterministic term $\propto S(\chi)$ comes in the form $-n\langle(\gamma-\langle\gamma\rangle)^{n-1}S(\chi)\rangle$, while the nth-order moment of the kernel w_χ comes in the form $(-1)^n\langle A_n(\chi,\gamma)\rangle$. This implies that, while the purely deterministic term always tends to decrease any moment of the electron distribution function (except in special geometries where the external field would take small enough values whenever[4] $\gamma > \langle\gamma\rangle$, the purely quantum term $\propto A_n(\chi,\gamma)$ tends to increase even moments and decrease odd moments. Concerning the other terms, it is in general difficult to state the behavior of the nth moment since it will involve all moments from 0 to n.

---

[4] Some pathological situations can be found where it is actually positive, their physical significance is however questionable and these scenarios will not be considered here.

5.5.3 The Perturbative Expansion

The equations of evolution of the successive moments of the distribution function, Eqs. (5.62) to (5.68), are exact. Yet, to evaluate the rhs of these equations, one needs to consider a particular distribution function. This would in general require to solve the full linear Boltzmann Eq. (5.28a). Throughout this work, we have often adopted the approach that consists in developing the various functions of χ around the average value $\langle\chi\rangle$. In what follows, we list the resulting (approximated) equations of evolution [introducing $\bar{\tau} = 3\tau_e/(2\alpha^2)$]. The equation of evolution of the mean energy is approximately given at first order by [here α = cLL, FP, MC]:

$$d_t\langle\gamma\rangle_\alpha \simeq -S(\langle\chi\rangle_\alpha) - \frac{1}{2}\hat{\sigma}_\gamma^2\,\langle\chi\rangle_\alpha^2\,S''(\langle\chi\rangle_\alpha). \tag{5.69}$$

In this case, the first order is identically zero while the second order term only involves the variance and second derivative of $S(\chi)$. In the case of the variance, the first order expansion involves only the derivative of S for the cLL description, while it involves in addition h and its derivative for the FP and MC descriptions

$$d_t\sigma_\gamma^2\big|_{\mathrm{cLL}} \simeq -2\sigma_\gamma^2\frac{\langle\chi\rangle}{\langle\gamma\rangle}S'(\langle\chi\rangle), \tag{5.70a}$$

$$\bar{\tau}\;d_t\sigma_\gamma^2\big|_\alpha \simeq \bar{\tau}\;d_t\sigma_\gamma^2\big|_{\mathrm{cLL}} + \langle\gamma\rangle h(\langle\chi\rangle) + \sigma_\gamma^2\frac{\langle\chi\rangle}{\langle\gamma\rangle}h'(\langle\chi\rangle). \tag{5.70b}$$

More generally, at order n, the expansion of the equation of evolution of the nth order moment μ_n of the electron distribution function in the MC description involves all the moments μ_k from 0 to n as well as the functions a_k and their first order derivative a'_k for $k = 0 \ldots n$. For $n = 3$, we thus get

$$d_t\mu_3|_{\mathrm{cLL}} \simeq -3\mu_3\frac{\langle\chi\rangle}{\langle\gamma\rangle}S'(\langle\chi\rangle), \tag{5.71a}$$

$$\bar{\tau}\;d_t\mu_3|_{\mathrm{FP}} \simeq \bar{\tau}\;d_t\mu_3|_{\mathrm{cLL}} + 3\sigma_\gamma^2[h(\langle\chi\rangle) + \langle\chi\rangle h'(\langle\chi\rangle)] + 3\mu_3\frac{\langle\chi\rangle}{\langle\gamma\rangle}h'(\langle\chi\rangle), \tag{5.71b}$$

$$\bar{\tau}\;d_t\mu_3|_{\mathrm{MC}} \simeq \bar{\tau}\;d_t\mu_3|_{\mathrm{FP}} - \langle\gamma\rangle^2 a_3(\langle\chi\rangle) - \sigma_\gamma^2[a_3(\langle\chi\rangle) + 2\,\langle\chi\rangle\,a'_3(\langle\chi\rangle)] - \mu_3\frac{\langle\chi\rangle}{\langle\gamma\rangle}a'_3(\langle\chi\rangle), \tag{5.71c}$$

while for any n, we can write

$$d_t \mu_n \Big|_{\text{cLL}} \simeq -n\mu_n \frac{\langle\chi\rangle}{\langle\gamma\rangle} S(\langle\chi\rangle), \tag{5.72a}$$

$$\begin{aligned} \bar{\tau} d_t \mu_n \Big|_{\text{FP}} &\simeq \bar{\tau} d_t \mu_n \Big|_{\text{cLL}} + \frac{n(n-1)}{2} \mu_{n-2} \langle\gamma\rangle h(\langle\chi\rangle) \\ &+ \frac{n(n-1)}{2} \mu_{n-1} [h(\langle\chi\rangle) + \langle\chi\rangle h'(\langle\chi\rangle)] \\ &+ \frac{n(n-1)}{2} \mu_n \frac{\langle\chi\rangle}{\langle\gamma\rangle} h'(\langle\chi\rangle), \end{aligned} \tag{5.72b}$$

$$\begin{aligned} \bar{\tau} d_t \mu_n \Big|_{\text{MC}} &\simeq \sum_{k=0}^{n-1} \sum_{l=0}^{n-k} (-1)^{n-k} \binom{n}{k} \binom{n-k}{l} \langle\gamma\rangle^{n-k-l} \\ &\left[\mu_{k+l}\, a_n(\langle\chi\rangle) + \mu_{k+l+1} \frac{\langle\chi\rangle}{\langle\gamma\rangle} a_n'(\langle\chi\rangle) \right]. \end{aligned} \tag{5.72c}$$

5.5.4 Electron Mean Energy

Let us now further discuss the temporal evolution of the electron mean energy as inferred from all three descriptions. As previously discussed, we will not consider here the additional Vlasov term and focus on the effect of radiation reaction.

The fact that the quantum-corrected leading term of the LL friction force naturally appears by taking the FP limit of the linear Boltzmann description already leads us to stress that, whenever the diffusion term [in Eq. (5.42)] can be neglected, all three descriptions will lead to the same evolution of the distribution function, and therefore to the same mean energy prediction [as seen from Eq. (5.62)]. This can be easily understood as all three models have been proved to share the same deterministic (drift) term.

Yet, even in a regime where the diffusion term is not negligible and where the three models result in sensibly different distribution functions, simulations relying on either the quantum-corrected friction force only or full MC simulations have been shown to lead to similar predictions on the electron mean energy (see, e.g., Ref. [47]). In Sect. 6.4, we will actually see that all three approaches lead to very similar predictions on the electron mean energy even when the overall electron distribution is very different from one model to another. In what follows, we explain this result.

To do so, we need to quantify the difference on the mean electron energy predictions by the different approaches. Our first step is to formally expand $S(\chi)$ around the average value $\langle\chi\rangle_\alpha$ in the last term of the rhs of Eq. (5.62)

$$\langle S(\chi)\rangle_\alpha \simeq S\left(\langle\chi\rangle_\alpha\right) + \frac{1}{2}\sigma_\chi^2\, S''(\langle\chi\rangle_\alpha)\,, \tag{5.73}$$

where $S''(\chi)$ is the second derivative of $S(\chi)$ with respect to χ, and

$$\sigma_\chi^2 = \left\langle(\chi - \langle\chi\rangle_\alpha)^2\right\rangle_\alpha \tag{5.74}$$

measures the variance of the distribution in χ of the electron population. From this, one expects all three descriptions to predict similar average electron energies whenever the first term in Eq. (5.73) dominates.

Then, we introduce the error on the rate of change of the electron energy:

$$\frac{d_t\langle\gamma\rangle_{\rm MC} - d_t\langle\gamma\rangle}{d_t\langle\gamma\rangle} \simeq \frac{F^2\hat{\sigma}_\gamma^2\langle\chi\rangle^2\widetilde{S}''(\langle\chi\rangle)}{\widetilde{S}(\langle\chi\rangle)} \equiv \widetilde{\rm Er}(\langle\chi\rangle, \hat{\sigma}_\gamma, F), \tag{5.75}$$

where we have introduced $F = \hat{\sigma}_\chi/\hat{\sigma}_\gamma$, $\hat{\sigma}_\gamma = \sigma_\gamma/\langle\gamma\rangle$ and $\hat{\sigma}_\chi = \sigma_\chi/\langle\chi\rangle$. Considering situations where all particles radiate in a similar field (e.g. localized electron bunch and/or uniform electromagnetic field), $\hat{\sigma}_\chi \sim \hat{\sigma}_\gamma$ $(F = 1)$ and we introduce ${\rm Er}(\langle\chi\rangle, \hat{\sigma}_\gamma) \equiv \widetilde{\rm Er}(\langle\chi\rangle, \hat{\sigma}_\gamma, 1)$ that now depends only on $\langle\chi\rangle$ and $\hat{\sigma}_\gamma$, as presented in Fig. 5.3. We can see that, for a given $\hat{\sigma}_\gamma$, ${\rm Er}(\langle\chi\rangle, \hat{\sigma}_\gamma)$ is only weakly dependent on $\langle\chi\rangle$, while it depends more strongly on $\hat{\sigma}_\gamma$ at fixed $\langle\chi\rangle$.

In next Sect. 5.5.5, we will show that whenever the *initial* electron distribution is such that $\hat{\sigma}_\gamma < \hat{\sigma}_\gamma^{\rm thr}$, with $\hat{\sigma}_\gamma^{\rm thr}$ a threshold energy dispersion [given by Eq. (5.81)] that depends only on $\langle\chi\rangle$, the energy dispersion of the electron distribution can increase up to, but never exceed $\hat{\sigma}_\gamma^{\rm thr}$. Replacing $\hat{\sigma}_\gamma$ by this threshold value $\hat{\sigma}_\gamma^{\rm thr}$ in our previous estimate thus provides us with an estimate of the error on the rate of change Eq. (5.75) that depends on $\langle\chi\rangle$ only. It is plotted (solid line) in Fig. 5.4 for mean quantum parameters in the range $10^{-3} < \langle\chi\rangle < 2$, and does not exceed a few percents (at $\langle\chi\rangle \simeq 1$). It increases with $\langle\chi\rangle$ as the threshold energy dispersion increases with $\langle\chi\rangle$. We also report in Fig. 5.4 the relative discrepancy (at the end of the simulation) in between Monte-Carlo and deterministic modeling, measured as $\Delta\gamma = (\langle\gamma\rangle_{\rm MC} - \langle\gamma\rangle_{\rm cLL})/\gamma_0$ [with $\gamma_0 = \langle\gamma\rangle(t = 0)$]. This discrepancy follows the same trend as Er (solid line) and the latter is found to provide an upper-bound to $\Delta\gamma$. For the two cases presented here [initially narrow electron bunch interacting with a constant magnetic field (blue crosses) and linearly polarized plane-wave (green crosses)], the two methods predict similar average energies with a relative discrepancy of a few percents, maximum at initially large average quantum parameters. Note that the results derived in this work are on the average quantities (and in particular the average χ here) and not on the "typical" χ of a narrow electron bunch. The equations derived here therefore still work for a more complicated electron distribution function than a regular one narrowly centered around a typical value of χ.

5.5.5 *Variance in Energy: Radiative Cooling vs. Energy Spreading*

We now get back to the equation of evolution of the variance σ_γ^2 and discuss the relative importance of radiative cooling and 'stochastic' energy spreading. To do so, we rewrite (exactly) Eq. (5.63) in the form

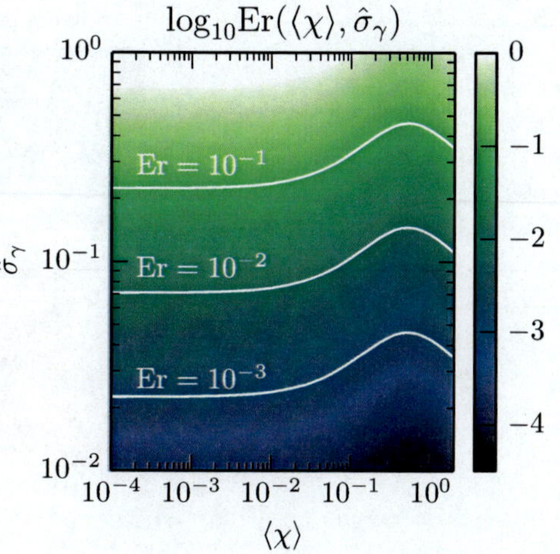

Fig. 5.3 Dependence with $\langle\chi\rangle$ and $\hat{\sigma}_\gamma$ of $\mathrm{Er}(\langle\chi\rangle, \hat{\sigma}_\gamma)$ which represents the relative difference between $d_t\langle\gamma\rangle_{\mathrm{MC}}$ and $d_t\langle\gamma\rangle_{\mathrm{cLL}}$. The curve $\mathrm{Er} = 10^{-3}$, 10^{-2} and 10^{-1} are plotted in white lines. [55]

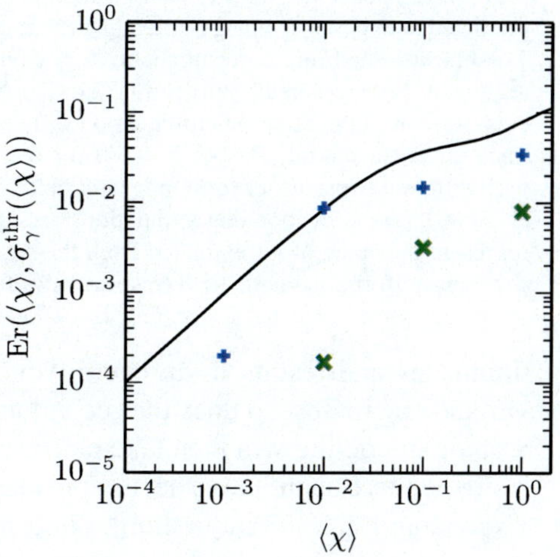

Fig. 5.4 Dependence with $\langle\chi\rangle$ of $\mathrm{Er}(\langle\chi\rangle, \sigma_\gamma^{\mathrm{thr}})$, which represents the relative difference between $d_t\langle\gamma\rangle_{\mathrm{MC}}$ and $d_t\langle\gamma\rangle_{\mathrm{cLL}}$ using $\hat{\sigma}_\gamma^{\mathrm{thr}}(\langle\chi\rangle)$. The green and blue crosses represent the value of $(\langle\gamma\rangle_{\mathrm{MC}}(t_{\mathrm{heat}}) - \langle\gamma\rangle_{\mathrm{cLL}}(t_{\mathrm{heat}}))/\langle\gamma\rangle_{\mathrm{cLL}}(t_{\mathrm{heat}})$ for the plane-wave field and the constant-uniform magnetic field (respectively) and for $\chi_0 = 10^{-2}$, 10^{-1} and 1, t_{heat} being the time at which σ_γ stops to increase [55].

$$\left(\frac{3\tau_e}{2\alpha^2}\right)\frac{d\sigma_\gamma^2}{dt} = \langle\gamma\rangle_\alpha\langle h(\chi)\rangle_\alpha - \left\langle(\gamma - \langle\gamma\rangle_\alpha)(2\widetilde{S}(\chi) - h(\chi))\right\rangle_\alpha, \tag{5.76}$$

where $\widetilde{S}(\chi) = \chi^2 g(\chi)$. There are now two possible situations : either (i) the energy distribution of the electron population is initially broad and the standard deviation σ_γ is of the same order than the average energy, or (ii) it is initially narrow and σ_γ is small with respect to the average energy.

In the first case, the second term in Eq. (6.1b) will be dominant at all times, and since for $\chi \lesssim 1$, $2\widetilde{S}(\chi) - h(\chi) > 0$ there will be cooling of the electron population even for initially large values of χ. In the second case, the first term in Eq. (6.1b)

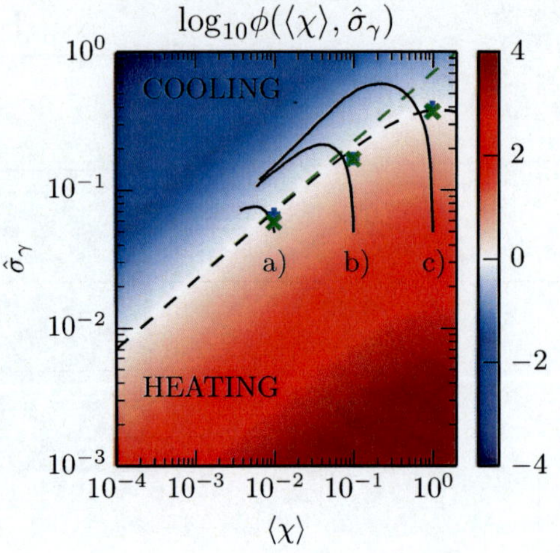

Fig. 5.5 Dependence with $\langle\chi\rangle$ and $\hat{\sigma}_\gamma$ of $\phi(\langle\chi\rangle, \hat{\sigma}_\gamma)$. When $\phi > 1$, the electron population is predicted to experience energy-spreading (heating), while it is predicted to cool down when $\phi < 1$. The black dashed line corresponds to $\phi = 1$ and represents the threshold $\hat{\sigma}_\gamma^{\mathrm{thr}}(\langle\chi\rangle)$ between the regions of energy spreading (heating) and cooling. The green dashed line shows the first order expansion in $\langle\chi\rangle$ of the previous equation and corresponds to the prediction of Ref. [42]. The black lines represent the trajectories $\hat{\sigma}_\gamma(\langle\chi\rangle)$ for the interaction of an ultra-relativistic electron bunch with different constant-uniform magnetic fields corresponding to a) $\chi_0 = 10^{-2}$, b) $\chi_0 = 10^{-1}$ and c) $\chi_0 = 1$ (the corresponding simulations are discussed in Sect. 6.4). The green and blue crosses represent the value of $\hat{\sigma}_\gamma$ extracted from the simulations considering the plane-wave field and the constant-uniform magnetic field (respectively) for $\chi_0 = 10^{-2}$, 10^{-1} and 1. [55]

dominates and results in an energy spreading of the electron population. As the variance increases, so does the second term that will eventually become dominant: a phase of cooling will then take place.

To be more quantitative, let us consider the latter case ($\sigma_\gamma \ll \langle\gamma\rangle$) in more details. Expanding Eq. (5.63) at first order in χ around $\langle\chi\rangle_\alpha$, we get:

$$\left(\frac{3\tau_e}{2\alpha^2}\right)\frac{d\sigma_\gamma^2}{dt} \simeq \langle\gamma\rangle_\alpha h(\langle\chi\rangle_\alpha) - \mathrm{Cov}(\gamma,\chi)_\alpha\big[2\widetilde{\mathrm{S}}'(\langle\chi\rangle_\alpha) - \mathrm{h}'(\langle\chi\rangle_\alpha)\big], \tag{5.77}$$

where $\mathrm{Cov}(\gamma,\chi)_\alpha = \langle(\gamma - \langle\gamma\rangle_\alpha)(\chi - \langle\chi\rangle_\alpha)\rangle_\alpha$.

Whether one should expect heating or cooling depends on the sign of the rhs of the previous equation. In particular, heating is expected whenever this rhs is positive, which arises for

$$\frac{\langle\gamma\rangle_\alpha\, h(\langle\chi\rangle_\alpha)}{\mathrm{Cov}(\gamma,\chi)_\alpha\big[2\widetilde{S}'(\langle\chi\rangle_\alpha) - h'(\langle\chi\rangle_\alpha)\big]} > 1, \tag{5.78}$$

with $2\widetilde{S}'(\chi) - h'(\chi) > 0$ whenever $\chi \lesssim 1$.

Introducing $F = \hat{\sigma}_\chi/\hat{\sigma}_\gamma$ and the correlation factor $\rho_{\gamma\chi} = \mathrm{Cov}(\gamma, \chi)/(\sigma_\gamma\sigma_\chi)$, where both coefficients F and $\rho_{\chi\gamma}$ depend on the geometry of the interaction, one can rewrite $\mathrm{Cov}(\gamma, \chi)_\alpha = \rho_{\gamma\chi} F\langle\chi\rangle\langle\gamma\rangle\hat{\sigma}_\gamma^2$. The previous condition Eq. (5.78) can then be rewritten in the form:

$$\tilde{\phi}(\langle\chi\rangle_\alpha, \hat{\sigma}_\gamma, F, \rho_{\gamma\chi}) = \frac{h(\langle\chi\rangle_\alpha)/\langle\chi\rangle_\alpha}{\rho_{\gamma\chi} F\, \hat{\sigma}_\gamma^2 \left[2\widetilde{S}'(\langle\chi\rangle_\alpha) - h'(\langle\chi\rangle_\alpha)\right]} > 1, \quad (5.79)$$

with $\hat{\sigma}_q = \sigma_q/\langle q\rangle_\alpha$.

This new functional parameter $\tilde{\phi}$ allows to account for the overall properties (mean energy and energy spread) of the electron population, and shows that electron heating is *not only* correlated to large values of the quantum parameter χ, as will be further demonstrated in Sect. 6.4.

Let us now focus on the case where all particles radiate in a similar external field (e.g. a localized electron bunch and/or uniform external field) for which $\hat{\sigma}_\chi \simeq \hat{\sigma}_\gamma$ ($F \simeq 1$) and $\rho_{\gamma\chi} \simeq 1$. Equation (5.79) then reduces to:

$$\phi(\langle\chi\rangle_\alpha, \hat{\sigma}_\gamma) = \frac{h(\langle\chi\rangle_\alpha)/\langle\chi\rangle_\alpha}{\hat{\sigma}_\gamma^2 \left[2\widetilde{S}'(\langle\chi\rangle_\alpha) - h'(\langle\chi\rangle_\alpha)\right]} > 1. \quad (5.80)$$

Comparing the value of $\phi\big(\langle\chi\rangle_\alpha, \hat{\sigma}_\gamma\big)$ with respect to 1, one can deduce whether heating ($\phi > 1$) and by extension cooling ($\phi < 1$) will take place. A threshold value $\hat{\sigma}_\gamma^{\mathrm{thr}}$ for the standard deviation can be derived considering $\phi = 1$, that reads:

$$\hat{\sigma}_\gamma^{\mathrm{thr}} = \sqrt{\frac{h(\langle\chi\rangle_\alpha)}{\langle\chi\rangle_\alpha[2\widetilde{S}'(\langle\chi\rangle_\alpha) - h'(\langle\chi\rangle_\alpha)]}}. \quad (5.81)$$

This threshold value gives the maximal standard deviation (in energy) that can be reached starting from an initially narrow electron energy distribution. Once the energy spread has reached this threshold, the cooling phase will take over.

Figure 5.5 presents, in color scale, $\phi\big(\langle\chi\rangle_\alpha, \hat{\sigma}_\gamma\big)$ as a function of the normalized standard deviation in energy $\hat{\sigma}_\gamma$ and the average quantum parameter $\langle\chi\rangle$. The dashed line corresponds to the threshold value $\sigma_\gamma^{\mathrm{thr}}$ [Eq. (5.81)] delimiting the regions in parameter space where cooling and heating are expected. The first order expansion of Eq. (5.81) is plotted as a green dashed line and corresponds to the prediction (derived in the limit $\chi \ll 1$) by Vranic *et al.* [42]. We plot in the same figure the measures of the maximal standard deviation extracted from simulations (discussed in Sect. 6.4) of an initially narrow electron bunch interacting with a constant magnetic field (blue crosses) and linearly polarized plane-wave (green crosses). These measures show the maximum (normalized) standard deviation $\hat{\sigma}_\gamma^{\mathrm{max}}$ obtained in the simulations as a function of the initial average quantum parameter $\chi_0 = \langle\chi\rangle(t = 0)$. Values reported here correspond to the results of either Monte-Carlo or Fokker-Planck simulations that lead to the same predictions, even for $\chi_0 = 1$. The evolution of the normalized standard deviation as a function of $\langle\chi\rangle$ (as it evolves with time) in the simulations considering a constant uniform magnetic field is also reported (solid lines). During the energy spreading (heating) phase (for $t < t_{\text{heating}}$ as defined in Sect. 6.4), the average χ is approximately constant, justifying the fact that in this phase, we can make the approximation $\langle \chi \rangle (t_{\text{heat}}) \simeq \chi_0$ when plotting the maximum standard deviation $\hat{\sigma}_\gamma^{\text{thr}}$ (blue and green crosses in Fig. 5.5). Moreover, we note that all curves end on the same line which acts as an attractor.

Note that, when the electron distribution function is initially broad, the hypothesis leading to the calculation of $\sigma_\gamma^{\text{thr}}$ are not valid. The general reasoning nevertheless holds and simulations presented in Sect. 6.4 indicate that $\hat{\sigma}_\gamma^{\text{thr}}$ can still be interpreted as a threshold: whenever the standard deviation of the considered electron distribution initially exceeds this threshold, only cooling of the electron population will be observed.

5.5.6 *Third Order Moment and Link to the Quenching of Radiation Losses*

In this section, we wish to show that the evolution of the third order moment can be used in order to interpret some interesting phenomenon associated to the quantum behavior of the system. As previously mentioned, we wish to link μ_3 to the so-called *quenching* of radiation losses. This process follows from the discrete nature of photon emission by the radiating electron in a quantum framework. As shown in Ref. [43] for an electron bunch interaction with a high-amplitude electromagnetic wave, a deformation of the distribution function can appear on short time scales as some electrons did radiate high-energy photons away thus decreasing their energy, while other electrons have not yet emitted any photon thus conserving their initial energy. As a result, the electron distribution function becomes asymmetric, showing a low energy tail, associated to a negative third order moment (negative skewness).

It is therefore interesting to investigate under which conditions the third order moment assumes negative values. As its evolution is correctly described in general in the linear Boltzmann (MC) approach *only*, we will consider Eq. (5.67) and derive the conditions under which $d\mu_3/dt < 0$. This would require to compute the average over the electron distribution function of the different quantities appearing in Eq. (5.67), which implies solving the full linear Boltzmann equation. Instead, we proceed as previously done and expand at first order the different functions in the rhs of Eq. (5.67) that depend on χ around $\langle \chi \rangle$. We find that $d\mu_3/dt < 0$ whenever

$$\tilde{\psi}(\langle \chi \rangle_\alpha, \hat{\sigma}_\gamma, F, \rho_{\gamma,\chi}, \tilde{C}_{\gamma,\chi}) < 1 \tag{5.82}$$

where we have introduced:

$$\tilde{C}_{\gamma,\chi} = \left\langle (\gamma - \langle \gamma \rangle_\alpha)^2 (\chi - \langle \chi \rangle_\alpha) \right\rangle_\alpha / \langle \gamma \rangle_\alpha^2 \tag{5.83}$$

and

$$\tilde{\psi} = \hat{\sigma}_\gamma^2 \left[f_1(\langle\chi\rangle) + F\rho_{\gamma,\chi} \langle\chi\rangle\, f_2(\langle\chi\rangle) \right] - \tilde{C}_{\gamma,\chi}\, f_3(\langle\chi\rangle) \tag{5.84}$$

with:

$$f_1(\chi) = \frac{3\,h(\chi) - a_3(\chi)}{a_3(\chi)}, \tag{5.85a}$$

$$f_2(\chi) = \frac{3\,h'(\chi) - 2a_3'(\chi)}{a_3(\chi)}, \tag{5.85b}$$

$$f_3(\chi) = \frac{3\,S'(\chi) + a_3'(\chi) - 3\,h'(\chi)}{a_3(\chi)}. \tag{5.85c}$$

Interestingly, Eq. (5.82) simplifies when considering a well localized (in space) electron population (e.g. narrow electron bunch and/or uniform electromagnetic field) so that $F \simeq 1$ and $\tilde{C}_{\gamma,\chi} \simeq \langle\chi\rangle\, \hat{\mu}_3$. The condition given by Eq. (5.82) then reduces to:

$$\psi(\langle\chi\rangle_\alpha, \hat{\sigma}_\gamma, \hat{\mu}_3) < 1, \tag{5.86}$$

where:

$$\psi = \hat{\sigma}_\gamma^2 \left[f_1(\langle\chi\rangle) + \langle\chi\rangle\, f_2(\langle\chi\rangle) \right] - \hat{\mu}_3 \langle\chi\rangle\, f_3(\langle\chi\rangle). \tag{5.87}$$

Considering the situation of an initially symmetric (in energy) electron beam, $\mu_3 = 0$, the condition for μ_3 to decrease (and thus assume negative values) can then be reduced to a condition on the normalized energy variance $\hat{\sigma}_\gamma^2$ and average quantum parameter only $\langle\chi\rangle$. Figure 5.6 presents, in color scale, the function $\psi(\langle\chi\rangle_\alpha, \hat{\sigma}_\gamma, \hat{\mu}_3 = 0)$ as a function of both $\langle\chi\rangle$ and $\hat{\sigma}_\gamma$. The solid line corresponds to the condition $\psi = 1$ and defines, for a given initial value of $\langle\chi\rangle$, a limiting energy variance for the electron bunch:

$$\hat{\sigma}_{\gamma,0}^{\lim}(\langle\chi\rangle) = \left[f_1(\langle\chi\rangle) + \langle\chi\rangle\, f_2(\langle\chi\rangle) \right]^{-1/2}. \tag{5.88}$$

This value of the initial electron bunch energy dispersion delimits the regions in parameter space where μ_3 is expected to increase or decrease.

Also reported are the measures of the standard deviation extracted from the simulations (discussed in Sect. 6.4) of an initially spatially narrow and energy symmetric $[\mu_3(t = 0) = 0]$ electron bunch interacting with a constant magnetic field (blue crosses) and linearly polarized plane-wave (green crosses) considering different initial values for the average quantum parameter $\langle\chi\rangle$. Vertical (+) crosses report the corresponding values of $\langle\chi\rangle$ and $\hat{\sigma}_\gamma$ at time $t = 0$. Crosses for $\langle\chi\rangle = 10^{-3}$ and 10^{-2} are either above or close to the delimiting solid line. In this simulation, one could thus expect an increase of μ_3. In our simulations, however, (and as expected from

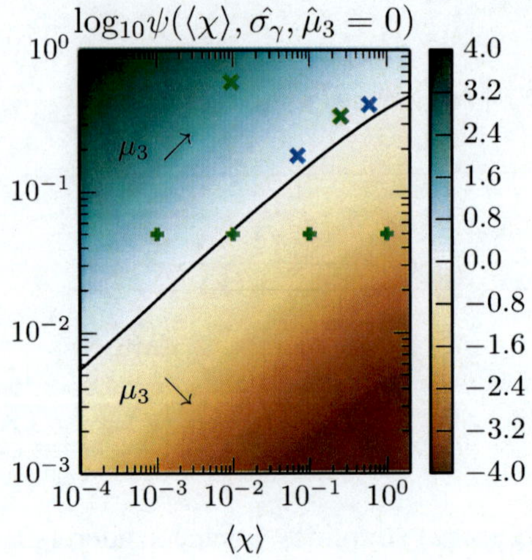

Fig. 5.6 Dependence with $\langle\chi\rangle$ and $\hat{\sigma}_\gamma$ of $\psi(\langle\chi\rangle, \hat{\sigma}_\gamma, \hat{\mu}_3 = 0)$. When $\psi > 1$, the electron population is predicted to acquire an asymmetry toward high energies (μ_3 increases, positive skewness) while it is predicted to acquire an asymmetry toward low energies (μ_3 decreases, negative skewness) when $\psi < 1$. The black dashed line correspond to $\psi = 1$ and represents the threshold $\hat{\sigma}_{\gamma,0}^{\text{lim}}(\langle\chi\rangle)$ [Eq. (5.88)]. The blue and green crosses represent the value of $\hat{\sigma}_\gamma$ when $\mu_3 = 0$ in the interaction of an ultra-relativistic electron bunch with different constant-uniform magnetic field and plane-waves (respectively) and initial quantum parameter a) $\chi_0 = 10^{-2}$, b) $\chi_0 = 10^{-1}$ and c) $\chi_0 = 1$. Vertical crosses correspond to the initial values of $\hat{\sigma}_\gamma$ and χ. Diagonal crosses correspond to the values of $\hat{\sigma}_\gamma$ and χ when μ_3 goes back to 0 (see, the 6th panel of Figs. 6.2 and 6.5). These simulations are discussed in Sect. 6.4. [55]

the scaling in χ^n of the successive moments discussed in Sect. 5.3.3), μ_3 assumes very small values and it was not possible to confirm this prediction. Nevertheless, for the case $\langle\chi\rangle = 10^{-1}$ and 1, where $|\mu_3|$ reaches larger values, all crosses are found to be below the limiting $\hat{\sigma}_{\gamma,0}^{\text{lim}}$ (solid) line, and the third order moment is found to decrease in the corresponding simulations, consistently with the present analysis. Also reported as "diagonal" ($\times$) crosses are the corresponding values of $\langle\chi\rangle(t^\star)$ and $\hat{\sigma}_\gamma(t^\star)$ extracted from the simulations at the first time $t^\star$ for which μ_3 is found to vanish. In every cases, μ_3 crosses 0 again when increasing, and finding all these points above the $\hat{\sigma}_{\gamma,0}^{\text{lim}}$ limit further confirms the present analysis.

For completeness, we study the influence of the initial value of $\hat{\mu}_3$ on its evolution, and we compute the limiting value of $\hat{\sigma}_\gamma$ at which $d\mu_3/dt$ changes sign for an initial non-zero value of $\hat{\mu}_3$

$$\hat{\sigma}_\gamma^{\text{lim}}(\langle\chi\rangle, \hat{\mu}_3) = \hat{\sigma}_{\gamma,0}^{\text{lim}}(\langle\chi\rangle)\sqrt{1 + \hat{\mu}_3\langle\chi\rangle f_3(\langle\chi\rangle)}. \tag{5.89}$$

Figure 5.7 presents, in color scale $\hat{\sigma}_\gamma^{\text{lim}}(\langle\chi\rangle, \hat{\mu}_3)$ as a function of the average quantum parameter $\langle\chi\rangle$ and for different values of $\hat{\mu}_3$. The black dashed line represents the value of $\hat{\sigma}_\gamma^{\text{thr}}$ [Eq. (5.81)], while the straight black line represents the value of

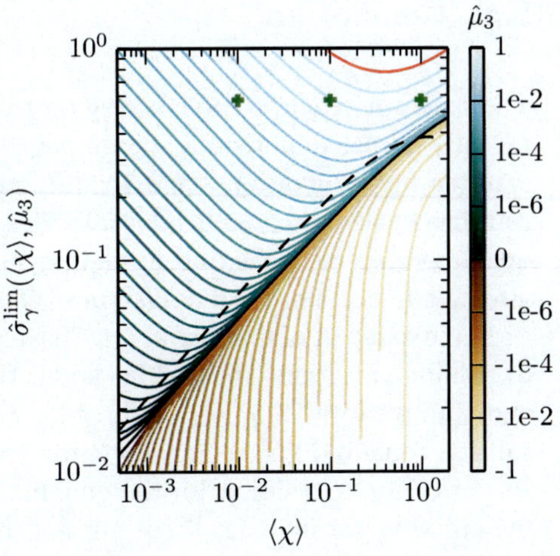

Fig. 5.7 Dependence of $\hat{\sigma}_\gamma^{\lim}$ [Eq (5.89)] with $\langle\chi\rangle$ for different values of $\hat{\mu}_3$. The red line represents $\hat{\sigma}_\gamma^{\lim}$ for the case $\hat{\mu_3} = 0.2$ used in Sect. 6.4.3. The black solid line corresponds to $\hat{\mu}_3 = 0$ while the black dotted line corresponds to $\hat{\sigma}_\gamma^{\text{thr}}$ [Eq. (5.81)]. The green crosses correspond to the three initial situations considered in Sect. 6.4.3 considering a broad Maxwell-Jüttner energy distribution [55].

$\hat{\sigma}_{\gamma,0}^{\lim}$ [Eq. (5.88)]. The other lines following the colorscale are the values of $\hat{\sigma}_\gamma^{\lim}$ for different values of $\hat{\mu}_3$. For a given initial $\hat{\mu}_3$, as for $\hat{\sigma}_{\gamma,0}^{\lim}$, if the initial values of $\hat{\sigma}_\gamma$ and $\langle\chi\rangle$ are above the limiting curve, the skewness will increase, and if below it will decrease.

We recall the meaning of the quantity $\hat{\sigma}_\gamma^{\text{thr}}$ (dashed line): if $\hat{\sigma}_\gamma$ is above this curve there will be cooling, and below, heating. For a system starting from initial non zero positive $\hat{\mu}_3$ (blue-shaded curves), the higher the value of $\hat{\mu}_3$, the broader the range of parameter in which we have cooling and decrease of μ_3, especially for small $\langle\chi\rangle$. On the contrary, for a system starting from an initially negative $\hat{\mu}_3$ (brown-shaded curves), we have the possibility of a range of parameters for which we can have both heating and an increase of μ_3.

The situation is different if we start from a symmetrical distribution function (or quasi-symmetrical distribution function, with very small values of μ_3). As we can see the lines $\hat{\sigma}_{\gamma,0}^{\lim}$ and $\hat{\sigma}_\gamma^{\text{thr}}$ are very close, and in this case values of $\hat{\sigma}_\gamma$ above this line correspond to an electron beam acquiring an asymmetry toward the left (negative skewness). Because of the proximity of these two lines, we can see in Fig. 5.7, that in most cases (for $\chi < 1$), cooling is accompanied by an increase of $\hat{\mu}_3$ and heating by a decrease of $\hat{\mu}_3$. In Sect. 6.4.3, we will examine in more detail the case of a Maxwell-Jüttner distribution that has a non-zero initial value of μ_3. The corresponding limiting line is marked in red in Fig. 5.7 and the initial conditions in $\langle\chi\rangle$ and $\hat{\sigma}_\gamma$ of the three simulations considered in that Section are reported in the figure as green crosses. According to our prediction, the system will cool down while its skewness diminishes (all points are below the red line), as will indeed be demonstrated in all these simulations.

5.6 Conclusion

The radiation reaction force acting on particles in strong electromagnetic fields has recently attracted increased attention as it may affect laser-plasma interaction under extreme light conditions, and for its impact on the particle dynamics in extreme astrophysical scenarios. It is usually described using either a classical (potentially quantum corrected) friction force or a full Monte-Carlo (MC) treatment, the latter allowing to account for the quantum and discrete nature of the photon emission.

In the first part of this chapter, we have revisited the basis of the classical treatment of radiation reaction. The Landau and Lifshitz (LL) force was rewritten in the simple and intuitive form of a friction force for radiating ultra-relativistic particles. Like the full LL equation, this reduced friction force also conserves the on-shell condition. Its correction to account for the quantum reduction of the power radiated away by the emitting particle was then introduced heuristically (as previously suggested in other works).

After briefly presenting the properties of high-energy photon emission as inferred from quantum electrodynamics (QED) in the non-linear moderately quantum regime, we then focused on a statistical description of photon emission and its back-reaction considering a population of ultra-relativistic electrons. Starting from a linear Boltzmann (lB) equation with a *collision* operator describing incoherent photon emission in a quantum description, we performed a Fokker-Planck (FP) expansion of the *collision* operator in the parameter γ_γ/γ, the limit of which were discussed in details. The resulting FP description, derived here for the first time under arbitrary particle and electromagnetic field configurations, is interesting for several reasons.

- (i) First, it takes the simple form of a partial differential equation, more easily handled in theoretical models than the lB description that relies on an integro-differential equation.
- (ii) The derived FP description is equivalent to a stochastic differential equation (SDE) for the electron momentum. In addition to the standard Lorentz force, this equation contains a deterministic drift term which is found to be leading term of the LL friction force with the quantum correction discussed above, hence justifying the heuristic treatment that consists in systematically correcting the LL friction force to account for the quantum reduction of the power-radiated away by the ultra-relativistic electron. Note that throughout this work, this quantum-corrected LL friction force provides the deterministic treatment of radiation reaction. Most importantly, this kinetic treatment also fully justifies - for the first time - the use of the quantum-corrected LL friction force to account for radiation reaction in Particle-In-Cell (PIC) codes.
- (iii) An additional diffusion term in the SDE is also derived for any $\chi \lesssim 1$. It accounts for the stochastic nature of photon emission inherent to its quantum nature.

Special attention was paid to the study of the successive moments of the electron distribution function. Using both analytical and numerical modelling, we evidence that the equation of evolution for the average energy of the particle ensemble is formally the same in all three models (quantum-corrected LL/deterministic, FP, lB/MC). An estimate on the discrepancy on the mean electron energy predicted by the different models is derived, and it is shown to be small for most cases: that is all three descriptions lead to the same prediction for the electron mean energy. This has serious implications for experiments, as the mean electron energy proves not to be, in general, a relevant measure to assess the importance of quantum effects beyond the simple reduction of the power radiated away by the particles.

The situation is different when considering the evolution of the higher order moments of the particles' energy distribution. In particular, only the FP and lB/MC descriptions provide the correct equation of evolution for the variance (energy dispersion). Instead the third and higher order moment equations of evolution differ from one approach to another.

The study of the equation of evolution of the energy variance allows us to define an energy spreading (*heating*) time, when the variance of an initially narrow beam reaches a maximum value. The existence of this maximum energy spread follows from the competition between the deterministic part of the radiation reaction force and the stochastic nature of the quantum process of high-energy photon emission. The former acts as a friction term and results in a cooling of the electron population, while the former leads to a natural spreading of the electron energy distribution function, associated to an increased effective temperature of the electron population. This spreading can set in the moderately quantum regime of radiation reaction ($\chi \lesssim 1$), and is shown to be correctly modelled by both the FP and lB/MC treatments. This has also interesting implications for experiments, as the stochastic nature of high-energy photons on the electron dynamics can then be diagnosed by a careful analysis of the resulting spread of the particle energy distribution.

Finally, the study of the evolution of the third order moment proved also to be particularly interesting as it reveals the discrepancy between all three models. In particular, this study allows us to link negative values of the third moment to the so-called quenching of radiation losses. This process is intimately linked to the discrete nature of high-energy photon emission and can be accounted for only using the lB/MC procedure. We show that this quenching process can be more easily observed, and impact the electron distribution, when the electron initial distribution is quite narrow. This has important implications for future experiments on radiation reaction, as it is here shown that "quenching", a purely quantum effect, can be diagnosed as a negative skewness of the resulting electron distribution function. Let us further note that, unlike previously claimed, even though quenching affects the overall shape of the electron distribution function, it does not impact the mean electron energy that is correctly modelled using the deterministic radiation reaction force, provided it is corrected to account for the quantum reduction on the radiated power.

References

1. Extreme light infrastructure (ELI). https://eli-laser.eu/
2. Cros B, Paradkar B, Davoine X, Chancé A, Desforges F, Dobosz-Dufrénoy S, Delerue N, Ju J, Audet T, Maynard G, Lobet M, Gremillet L, Mora P, Schwindling J, Delferrière O, Bruni C, Rimbault C, Vinatier T, Piazza AD, Grech M, Riconda C, Marquès J, Beck A, Specka A, Martin P, Monot P, Normand D, Mathieu F, Audebert P, Amiranoff F (2014) Laser plasma acceleration of electrons with multi-PW laser beams in the frame of CILEX. Nuclear Instrum Methods Phys Res Sect A Accel Spectrom Detect Assoc Equip 740:27 – 33. 2013 Proceedings of the first European advanced accelerator concepts workshop
3. Piazza D, Müller C, Hatsagortsyan K, Keitel CH (2012) Extremely high-intensity laser interactions with fundamental quantum systems. Rev Mod Phys 84:1177
4. Tamburini M, Pegoraro F, Piazza AD, Keitel CH, Macchi A (2010) Radiation reaction effects on radiation pressure acceleration. New J Phys 12(12):123005
5. Nerush EN, Kostyukov IY, Fedotov AM, Narozhny NB, Elkina NV, Ruhl H (2011) Laser field absorption in self-generated electron-positron pair plasma. Phys Rev Lett 106:035001
6. Gonoskov A, Bastrakov S, Efimenko E, Ilderton A, Marklund M, Meyerov I, Muraviev A, Sergeev A, Surmin I, Wallin E (2015) Extended particle-in-cell schemes for physics in ultrastrong laser fields: review and developments. Phys Rev E 92:023305
7. Ridgers CP, Brady CS, Duclous R, Kirk JG, Bennett K, Arber TD, Robinson APL, Bell AR (2012) Dense electron-positron plasmas and ultraintense γ rays from laser-irradiated solids. Phys Rev Lett 108:165006
8. Capdessus R, d'Humières E, Tikhonchuk VT (2013) Influence of ion mass on laser-energy absorption and synchrotron radiation at ultrahigh laser intensities. Phys Rev Lett 110:215003
9. Blackburn TG, Ridgers CP, Kirk JG, Bell AR (2014) Quantum radiation reaction in laser electron-beam collisions. Phys Rev Lett 112:015001
10. Lobet M, Ruyer C, Debayle A, d'Humières E, Grech M, Lemoine M, Gremillet L (2015) Ultrafast synchrotron-enhanced thermalization of laser-driven colliding pair plasmas. Phys Rev Lett 115:215003
11. Lobet M, Davoine X, d'Humières E, Gremillet L (2017) Generation of high-energy electron-positron pairs in the collision of a laser-accelerated electron beam with a multipetawatt laser. Phys Rev Accel Beams 20:043401
12. Grismayer T, Vranic M, Martins JL, Fonseca RA, Silva LO (2017) Seeded QED cascades in counterpropagating laser pulses. Phys Rev E 95:023210
13. Chen H, Wilks SC, Bonlie JD, Liang EP, Myatt J, Price DF, Meyerhofer DD, Beiersdorfer P (2009) Relativistic positron creation using ultraintense short pulse lasers. Phys Rev Lett 102:105001
14. Sarri G, Poder K, Cole JM, Schumaker W, Di Piazza A, Reville B, Dzelzainis T, Doria D, Gizzi LA, Grittani G, Kar S, Keitel CH, Krushelnick K, Kuschel S, Mangles SPD, Najmudin Z, Shukla N, Silva LO, Symes D, Thomas AGR, Vargas M, Vieira J, Zepf M (2015) Generation of neutral and high-density electron–positron pair plasmas in the laboratory. Nat Commun 6:6747 EP
15. Cole JM, et al (2017) Experimental observation of radiation reaction in the collision of a high-intensity laser pulse with a laser-wakefield accelerated electron beam. arXiv:1707.06821
16. Poder K, et al (2017) Evidence of strong radiation reaction in the field of an ultra-intense laser. arXiv:1709.01861
17. Jaroschek CH, Hoshino M (2009) Radiation-dominated relativistic current sheets. Phys Rev Lett 103:075002
18. Cerutti B, Werner GR, Uzdensky DA, Begelman MC (2014) Gamma-ray flares in the Crab Nebula: a case of relativistic reconnection? Phys Plasmas 21(5):056501
19. Cerutti B, Philippov AA, Spitkovsky A (2016) Modelling high-energy pulsar light curves from first principles. Mon Not Roy Astron Soc 457(3):2401
20. Spohn H (2000) The critical manifold of the Lorentz-Dirac equation. EPL (Europhys Lett) 50(3):287

21. Rohrlich F (2008) Dynamics of a charged particle. Phys Rev E 77:046609
22. Sokolov IV, Naumova NM, Nees JA, Mourou GA, Yanovsky VP (2009) Dynamics of emitting electrons in strong laser fields. Phys Plasmas 16(9):093115
23. Sokolov IV, Naumova NM, Nees JA, Mourou GA (2010) Pair creation in QED-strong pulsed laser fields interacting with electron beams. Phys Rev Lett 105:195005
24. Bulanov SV, Esirkepov TZ, Kando M, Koga JK, Bulanov SS (2011) Lorentz-Abraham-Dirac versus Landau-Lifshitz radiation friction force in the ultrarelativistic electron interaction with electromagnetic wave (exact solutions). Phys Rev E 84:056605
25. Burton DA, Noble A (2014) On the entropy of radiation reaction. Phys Lett A 378(14):1031–1035
26. Capdessus R, Noble A, McKenna P, Jaroszynski DA (2016) Publisher's Note: Role of momentum and velocity for radiating electrons [Phys. Rev. D 93, 045034 (2016)]. Phys Rev D 93:089908
27. Moniz EJ, Sharp DH (1977) Radiation reaction in nonrelativistic quantum electrodynamics. Phys Rev D 15:2850–2865
28. Krivitskiĭ VS, Tsytovich VN (1991) Average radiation-reaction force in quantum electrodynamics. Sov Phys Usp 34(3):250
29. Di Piazza A, Hatsagortsyan KZ, Keitel CH (2010) Quantum radiation reaction effects in multiphoton compton scattering. Phys Rev Lett 105:220403
30. Meuren S, Di Piazza A (2011) Quantum electron self-interaction in a strong laser field. Phys Rev Lett 107:260401
31. Ilderton A, Torgrimsson G (2013) Radiation reaction from QED: lightfront perturbation theory in a plane wave background. Phys Rev D 88:025021
32. Ilderton A, Torgrimsson G (2013b) Radiation reaction in strong field QED. Phys Lett B 725(4):481–486
33. Duclous R, Kirk JG, Bell AR (2011) Monte Carlo calculations of pair production in high-intensity laser-plasma interactions. Plasma Phys Control Fusion 53(1):015009
34. Arber TD, Bennett K, Brady CS, Lawrence-Douglas A, Ramsay MG, Sircombe NJ, Gillies P, Evans RG, Schmitz H, Bell AR, Ridgers CP (2015) Contemporary particle-in-cell approach to laser-plasma modelling. Plasma Phys Control Fusion 57:113001
35. Lobet M, d'Humières E, Grech M, Ruyer C, Davoine X, Gremillet L (2016) Modeling of radiative and quantum electrodynamics effects in PIC simulations of ultra-relativistic laser-plasma interaction. J Phys: Conf Ser 688(1):012058
36. Landau LD, Lifshitz EM (1947) The classical theory of fields. Butterworth-Heinemann, Oxford
37. Uggerhøj UI (2005) The interaction of relativistic particles with strong crystalline fields. Rev Mod Phys 77:1131–1171
38. Walser MW, Keitel CH (2001) Geometric and algebraic approach to classical dynamics of a particle with spin. Lett Math Phys 55(1):63–70
39. Neitz N, Piazza AD (2013) Stochasticity effects in quantum radiation reaction. Phys Rev Lett 111:054802
40. Tamburini M, Pegoraro F, Piazza AD, Keitel C, Liseykina T, Macchi A (2011) Radiation reaction effects on electron nonlinear dynamics and ion acceleration in laser–solid interaction. Nuclear Instrum Methods Phys Res Sect A Accel Spectrom Detect Assoc Equip 653(1):181–185. Superstrong 2010
41. Lehmann G, Spatschek KH (2012) Phase-space contraction and attractors for ultrarelativistic electrons. Phys Rev E 85:056412
42. Vranic M, Grismayer T, Fonseca RA, Silva LO (2016) Quantum radiation reaction in head-on laser-electron beam interaction. New J Phys 18(7):073035
43. Harvey CN, Gonoskov A, Ilderton A, Marklund M (2017) Quantum quenching of radiation losses in short laser pulses. Phys Rev Lett 118:105004
44. Elkina NV, Fedotov AM, Herzing C, Ruhl H (2014) Accurate numerical simulation of radiation reaction effects in strong electromagnetic fields. arXiv

45. Ritus VI (1979) Quantum effects of the interaction of elementary particles with an intense electromagnetic field. Ph.D. thesis
46. Erber T (1966) High-energy electromagnetic conversion processes in intense magnetic fields. Rev Mod Phys 38:626–659
47. Ridgers C, Kirk J, Duclous R, Blackburn T, Brady C, Bennett K, Arber T, Bell A (2014) Modelling gamma-ray photon emission and pair production in high-intensity laser-matter interactions. J Comput Phys 260:273–285
48. Lapeyre B, Pardoux E, Sentis R (1998) Méthode de Monte-Carlo pour les équations de transport et de diffusion. Springer
49. Schlegel T, Naumova N, Tikhonchuk VT, Labaune C, Sokolov IV, Mourou G (2009) Relativistic laser piston model: ponderomotive ion acceleration in dense plasmas using ultraintense laser pulses. Phys Plasmas 16(8):083103
50. Elkina NV, Fedotov AM, Kostyukov IY, Legkov MV, Narozhny NB, Nerush EN, Ruhl H (2011) QED cascades induced by circularly polarized laser fields. Phys Rev ST Accel Beams 14:054401
51. Ridgers CP (2017) Signatures of quantum effect on radiation reaction in laser-electron-beam collisions. arXiv:1708.04511
52. Touati M, Feugeas J-L, Nicolaï P, Santos JJ, Gremillet L, Tikhonchuk VT (2014) A reduced model for relativistic electron beam transport in solids and dense plasmas. New J Phys 16(7):073014
53. Pawula RF (1967) Approximation of the linear Boltzmann equation by the Fokker-Planck equation. Phys Rev 162:186–188
54. Kloeden PE, Platen E (1991) Numerical solution of stochastic differential equations. Springer
55. Niel F, Riconda C, Amiranoff F, Duclous R, Grech M (2018) From quantum to classical modeling of radiation reaction: a focus on stochasticity effects. Phys Rev E 97:043209

Chapter 6
Domain of Validity of the Different Descriptions and Numerical Simulations

6.1 Outline

In the previous Chapter, we introduced three models for radiation reaction, derived the equation of evolution for their successive energy moments and deduced the threshold for the increase or decrease of these moments. Here, we will use this hierarchy of moment equations to identify the domains of validity of the various descriptions of radiation reaction. These domains being intimately linked to the relative importance of different effects of radiation reaction, our analysis provides new physical insights into how to observe these different effects. Moreover, they will also help ourselves better determine the threshold parameters between the different methods in numerical codes.

This Chapter is structured as follows. In Sect. 6.2, we establish the limit between the different regimes of validity using the equations of evolution of the electron energy moments. Section 6.3 then presents - in addition to the two numerical algorithms described in Chap 4 to account for radiation reaction - the new particle pusher obtained from the FP description. Beyond the theoretical insights offered by the developed descriptions, this approach is also particularly interesting for numerical [in particular Particle-In-Cell (PIC)] simulations as a simple solver (so-called *particle pusher*) is obtained which can be easily implemented in kinetic simulation tools to account for the on-set of QED effects (namely the reduction of the power radiated away by the electron, so-called quantum correction, and the straggling following from the stochastic nature of photon emission) in the intermediate quantum regime, without having to rely on the computationally demanding Monte-Carlo approach. Section 6.4 then considers different physical configurations to validate both our theoretical analysis and numerical tools, and allows us to further investigate both the domains of validity of the different descriptions as well as the different aspects of radiation reaction.

F. Niel, *Classical and Quantum Description of Plasma and Radiation in Strong Fields*, Springer Theses, https://doi.org/10.1007/978-3-030-73547-0_6

6.2 Domain of Validity of the Three Descriptions

The analysis of the evolution of the successive moments allows us to infer more precisely the domain of validity of the three descriptions by taking into account the properties of the electron distribution function (in particular its first three moments). Most importantly, these domains of validity also provide us with a deeper insight into the various aspects of radiation reaction from which can be drawn new guidelines for designing experiments, as will be further discussed in Sect. 6.2.3.

In what follows, we discuss this, first considering an initially symmetric distribution function for the electron (most interesting when considering, e.g., an electron beam), then considering an asymmetric electron distribution function (as can be expected considering a hot relativistic electron plasma, see, e.g., Sect. 6.4.3).

6.2.1 *Case of an Initially Symmetric Energy Distribution*

In the case of an initially symmetric (or nearly symmetric) distribution function, the initial third order moment $\hat{\mu}_3$ is equal to zero. In this situation, the deterministic (cLL) description is expected to hold whenever:

- (i) the purely quantum terms in the equation of evolution for the variance [second and third terms in the rhs of Eq. (6.1b)] are negligible with respect to the classical one [Eq. (6.1a)], and
- (ii) the third order moment remains negligible compared to the second order one (i.e. it does not play any role in the description of the distribution function).

Let us recall for convenience the equations of evolution for the variance and third order moment of the electron distribution function as described by the cLL and by the FP/MC methods, at first order in perturbation (see Sect. 5.5.3)

$$d_t\sigma_\gamma^2\big|_{\rm cLL} \simeq -2\sigma_\gamma^2\frac{\langle\chi\rangle}{\langle\gamma\rangle}S'(\langle\chi\rangle)\,, \tag{6.1a}$$

$$\begin{aligned}\bar{\tau}\; d_t\sigma_\gamma^2\big|_\alpha \simeq\; & \bar{\tau}\; d_t\sigma_\gamma^2\big|_{\rm cLL} \\ & + \langle\gamma\rangle h(\langle\chi\rangle) + \sigma_\gamma^2\frac{\langle\chi\rangle}{\langle\gamma\rangle}h'(\langle\chi\rangle)\,,\end{aligned} \tag{6.1b}$$

$$d_t \mu_3|_{\text{cLL}} \simeq -3\mu_3 \frac{\langle \chi \rangle}{\langle \gamma \rangle} S'(\langle \chi \rangle), \tag{6.2a}$$

$$\begin{aligned} \bar{\tau}\, d_t \mu_3|_{\text{FP}} \simeq\ & \bar{\tau}\, d_t \mu_3|_{\text{cLL}} + 3\sigma_\gamma^2 [h(\langle \chi \rangle) \\ & + \langle \chi \rangle h'(\langle \chi \rangle)] + 3\mu_3 \frac{\langle \chi \rangle}{\langle \gamma \rangle} h'(\langle \chi \rangle), \end{aligned} \tag{6.2b}$$

$$\begin{aligned} \bar{\tau}\, d_t \mu_3|_{\text{MC}} \simeq\ & \bar{\tau}\, d_t \mu_3|_{\text{FP}} - \langle \gamma \rangle^2 a_3(\langle \chi \rangle) \\ & - \sigma_\gamma^2 [a_3(\langle \chi \rangle) + 2 \langle \chi \rangle a_3'(\langle \chi \rangle)] \\ & - \mu_3 \frac{\langle \chi \rangle}{\langle \gamma \rangle} a_3'(\langle \chi \rangle), \end{aligned} \tag{6.2c}$$

where $\alpha =$ [FP, MC]. The first condition (i) then reads:

$$\left| -2\sigma_\gamma^2 \frac{\langle \chi \rangle}{\langle \gamma \rangle} \tilde{S}'(\langle \chi \rangle) \right| \gg \left| \langle \gamma \rangle h(\langle \chi \rangle) + \sigma_\gamma^2 \frac{\langle \chi \rangle}{\langle \gamma \rangle} h'(\langle \chi \rangle) \right|, \tag{6.3}$$

and reduces to:

$$\hat{\sigma}_\gamma^2 \gg \frac{h(\langle \chi \rangle) + \langle \chi \rangle h'(\langle \chi \rangle)\, \hat{\sigma}_\gamma^2}{2\langle \chi \rangle\, S'(\langle \chi \rangle)}. \tag{6.4}$$

The second condition (ii) can be computed estimating the extremum value of $\hat{\mu}_3$ [obtained canceling the rhs of Eq. (6.2c)]. This yields (setting $\psi = 1$ in Eq. (6.5))

$$\hat{\mu}_3^{\text{extr}} = \frac{\hat{\sigma}_\gamma^2 [f_1(\langle \chi \rangle) + \langle \chi \rangle\, f_2(\langle \chi \rangle)] - 1}{\langle \chi \rangle\, f_3(\langle \chi \rangle)}, \tag{6.5}$$

where we used the functions $f_{1,2,3}$ defined in Sect. 5.5.6 by

$$f_1(\chi) = \frac{3\,h(\chi) - a_3(\chi)}{a_3(\chi)}, \tag{6.6a}$$

$$f_2(\chi) = \frac{3\,h'(\chi) - 2a_3'(\chi)}{a_3(\chi)}, \tag{6.6b}$$

$$f_3(\chi) = \frac{3\,S'(\chi) + a_3'(\chi) - 3\,h'(\chi)}{a_3(\chi)}. \tag{6.6c}$$

The condition (ii) then reads

$$\frac{\hat{\sigma}_\gamma^2 \langle \chi \rangle f_3(\langle \chi \rangle)}{|\hat{\sigma}_\gamma^2 [f_1(\langle \chi \rangle) + \langle \chi \rangle f_2(\langle \chi \rangle)] - 1|} \gg 1. \tag{6.7}$$

In contrast, the FP model providing the correct description for the energy variance, its validity is constrained by the second condition (ii) [Eq. (6.7)] only.

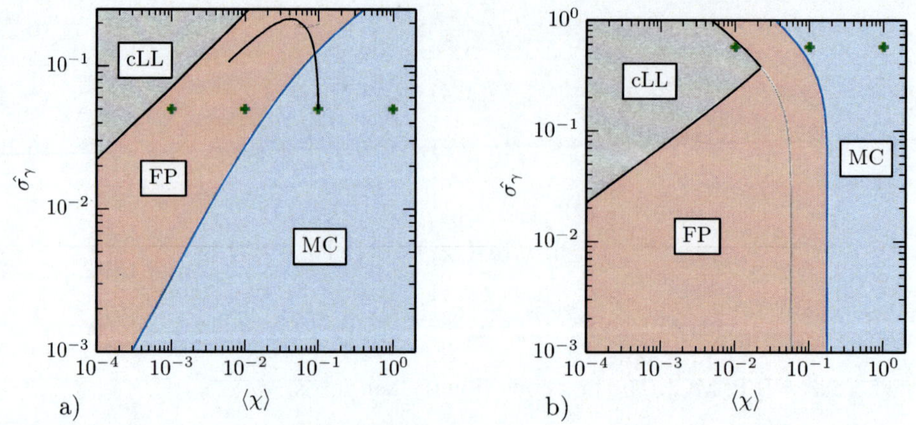

Fig. 6.1 Domain of validity of the different approaches. **a** In the case of an initially symmetric electron energy distribution; as given by Eqs. (6.4) and (6.7) for a symmetric distribution function ($\hat{\mu}_3 = 0$). **b** In the case of an initially asymmetric distribution function; as given by Eqs. (6.9) and (6.11). Here presented for $\hat{\mu}_3 \simeq 0.2$ corresponding to the Maxwell-Jüttner distribution presented in Sect. 6.4.3 [1]

The derived conditions, Eqs. (6.4) and (6.7), depend on both, the average quantum parameter $\langle\chi\rangle$ and the electron distribution normalized energy variance $\hat{\sigma}_\gamma$. They provide us with a path to define the domain of validity of the different approaches (deterministic/cLL, FP or MC) in the $(\langle\chi\rangle,\hat{\sigma}_\gamma)$ parameter-space. Our results are summarized in Fig. 6.1.

The blue shaded regions correspond to the parameter-space for which condition (ii) [Eq. (6.7)] is not satisfied (more exactly the lhs of Eq. (6.7) is ≤ 10[1]). In this region, the third order moment of the distribution function can assume large values, so that only the MC procedure provides a correct description.

Finally, outside of this domain (grey and red shaded regions), the FP description holds while the gray shaded region corresponds to the region where the deterministic (cLL) description is valid, i.e. both conditions (i) and (ii) [Eqs. (6.4) and (6.7), respectively] are satisfied [for condition (i), we specify here that the lhs of Eq. (6.4) is ≥ 10 which defines the region above the black-dashed line].

6.2.2 *Case of an Initially Non-symmetric Energy Distribution*

A similar analysis can be performed with an initially asymmetric distribution function. In that case however, even though condition (i) [Eq. (6.4)] is unchanged, one has to reconsider the condition on $\hat{\mu}_3$ that is now not negligible, but should be correctly handled by the various description. More precisely, for the deterministic (cLL) description to hold, one thus requires that all the purely quantum terms in the

[1]There is of course a degree of arbitrariness here, and the domain of validity can be adjusted depending on the required degree of precision

equation of evolution Eq. (6.2c) of μ_3 are negligible compared to the classical term appearing in Eq. (6.2a). This condition writes

$$\Big| d_t\mu_3|_{\rm cLL} \Big| \gg \Big| d_t\mu_3|_{\rm MC} - d_t\mu_3|_{\rm cLL} \Big|, \tag{6.8}$$

and simplifies to:

$$\frac{3\hat{\mu}_3\langle\chi\rangle S'(\langle\chi\rangle)/a_3(\langle\chi\rangle)}{\left|-\hat{\mu}_3\langle\chi\rangle f_3(\langle\chi\rangle) + 3\hat{\mu}_3\langle\chi\rangle\tilde{S}'(\langle\chi\rangle)/a_3(\langle\chi\rangle) + \hat{\sigma}_\gamma^2[f_1(\langle\chi\rangle) + \langle\chi\rangle f_2(\langle\chi\rangle)] - 1\right|} \gg 1. \tag{6.9}$$

Proceeding in a similar way for the FP description, we require that

$$\Big| d_t\mu_3|_{\rm FP} \Big| \gg \Big| d_t\mu_3|_{\rm MC} - d_t\mu_3|_{\rm FP} \Big|, \tag{6.10}$$

which leads condition (ii):

$$\frac{\left|3\hat{\sigma}_\gamma^2[h(\langle\chi\rangle) + \langle\chi\rangle h'(\langle\chi\rangle)] + 3\hat{\mu}_3\langle\chi\rangle[h'(\langle\chi\rangle) - \tilde{S}(\langle\chi\rangle)]\right|}{\left|a_3(\langle\chi\rangle) + \hat{\sigma}_\gamma^2[a_3(\langle\chi\rangle) + 2\langle\chi\rangle a_3'(\langle\chi\rangle)] + \hat{\mu}_3\langle\chi\rangle a_3'(\langle\chi\rangle)\right|} \gg 1, \tag{6.11}$$

The derived conditions of validity, Eqs. (6.4) and (6.9) for the deterministic (cLL) description and Eqs. (6.4) and (6.11) for the Fokker-Planck description now depend not only on $\langle\chi\rangle$ and $\hat{\sigma}_\gamma$, but also on the normalized third order moment $\hat{\mu}_3$. In Fig. 5.7, we have projected them on the $(\langle\chi\rangle, \hat{\sigma}_\gamma)$ parameter-space assigning for $\hat{\mu}_3$ its initial value ($\hat{\mu}_3 = 0.2$) for the particular case (broad Maxwell-Jüttner distribution) studied in Sect. 6.4.3.

The regions of validity of the different models are shown in Fig. 6.1b following the same lines then for Fig. 6.1a. While the region of validity of the deterministic (cLL) description is mainly unchanged (note that the range of accessible $\hat{\sigma}_\gamma$ is larger for asymmetric functions), one finds that the region validity of the FP description is significantly increased. As will be discussed in Sect. 6.4.3, this is indeed what is observed in our simulations.

6.2.3 *Physical Implications*

Our analysis of the successive (and in particular second and third) moments has allowed us to identify more clearly the domain of validity of the different descriptions in terms of both the initial average quantum parameter and shape of the energy distribution of the electron population. As each of these descriptions account for different physical effects of radiation reaction, these domains of validity provide us with some insights on which processes play a dominant role under given conditions of the external field and properties of the electron population, as well as some guidelines for future experiments to observe various aspects of radiation reaction.

Indeed, should our physical conditions (e.g. experimental set-up or astrophysical scenario) lie in the regime where the deterministic (cLL) description is valid, radiation reaction acts as a friction force and one can expect to observe both a reduction of the electron population mean energy as well as a narrowing (cooling) of the electron energy distribution function. In the case where our physical conditions are characteristic of the regime where the deterministic description is not valid anymore but the FP approach is, the stochastic nature of high-energy photon emission starts to play an important role. Under such conditions, and in particular for short (with respect to the so-called *heating* time that will be further discussed in Sect. 6.4) times, one can expect to observed a measurable broadening of the electron energy distribution. Finally, should our experimental (or astrophysical) conditions lie in the regime where only the linear Boltzmann (MC) description is valid, not only the stochastic nature of photon emission, but its discrete nature will play a role. In particular, this defines the experimental regime for which the *quenching* of radiation losses can be observed, and diagnosed as a negative skewness of the electron distribution function. In this regime, and for short enough times, the diffusion approximation supporting the FP approach is not valid anymore and rare emission events, that can be described only as a Poisson process (i.e. by the MC procedure), are of outmost importance.

6.3 Stochastic (Fokker-Planck) Pusher

The stochastic pusher we now describe is based on the Fokker-Planck treatment developed in Sect. 5.3. It follows the very same step as described earlier for the deterministic pusher with the difference that the radiation reaction force in Eq. (4.27) now contains an additional stochastic term:

$$\mathbf{f}_{\rm rad}^{(n)} = \left[-P_{\rm rad}\Delta t + mc^2\sqrt{R(\chi,\gamma)}\,dW\right]\boldsymbol{\beta}/(c\boldsymbol{\beta}^2)\,, \tag{6.12}$$

where $\boldsymbol{\beta} = \mathbf{p}^{(n-\frac{1}{2})}/(mc\gamma)$, and dW is a random number generated using a normal distribution of variance Δt. Both functions $g(\chi)$ and $h(\chi)$ [the latter appearing when evaluating $R(\chi,\gamma)$ using Eq. (5.38b)] can be either tabulated or estimated from a fit.

Let us now note that Eq. (6.12) can in some cases (when its rhs is positive) lead to an electron gaining energy. This up-scattering is not physical, and is a well-known short-cut of the Fokker-Planck approach. It may become problematic only in cases where $\chi \to 1$ for which the stochastic term can become of the order of the drift term. However, if this may be a problem should one consider only a single particle dynamics, this problem is alleviated when using this kind of pusher in Particle-In-Cell (PIC) codes. In that case indeed, one deals not with real particles but with so-called macro-particles that actually represent discrete element of a distribution function (see, e.g., Ref. [2]), and up-scattering is then in average suppressed. This will be discussed into more details in the next Sect. 6.4.

For the sake of completeness, we also note that Wiener process involve sample paths that are non-differentiable [3]. This requires much care when the issue comes to the numerical treatment of these random discrete increments. Here, as a first tentative, we introduce the simplest possible scheme, know as Euler-Maruyama [3]. Of course more sophisticated and accurate schemes exist, that have not been tested in this work. An importance issue lies in the stiffness of the Stochastic Differential Equation (SDE). This stiffness can be quantified with use of the SDE Lyapunov exponents, which basically indicates the presence of different scales in the solution [3]. *A priori*, the stiffness of the rate of photon emission is avoided by the SDE because it is precisely the operating regime of the Monte-Carlo pusher.

6.4 Numerical Results

In this Section, we confront the various numerical algorithms (pushers) introduced above against each other as well as against our theoretical predictions (Sect. 5.4). We consider an electron beam, first in a constant magnetic field, then in a counter-propagating linearly polarized plane-wave. The case of an electron bunch with a broad (Maxwell-Jüttner) energy distribution in a constant magnetic field is also discussed. Note that, throughout this Section, the Monte-Carlo (MC) simulations will be used as a reference as they provide a more general description equivalent to the full linear-Boltzmann description.

6.4.1 Constant-Uniform Magnetic Field

We start by simulating the interaction of a Gaussian electron beam with mean energy $\gamma_0 = 1800$ and standard deviation $\sigma_0 = 90$ (corresponding to approximately 920 ± 46 MeV) with different constant-uniform magnetic fields of magnitude corresponding to $\chi_0 = \langle\chi\rangle(t = 0) = 10^{-3}$, 10^{-2}, 10^{-1} and 1 (correspondingly, $B = 2.5$ kT, 25 kT, 250 kT and 2.5 MT). The end of the simulation is taken when the energy decrease becomes very slow (i.e. we approach a regime in which radiation losses are not important) except for the case $\chi_0 = 10^{-3}$, where the energy loss is always small and we stop arbitrarily at $t_{\text{end}} = 20/\omega_c$, with $\omega_c = eB/(m\gamma)$ the synchrotron frequency. For $\chi_0 = 1$ the simulation ends at $t_{\text{end}} = 3/\omega_c$, for $\chi_0 = 10^{-1}$ at $t_{\text{end}} = 5/\omega_c$, and for $\chi_0 = 10^{-2}$ and 10^{-3} at $t_{\text{end}} = 20/\omega_c$. In all cases, we used 10 000 test particles.

The results are summarized in Fig. 6.2. The first row a) corresponds to $\chi_0 = 10^{-3}$, the second b) to $\chi_0 = 10^{-2}$, the third c) to $\chi_0 = 10^{-1}$ and the last one d) to $\chi_0 = 1$. The first three columns correspond to the evolution of the distribution function

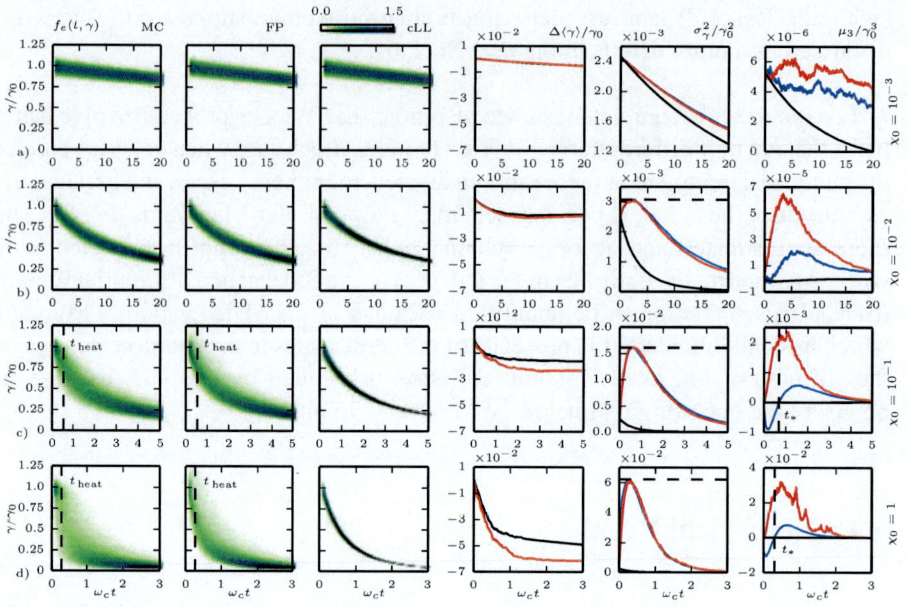

Fig. 6.2 Simulations of an ultra-relativistic electron beam in a constant uniform magnetic field for **a** $\chi_0 = 10^{-3}$, **b** $\chi_0 = 10^{-2}$, **c** $\chi_0 = 10^{-1}$ and **d** $\chi_0 = 1$. The first three panels of each row show the electron distribution functions from the Monte-Carlo simulations (MC, first panels), stochastic (Fokker-Planck) simulations (FP, second panels) and quantum-corrected deterministic simulations (cLL, third panels). The fourth panel shows the difference in the prediction of the mean electron energy in between the MC simulation and the deterministic (black line) and FP (red line) simulations. The two last panels (in each row) correspond to the moments of order 2 (energy variance) and 3 for the MC (blue line), FP (red line) and deterministic (black line) simulations [1]

$f_e(t, \gamma)$ respectively in the case of the Monte-Carlo simulation (MC), the stochastic (Fokker-Planck) pusher (FP) and the deterministic (cLL) radiation reaction pusher [including the quantum correction $g(\chi)$]. The fourth column corresponds to the (normalized) difference between the average energy extracted from the Monte-Carlo simulations and the average energy obtained from the stochastic pusher (red line), and that obtained using the deterministic pusher (black line). Both are normalized to the initial mean energy γ_0: $\Delta\gamma_\alpha/\gamma_0 = (\langle\gamma\rangle_{MC} - \langle\gamma\rangle_\alpha)/\gamma_0$, with $\alpha = cLL$ or FP. Finally the last two rows correspond to the normalized variance $\sigma_\gamma^2/\gamma_0^2 = \langle(\gamma - \langle\gamma\rangle)^2\rangle/\gamma_0^2$ and to the normalized moment of order 3, $\mu_3/\gamma_0^3 = \langle(\gamma - \langle\gamma\rangle)^3\rangle/\gamma_0^3$ (in all plots, the blue line corresponds to the Monte-Carlo simulation, the red line to the stochastic pusher and the black line to the deterministic pusher [with the quantum correction $g(\chi)$].

Let us first consider the case $\chi_0 = 10^{-3}$. As $\chi_0 < \chi_{\rm cl}$, one could expect all three descriptions to lead to similar results, and indeed, there is a very good agreement in the evolution of the distribution function as calculated by the three models [see first three panels of Fig. 6.2a]. Small differences eventually appear that are as predicted by the analysis performed in the previous sections. In particular we see in the variance [fifth

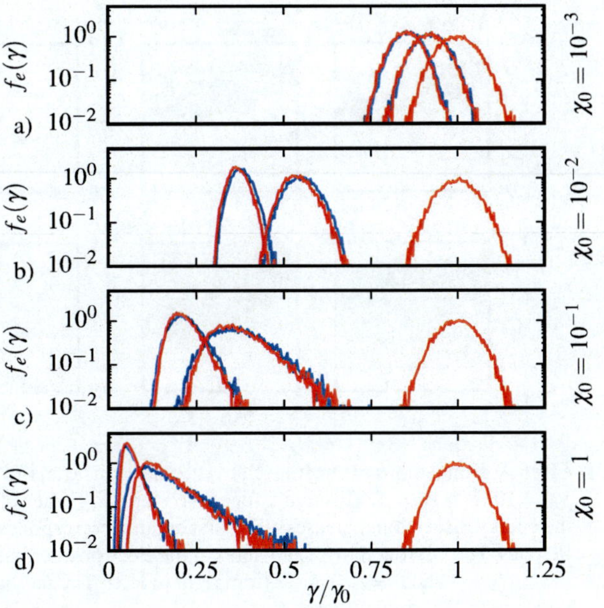

Fig. 6.3 Simulations of an ultra-relativistic electron beam in a constant uniform magnetic field for **a** $\chi_0 = 10^{-3}$, **b** $\chi_0 = 10^{-2}$, **c** $\chi_0 = 10^{-1}$ and **d** $\chi_0 = 1$ electron distribution functions at times $t = 0$, $t = t_{\text{end}}/2$ and $t = t_{\text{end}}$ (from right to left). The red lines correspond to FP simulations, the blue one to MC simulations [1]

panel of Fig. 6.2a] that cooling is slightly overestimated by the deterministic (cLL) model. This is expected as we are close but not exactly into the deterministic domain of validity (see Fig. 6.1a) so that the quantum terms leading to energy spreading are not completely negligible. Moreover, there is a small difference between models in the third order moment [last panel of Fig. 6.2a], as expected from Sect. 5.5. Yet this moment remains 3 orders of magnitude smaller than the variance, as expected from the scaling $\propto \chi$ of the various moments (see Sect. 5.3.3), and it is thus negligible (see also Fig. 6.3a).

We now examine the cases $\chi_0 = 10^{-2}$ and 10^{-1}. We are now in what we called the intermediate quantum regime ($\chi_{\text{cl}} < \chi < \chi_{\text{qu}}$), and one expects the deterministic (cLL) description to fail in describing the evolution of the electron population. It turns out to be the case as, while the evolution of the distribution function obtained from the stochastic (FP) pusher is in good agreement with the one obtained using the MC procedure, both are very different from the deterministic one. Still, and as expected from Sect. 5.5, all three models give similar results concerning the evolution of the average energy which is found to be very close to the evolution of a single classical particle with initial energy equal to the average energy of the initial population (see the fourth panels of Fig. 6.2b and c).

The main difference between the deterministic model and the quantum (FP and MC) models is in the variance (fifth panels of Fig. 6.2b and c). The quantum models exhibit an energy spreading (effective heating) phase, with σ_γ increasing up to a maximum value, and a later phase of cooling, while the deterministic model predicts only cooling. As predicted, both quantum models (MC and FP) predict the same

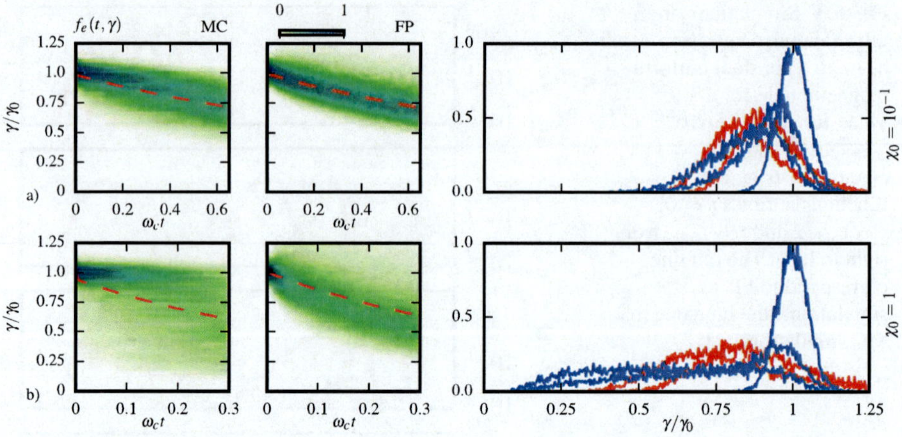

Fig. 6.4 Simulations of an ultra-relativistic electron beam in a constant, uniform magnetic field for **a** $\chi_0 = 10^{-1}$, **b** $\chi_0 = 1$. This figure focuses on the early times of interaction ($t \lesssim t_{\text{heating}}$ during which the energy dispersion increases). The first column corresponds to MC simulations, the second to the FP ones. The last row shows snap-shots of the electron distribution functions at different times $t = 0$, $t = t_{\text{heating}}/4$ and $t = t_{\text{heating}}/2$ (from right to left). The red lines correspond to FP simulations, the blue one to MC simulations [1]

evolution of the energy variance, and the maximum value of $\hat{\sigma}_\gamma$ is found to be in perfect agreement with the predicted value $\hat{\sigma}_\gamma^{\text{thr}}$ given by Eq. (5.81). The good agreement between both quantum models with respect to the temporal evolution of the distribution function can be clearly seen in Figs. 6.3b and c, where we superimposed the electron distribution functions obtained from the stochastic (FP) pusher (blue line) and Monte Carlo approach (red line) at different times $t = 0$, $t = t_{\text{end}}/2$ and $t = t_{\text{end}}$.

As expected, differences in between the FP and MC models only appear in the evolution of the third order moment (last panels in Figs. 6.2b and c). Yet, for both cases, this third order moment is found to be χ times smaller than the second order moment. For the case $\chi_0 = 10^{-2}$, this discrepancy is negligible and both FP and MC are equivalent. For the case $\chi_0 = 10^{-1}$, one can argue that for short times, this difference of the order of 10% is not so negligible. This is exactly what one could have expected from the extended analysis of the domain of validity presented in Sect. 6.2, and Fig. 6.1a, where the case $\chi_0 = 10^{-1}$ is found to lie close to, yet outside of the domain of validity of the FP description. Furthermore, at longer times ($t > t_{\text{heat}}$), the slight decrease of $\langle\chi\rangle$ and significant increase of $\hat{\sigma}_\gamma$ bring the simulation conditions back into the FP domain of validity as shown by the solid black line in Fig. 6.1 Note that this finding supports our claim that not only $\langle\chi\rangle$ but also $\hat{\sigma}_\gamma$ are relevant to discriminating which physical processes of radiation reaction are important, in particular at short times.

Finally, for $\chi_0 = 1 > \chi_{qu}$, we are in the quantum regime and we start to see some differences in the global shape of the distribution function among the two different quantum models (see first two panels of Figs. 6.2d and 6.3d), in particular during the early stage of interaction. This is expected as, for $\chi \sim 1$, the higher order moment ($n \geq 3$) contributions are not negligible anymore (see Sect. 5.3.3) and leads to different predictions whether one considers the FP or linear Boltzmann approach (see Sect. 5.4), as clearly seen in the last panel of Fig. 6.2d. This particular case also clearly lies outside of the deterministic and FP domains of validity (as shown in Fig. 6.1a and discussed in Sect. 6.2). Let us note however that the prediction of the average electron energy (fourth panel of Fig. 6.2d) and energy dispersion (fifth panel of Fig. 6.2d) is consistent in between all three approaches.

In addition, the clear difference between the FP and MC description is in the third order moment. Figure 6.4 shows the evolution of the distribution function in the FP and MC models focusing on the initial stage of interaction (heating phase, corresponding to an increase of the variance) at three different times $t = 0$, $t = t_{\text{heating}}/4$ and $t = t_{\text{heating}}/2$. Let us first note that the FP simulation, in contrast with the MC one, exhibits a non negligible amount of particles gaining energy. This unphysical behavior follows from what we earlier introduced as particle up-scattering (see Sect. 6.3). As $\chi \to 1$, the contribution of the diffusion term becomes of the same order of the drift term, and clearly the FP model reaches his limit.

Furthermore, the energy spreading in the MC simulation is, for such large values of χ, strongly asymmetric (see also Fig. 6.3d). As the variance increases, the moment of order 3 (that becomes of the same order than the variance as $\chi \to 1$, see Fig. 6.3d last two panels) is negative. This corresponds to a tail towards the low energies, with the distribution still being peaked at high-energy (similar to the time $t = 0$ peak). As the variance reaches its threshold value [still correctly predicted by Eq. (5.81), see also Fig. 5.5 (blue crosses)], we reach $t = t_{\text{heating}}$ and our simulation shows that the sign of the third order moment changes at this time. This is coherent with the fact that $\hat{\sigma}_{\gamma}^{\text{thr}} \simeq \hat{\sigma}_{\gamma_0}^{\text{lim}}$ as explained in more detail in Sect. 5.5.6.

This function peaked at high (close to initial) energy can be interpreted as the result of the *quenching* of radiation losses [4]. This quantum quenching, which is not accounted for in the deterministic (even quantum corrected) and FP approaches, follows from the discrete nature of photon emission. As a result, in the regime where *quenching* is important, each electron trajectory can be modelled only considering the discrete nature of the emission process, i.e. it requires the use of the full Monte-Carlo approach.

Yet, when one follows the mean energy only, all three descriptions provide similar results. That is, even in this regime of quantum quenching, the mean energy of the overall electron population is reduced and still closely follows deterministic radiation reaction (quantum corrected) and FP predictions. This has consequences on future experiments, where only a careful measurement of the electron energy spectra (and in particular their symmetry) will allow to observe this quenching process. We will get back to this particular point in more detail at the end of next Sect. 7.3.3.

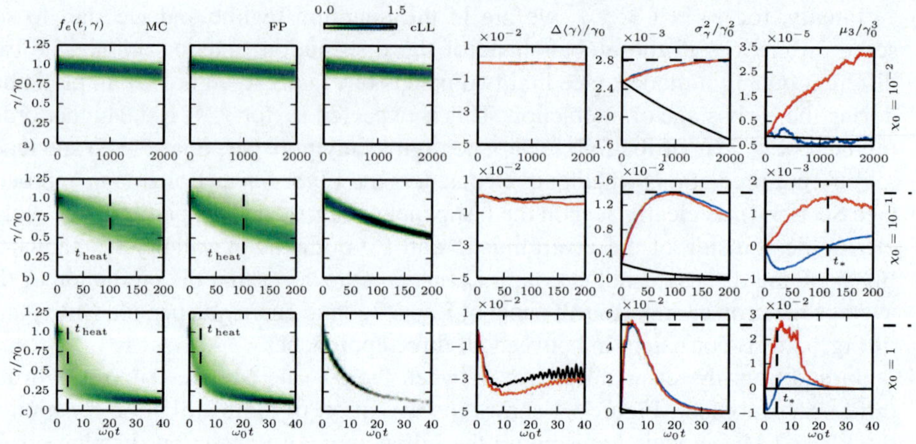

Fig. 6.5 Simulations of an ultra-relativistic electron beam in a counter-propagating linearly polarized plane wave for **a** $\chi_0 = 10^{-2}$, **b** $\chi_0 = 10^{-1}$, and **c** $\chi_0 = 1$. The first three panels of each row show the electron distribution functions from the Monte-Carlo simulations (MC, first panels), stochastic (Fokker-Planck) simulations (FP, second panels) and quantum-corrected deterministic simulations (cLL, third panels). The fourth panel shows the difference in the prediction of the mean electron energy in between the MC simulation and the deterministic (black line) and FP (red line) simulations. The two last panels (in each row) correspond to the moments of order 2 (energy variance) and 3 for the MC (blue line), FP (red line) and deterministic (black line) simulations [1]

6.4.2 *Linearly Polarized Plane-Wave*

We now consider the interaction of this same Gaussian electron beam (mean energy $\gamma_0 = 1800$ and standard deviation $\sigma_0 = 90$) with counter-propagating (linearly polarized) electromagnetic plane-waves with different amplitudes with $\chi_0 = \langle\chi\rangle(t = 0) = 10^{-2}$, 10^{-1} and 1 (corresponding to the wave normalized vector potentials $a_0 = 1.14$, 11.4 and 114, respectively). The duration of each simulation is chosen so that we get the interesting features of the interaction. For $\chi_0 = 10^{-2}$, the duration of the simulation is $t_{\rm end} = 2000/\omega_0$, for $\chi_0 = 10^{-1}$ $t_{\rm end} = 200/\omega_0$ and for $\chi_0 = 1$ $t_{\rm end} = 40/\omega_0$, where ω_0 is the electromagnetic wave angular frequency ($\omega_0 = 2\pi c/\lambda_0$, where we have considered $\lambda_0 = 1\ \mu$m). In all cases, 10 000 test particles were used.

The simulation results are summarized in Fig. 6.5, following the same presentation than Fig. 6.2. The first row a) corresponds to $\chi_0 = 10^{-2}$, the second b) to $\chi_0 = 10^{-1}$, the third c) to $\chi_0 = 1$. The interpretation of these simulations follows the same lines than for the case of a constant magnetic field and, as a result, the same conclusions can be drawn.

Let us start by considering the case $\chi_{\rm cl} < \chi_0 = 10^{-2} < \chi_{\rm qu}$. As $\chi_0 \ll 1$ the $n \leq 2$ higher order moments ($\propto \chi^{n+1}$) are small, and the overall evolution of the electron distribution function f_e is well reproduced, as shown in Fig. 6.5a should one focus on the first four panels in Fig. 6.5a only. Looking into more details, and in particular

Fig. 6.6 Simulations of an ultra-relativistic electron beam in a counter-propagating linearly polarized plane wave for **a** $\chi_0 = 10^{-2}$, **b** $\chi_0 = 10^{-1}$ and **c** $\chi_0 = 1$ electron distribution functions at times $t = 0$, $t = t_{\text{end}}/2$ and $t = t_{\text{end}}$ (from right to left). The red lines correspond to FP simulations, the blue one to MC simulations [1]

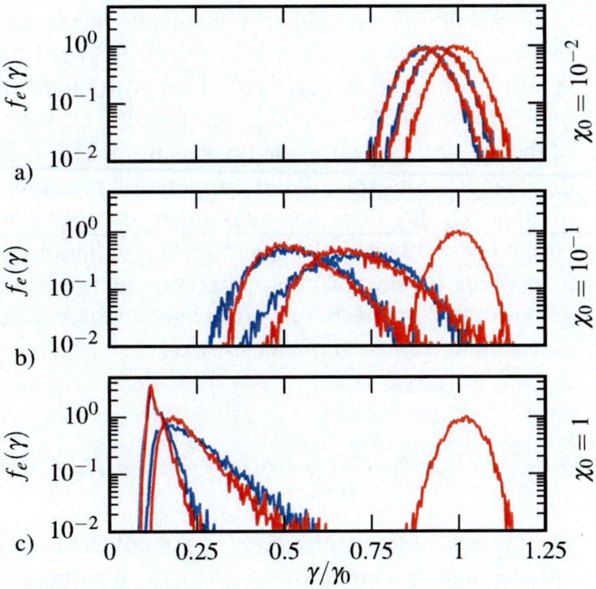

to the fifth panel (energy variance), one sees that here again the deterministic (cLL) description strongly overestimate the electron beam cooling. This is explained again by this particular case lies outside of the regime of applicability of the deterministic (cLL) description (Fig. 6.1a), and well into that of the FP description (Fig. 6.6).

Similarly, for $\chi_{\text{cl}} < \chi_0 = 10^{-1} < \chi_{\text{qu}}$, the stochastic (FP) scheme reproduces correctly the evolution of both the electron mean energy *and* variance, and the global properties of the electron distribution function f_e considering the two quantum approaches are similar. In particular, the existence of a heating to cooling transition is recovered, and analytical predictions of Eq. (5.81) are found to be in excellent agreement with our simulation results, see also Fig. 5.5 (green crosses). As discussed previously for the case of an electron beam radiating a constant magnetic field, and as predicted by our theoretical analysis (this particular simulation lies slightly outside of the regime of validity of the FP description, see Fig. 6.1a), discrepancies in the third order moment appear on short time. For this case however, this error remains quite small, of the order of $\chi_0 = 10^{-1}$ times the normalized variance (second order moment). Here again, our numerical results support the extended analysis of the domains of validity of the different approaches (Sect. 6.2).

Finally, when $\chi_0 = 1 > \chi_{\text{qu}}$, important differences in the different approaches are observed, in particular visible in the third momentum, and in the overall shape of the distribution function at early times as μ_3 then assumes values of the same order than σ_γ. Once more, μ_3 is negative during the heating phase, and quantum quenching sets in. Here again, this third order moment flip signs at $t_{\text{heating}} \simeq 2\pi/\omega_0$, which thus provides a good measure for the time up to which the discrete nature of photon emission has a noticeable effect on the overall shape of the distribution function.

As previously stressed, this quenching is intimately linked to the discrete nature of photon emission, and is here shown to greatly impact the physics at $\langle\chi\rangle \to 1$. As it is associated to negative values of the third order moment μ_3, the impact of *quenching* can be measured in forthcoming experiments by analysing the skewness of the measured electron energy spectrum. This would provide a clear signature of the quantum (discrete) nature of radiation reaction. Yet, unlike previously claimed in Ref. [4], this does not translate in the electron bunch mean energy which is correctly described by the deterministic (radiation friction) description provided that it accounts for the quantum correction as previously underlined in Sect. 5.4, and demonstrated in Fig. 6.5 (fourth panels) where all three descriptions are shown to give similar results on the mean energy.

6.4.3 Electron Population with a Broad Energy Dispersion

For the sake of completeness, we finally consider the evolution of an electron population with an initially broad energy distribution radiating in a constant uniform external magnetic field. The electron energy distribution at the beginning of the simulation follows a (zero-drift) 3D Maxwell-Jüttner distribution:

$$f_e(t=0,\gamma) = \frac{\gamma\sqrt{\gamma^2-1}}{\theta K_2(1/\theta)} \exp\left(-\frac{\gamma}{\theta}\right), \tag{6.13}$$

where K_2 is the modified Bessel function of second kind, and $\theta = T/(mc^2) = 600$ is the normalized temperature corresponding to an initial electron mean Lorentz factor $\gamma_0 = \langle\gamma\rangle \simeq 1800$, and initial energy standard deviation $\sigma_\gamma \simeq 0.57\,\gamma_0$. Note also that such a broad distribution also presents a large initial asymmetry and $\hat{\mu}_3 \simeq 0.2$ at the beginning of the simulation. Three magnetic field strengths have been considered corresponding to $\chi_0 = \langle\chi\rangle(t=0) = 10^{-2}$, $\chi_0 = 10^{-1}$ and $\chi = 1$.

The simulation results are summarized in Fig. 6.7 following the same presentation than Figs. 6.2 and 6.5.

Let us first briefly focus on the case $\chi_0 = 10^{-2}$. This particular case is indeed particularly interesting as for such χ_0, both our extended analysis of the domain of validity (Sect. 6.2) and previous simulations considering an initially narrow and symmetric electron beam showed that the on-set of stochasticity effects translate in an overestimated cooling by the deterministic (cLL) description. In contrast, for this particular case of a broad energy distribution, and as predicted in Fig. 6.1b, this particular case $\chi_0 = 10^{-2}$ now lies exactly into the regime of the deterministic (cLL) approach. This is indeed confirmed by our simulations results (Fig. 6.7a) where all three models predict the same evolution for all three moments. This finding further confirms our analysis that not only the average quantum parameter, but also the shape of the energy distribution of the electron population is important to assess the relative importance, and measurable aspect, of various effects of radiation reaction.

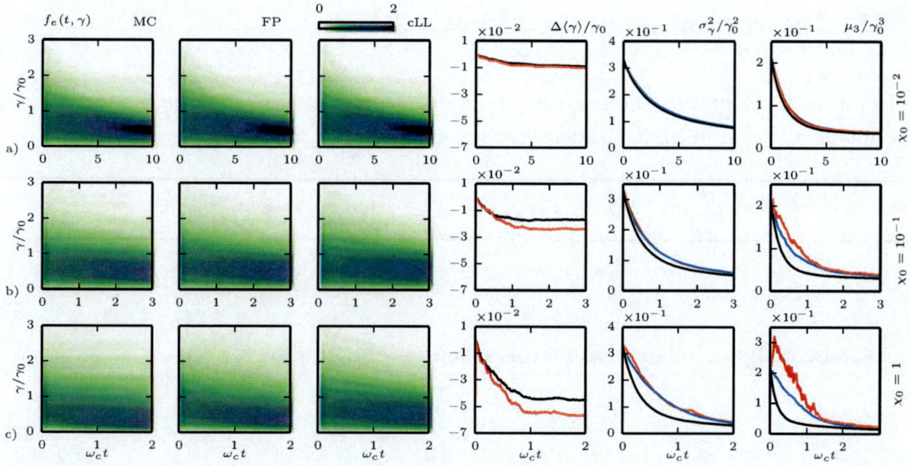

Fig. 6.7 Simulations of an electron bunch with initially broad (Maxwell-Jüttner) energy distribution in a constant, uniform magnetic field for **a** $\chi_0 = 10^{-3}$, **b** $\chi_0 = 10^{-2}$, **c** $\chi_0 = 10^{-1}$ and **d** $\chi_0 = 1$. The first three panels of each row shows the electron distribution functions from the Monte-Carlo simulations (MC, first panels), stochastic (Fokker-Planck) simulations (FP, second panels) and quantum-corrected deterministic simulations (cLL, third panels). The fourth panels show the difference in the prediction of the mean electron energy in between the MC simulation and the deterministic (black line) and FP (red line) simulations. The two last panels (in each row) correspond to the moments of order 2 (energy variance) and 3 for the MC (blue line), FP (red line) and deterministic (black line) simulations [1]

Furthermore, and for all values of χ_0, the very large initial standard deviation, $\sigma_\gamma/\gamma_0 \simeq 0.57$, exceeds the predicted threshold $\sigma_\gamma^{\rm thr}/\gamma_0 \simeq 0.37$ [here computed for $\chi = 1$ using Eq. (5.81)]. As a result, and even at large χ, all simulations (and in particular all quantum ones) predict only a cooling of the electron distribution with σ_γ continuously decreasing with time.

Another remarkable point is that the FP description reproduces the MC results for $\chi_0 = 10^{-1}$ (and to some extent for $\chi_0 = 1$) much more closely than for the initially narrow electron beam. Again, this is in good agreement with our extended analysis of the domain of valid of the different descriptions of radiation reaction and is explained by the initially broad energy dispersion (variance) so that the deterministic terms in the equation of evolution of the moments are dominant, while the quantum terms only give a correction.

Let us finally note that, beyond demonstrating the validity our approach and its capacity at being applied to general electron populations, this particular case of a Maxwell-Jüttner distribution is also interesting to broaden the spectrum of applications of our study. Indeed, such relativistic distribution functions are typical of laser-solid interaction in the ultra-high intensity regime, as well as in relativistic astrophysics.

6.5 Interpretation of the "Vlasov Terms"

Let us analyze in more details the so-called "Vlasov terms" obtained in the previous chapter by computing the successive moments of the operator

$$\frac{d}{dt} f_e = \nabla \cdot [c\beta \mathbf{\Omega} f_e] - \frac{1}{mc^2}\, \partial_\gamma [ec\beta(\mathbf{\Omega}\cdot\mathbf{E}) f_e] \\ - \frac{e}{p} \nabla_{\mathbf{\Omega}} \cdot [(\mathbb{1} - \mathbf{\Omega}\otimes\mathbf{\Omega})\cdot(\mathbf{E} + \beta\mathbf{\Omega}\times\mathbf{H}) f_e]\,. \quad (6.14)$$

As indicated previously, these terms take the following form

$$\frac{n\,e}{mc} \big\langle (\gamma - \langle\gamma\rangle_\alpha)^{n-1}\, (\boldsymbol{\beta}\cdot\mathbf{E}) \big\rangle_\alpha\,, \quad (6.15)$$

and they are due to the action of Lorentz force alone on the electron distribution function. Indeed, considering a particle subject to the Lorentz force only, we have

$$\frac{d\gamma}{dt} = -\frac{e}{mc} \boldsymbol{\beta}\cdot\mathbf{E}\,, \quad (6.16)$$

from which we easily deduce for any integer n

$$\frac{d\gamma^n}{dt} = -\frac{ne}{mc} \gamma^{n-1} \boldsymbol{\beta}\cdot\mathbf{E}\,, \quad (6.17)$$

which leads, after averaging, the different terms (6.15). In the work by [5], the considered situation was that of an electron bunch of energy γ_0 counterpropagating with a laser with normalized strength $a_0 \ll \gamma_0$. In this situation, $\boldsymbol{\beta} \perp \mathbf{E}$ and all the Vlasov terms vanish. Here we developed the formalism keeping these term. This allows us to see how the different equations developed in this chapter and the previous (in particular for the maximum variance $\hat{\sigma}_\gamma^2$) are changed by the Vlasov terms. Moreover, we will see in Chap. 8 that these terms are negligible in other situations than that considered in this chapter. From a conceptual point of view, our goal here was to investigate RR. The stochastic heating coming from the Lorentz force only is therefore out of the scope of this work. The interested reader is referred to [6–9].

6.6 Conclusion

The formalism developed in the previous Chapter allowed to identify the limit of validity of the deterministic approach and to propose a first criterion, in terms of the quantum parameter χ, for the transition between FP and IB/MC treatments, namely $\chi < \chi_{qu} \simeq 0.25$. This criterium is not strict and has been extended in this

Chapter, following considerations on the evolution of the successive moments of the distribution function, to account for the general shape of the electron distribution function.

The analytical study of the successive moments of the electron energy distribution function allowed us to gain a deeper insight into the regimes of validity of the various approaches considering more general distribution functions than usually investigated. While previous studies have considered the average quantum parameter $\langle \chi \rangle$ as the only relevant parameter, we show that the initial electron spread in energy is also a key-parameter for determining the validity of the different (deterministic, FP or IB/MC) models. Our moment analysis thus allowed us to identify more rigorously the various regimes of validity of the different models as a function of both, the initial quantum parameter $\langle \chi \rangle$ and normalized electron energy spread $\sigma_\gamma / \langle \gamma \rangle$. By doing so, we could also highlight under which conditions the various physical effects, and in particular the stochastic and/or discrete nature of high-energy photon emission may be more easily accessible in forthcoming experiments.

The numerical implementation of the SDE accounting for this diffusion term is discussed, leading to the development of a new stochastic pusher for PIC codes that can be easily implemented by modifying the pusher accounting for the (quantum-corrected) radiation-reaction force. Its ability to correctly address various physical configurations is demonstrated, and it can be used in an intermediate regime where the lowest order term (deterministic friction force) is not accurate enough, and the full MC procedure (equivalent to the IB description) is not yet necessary.

The theoretical findings of this work as well as the numerical tools developed here are important for future experiments on extreme light facilities. They are also important to the relativistic astrophysics community as radiation reaction is known to strongly affect the particle dynamics in various extreme scenarios.

References

1. Niel F, Riconda C, Amiranoff F, Duclous R, Grech M (2018a) From quantum to classical modeling of radiation reaction: A focus on stochasticity effects. Phys. Rev. E 97:
2. Derouillat, J., Beck, A., Pérez, F., Vinci, T., Chiaramello, M., Grassi, A., Flé, M., Bouchard, G., Plotnikov, I., Aunai, N., Dargent, J., Riconda, C., and Grech, M. (2018). Smilei : A collaborative, open-source, multi-purpose particle-in-cell code for plasma simulation. *Computer Physics Communications*, 222(Supplement C):351 – 373
3. Kloeden PE, Platen E (1991) Numerical solution of stochastic differential equations. Springer
4. Harvey CN, Gonoskov A, Ilderton A, Marklund M (2017) Quantum Quenching of Radiation Losses in Short Laser Pulses. Phys. Rev. Lett. 118:
5. Neitz N, Piazza AD (2013a) Stochasticity Effects in Quantum Radiation Reaction. Phys. Rev. Lett. 111(054802):
6. Doveil F (1981) Stochastic plasma heating by a large-amplitude standing Phys. Rev. Lett. 46(8):
7. Bauer D (1995) Relativistic ponderomotive force, uphill acceleration, and transition to chaos. Phys. Rev. Lett. 75(25):

8. Esirkepov T (2015) Attractors and chaos of electron dynamics in electromagnetic standing waves. Physics Letters A 379(2044–2054):
9. Esirkepov T (2017) Paradoxical stabilization of forced oscillations by strong nonlinear friction. Physics Letters A 381(2559–2564):

Chapter 7
Photon Distribution Function

7.1 Outline

Radiation reaction (RR) is the influence of the electromagnetic field emitted by a charged particle on its own dynamics. It is one of the oldest problems in electrodynamics [1–3] and it has recently been at the center of various studies motivated either by relativistic astrophysics studies [4–6], or by the advent of multi-petawatt laser systems [7–19]. Among the latter, the Apollon [20], Gemini [21, 22] and ELI [23] facilities provide us with test beds for our models of high-energy photon emission and its back-reaction on the particle dynamics.

In order to provide a link between theoretical/numerical modelling and experiments, many authors have studied the signatures of RR on either the electron or photon distributions and proposed accordingly possible experimental setups in order to observe such signatures. Among the signatures in the electron distribution function, one can mention the (quantum) stochastic heating first predicted by Neitz and Di Piazza [24] and further investigated in Refs. [16, 25, 26] or the quenching of radiation losses first introduced in Ref. [27] and more recently revisited as an asymmetry (or skewness) of the electron distribution function in Ref. [25].

Signatures in the emitted photon spectrum have also been studied. RR was identified with the multiple photon recoils experienced by electrons emitting consecutively and incoherently high-energy photons in Ref. [28], a study in which the authors also highlighted the impact of RR on the multi-photon Compton spectra. More recently, Neitz and Di Piazza also used a kinetic approach to study the interplay of RR and quantum electrodynamics (QED) effects on both the electron and emitted photon spectra, suggesting that such effects could be observed at already available laser intensities [29]. Such findings were further supported by Blackburn *et al.*, who used a Monte-Carlo based Particle-In-Cell (PIC) approach to study high-energy photon emission in the head-on collision of an electron-beam with an ultra-intense laser pulse in a regime where QED effects are important [13].

F. Niel, *Classical and Quantum Description of Plasma and Radiation in Strong Fields*, Springer Theses, https://doi.org/10.1007/978-3-030-73547-0_7

In the present Chapter, we investigate RR effects on the spectrum of high-energy photons focusing on the comparison of the three different models discussed in Ref. [25]. A remarkable result is that the emitted radiation spectrum is weakly sensitive to the details of the electron energy distribution function as long as the average energy is the same and the energy spreads are comparable. The chapter is structured as follows. First, Sect. 7.2 lays the basis for this study. We summarize previous results on the radiation spectrum considering a single radiating electron in the local constant field approximation. We then remind briefly the three complementary approaches for treating RR in the moderately quantum regime which have been recently presented in Ref. [25] and discuss how the emitted high-energy photon properties can be extracted using these approaches. Then, in Sect. 7.3, we present the results of PIC simulations performed with the open-source code SMILEI [30]. These simulations shed a new light on how the three models for RR mentioned above perform in predicting the spectral properties of the emitted high-energy photons. Finally, Sect. 7.4 summarizes our findings.

7.2 High-Energy Photon Emission and Its Back Reaction

Throughout this work, we place ourselves within the local constant field approximation (LCFA) which relies on the possibility of neglecting the space and time variation of the external field over the region of formation of a high-energy photon emitted by an ultra-relativistic electron. The LCFA validity has been investigated in recent works [31], but its discussion remains beyond the scope of this work.

In what follows, we first recall the properties of high-energy photon emission under the LCFA, focusing on the single electron emission rate and spectral properties of the emitted photons (Sect. 7.2.1). We then summarize previous findings on the modelling of the back-reaction of high-energy photon emission on the radiating electron dynamics, a.k.a. radiation reaction, in the moderately quantum regime (Sect. 7.2.2). Finally, in Sect. 7.2.3, we take a closer look at the high-energy photon emission by an ensemble of ultra-relativistic electrons.

7.2.1 Photon Emission Rate and Energy Spectrum

Let us consider an electron (with charge $-e$ and mass m) radiating in an external electromagnetic field with relativistic strength:

$$a_0 = \frac{e|A^{\mu}|}{mc^2} \gg 1, \tag{7.1}$$

where c is the speed of light [SI units will be used throughout this work unless specified otherwise], and A_μ is the four potential associated to the electromagnetic field tensor $F^{\mu\nu} = \partial^\mu A^\nu - \partial^\nu A^\mu$. We will further consider undercritical (otherwise arbitrary) electromagnetic fields, that is both field invariants are small with respect to the square of the Schwinger field $E_s = m^2c^3/(e\hbar) \simeq 1.3 \times 10^{18}$ V/m. The Lorentz invariant rate of photon emission by the electron has been first derived in the LCFA by Nikishov and Ritus [32]. It reads:

$$\frac{d^2N_\gamma}{d\tau d\chi_\gamma} = \frac{2}{3}\frac{\alpha^2}{\tau_e}\frac{G(\chi,\chi_\gamma)}{\chi_\gamma}, \tag{7.2}$$

where τ is the electron proper time, $\alpha = e^2/(4\pi\epsilon_0\hbar c)$ is the fine structure constant (ϵ_0 and $\hbar$ being the permittivity of vacuum and reduced Planck constant, respectively), $\tau_e = r_e/c$ is the time for light to cross the classical radius of the electron $r_e = e^2/(4\pi\epsilon_0 mc^2)$, and $G(\chi, \chi_\gamma)$ is the quantum emissivity:

$$G(\chi,\chi_\gamma) = \frac{\sqrt{3}}{2\pi}\frac{\chi_\gamma}{\chi}\left[\int_\nu^{+\infty} K_{5/3}(y)\,dy + \frac{3}{2}\chi_\gamma\,\nu\,K_{2/3}(\nu)\right] \tag{7.3}$$

with $\nu = 2\chi_\gamma/[3\chi(\chi - \chi_\gamma)]$. Both the photon emission rate and the quantum emissivity depend only on the electron and photon parameters at the moment of emission:

$$\chi = \left|\frac{F^{\mu\nu}}{E_s}\frac{p_\nu}{mc}\right| \quad \text{and} \quad \chi_\gamma = \left|\frac{F^{\mu\nu}}{E_s}\frac{\hbar k_\nu}{mc}\right| \tag{7.4}$$

with p^μ and $\hbar k^\mu$ the electron and photon four-momentum, respectively.

Let us now consider a reference frame, henceforth referred to as the laboratory-frame, in which the electron is ultra-relativistic (that is the electron Lorentz factor is $\gamma \gg 1$). The photon quantum parameter at the time of emission can be directly linked to the electron quantum parameter at this time as $\chi_\gamma = \xi\,\chi$, with $\xi = \gamma_\gamma/\gamma$ the ratio of the photon normalized energy $\gamma_\gamma = \hbar\omega/(mc^2)$ (ω denoting the photon angular frequency) by the electron Lorentz factor. The instantaneous power spectrum of the emitted high-energy photons (in this frame, and for a single radiating electron) is obtained from Eq. (7.2) and reads:

$$\frac{dP_{\text{inst}}}{d\gamma_\gamma} = \frac{\sqrt{3}}{2\pi}P_\alpha\,\frac{\xi}{\gamma}\left[\int_\nu^{+\infty} K_{5/3}(y)\,dy + \frac{\xi^2}{1-\xi}\,K_{2/3}(\nu)\right], \tag{7.5}$$

with $P_\alpha = 2\alpha^2mc^2/(3\tau_e)$. Let us note that, in the classical limit $\chi \ll 1$, Eq. (7.5) reduces to the standard synchrotron energy spectrum (see, e.g., Ref. [33]), and integration of Eq. (7.5) over all photon energies leads to the *quantum corrected* instantaneous power radiated away by the electron:

$$P_{\text{rad}} = P_{\text{cl}}\,g(\chi), \tag{7.6}$$

with $P_{\rm cl} = P_\alpha \chi^2$ the classical Larmor power and

$$g(\chi) = \frac{9\sqrt{3}}{8\pi}\int_0^{+\infty} d\nu \left[\frac{2\nu^2\,\mathrm{K}_{5/3}(\nu)}{(2+3\nu\chi)^2} + \frac{4\nu\,(3\nu\chi)^2}{(2+3\nu\chi)^4}\,\mathrm{K}_{2/3}(\nu)\right] \tag{7.7}$$

the quantum correction introduced in various works [34, 35] (see also Ref. [25] for more details). Let us finally stress that considering ultra-relativistic electrons further allows us to assume that all photons are emitted in the direction $\mathbf{\Omega}$ of the electron velocity at the moment of emission.

7.2.2 *Radiation Reaction*

In the moderately quantum regime [3] ($a_0 \gg 1$ and $\chi \lesssim 1$), electron-positron pair production can be neglected and the problem of radiation reaction acting on a population of ultra-relativistic electron can be casted in the form of a Master equation for the electron and photon distribution functions $f_e(t, \mathbf{x}, \gamma, \mathbf{\Omega})$ and $f_\gamma(t, \mathbf{x}, \gamma, \mathbf{\Omega})$, respectively [8, 25, 26, 36, 37]:

$$\frac{d}{dt} f_e = \int_0^{+\infty} d\gamma_\gamma\; w_\chi(\gamma+\gamma_\gamma, \gamma_\gamma)\, f_e(t, \mathbf{x}, \gamma+\gamma_\gamma, \mathbf{\Omega}) - f_e(t, \mathbf{x}, \gamma, \mathbf{\Omega}) \int_0^{+\infty} d\gamma_\gamma\, w_\chi(\gamma, \gamma_\gamma)\,, \tag{7.8}$$

$$\frac{d}{dt} f_\gamma = \int_1^{+\infty} d\gamma\; w_\chi(\gamma, \gamma_\gamma)\, f_e(t, \mathbf{x}, \gamma, \mathbf{\Omega}), \tag{7.9}$$

where $w_\chi(\gamma, \gamma_\gamma)$ is the rate of emission of photons with energy $mc^2\gamma_\gamma$ by an electron with a given quantum parameter χ and energy $mc^2\gamma$:

$$w_\chi(\gamma, \gamma_\gamma) = \frac{2}{3}\frac{\alpha^2}{\tau_e}\frac{G(\chi, \chi_\gamma)}{\gamma\gamma_\gamma}\,, \tag{7.10}$$

and where it has been assumed that radiation emission (and its back-reaction) is dominated by the contribution of ultra-relativistic electrons (for which $\mathbf{p} \simeq mc\gamma\mathbf{\Omega}$), and that such ultra-relativistic electrons emit radiation in the direction $\mathbf{\Omega}$ of their velocity. Note also that the time derivative d/dt in the left-hand-side of Eqs. (5.28a) and (5.28b) are total time derivatives and have to be handled carefully.

This Master equation describes the process of high-energy photon emission as a discontinuous jump process. Equation (5.28a) is a linear Boltzmann equation whose rhs acts as a *collision operator*, while the rhs of Eq. (5.28b) is a source term for the photon population. A discrete formulation of this Master equation can be rigorously derived in the form of the Monte-Carlo procedure [38] routinely implemented in

Particle-In-Cell (PIC) codes to account for high-energy photon emission and its back-reaction on the electron dynamics [9, 39].

In a recent work, Ref. [25], it was shown that for $\chi \lesssim 1$ (and arbitrary geometry of interaction), this discontinuous jump process is well approximated by a diffusion process. Performing a Fokker-Planck expansion of the *collision operator* in the limit $\gamma_\gamma \ll \gamma$, it was shown that the electron momentum obeys a stochastic differential equation with a deterministic term containing both the Lorentz force and *quantum corrected* Landau-Lifshitz radiation friction force [1] and a (stochastic) diffusion term accounting for the stochasticity of high-energy photon emission inherent to its quantum nature. This work allowed us to bridge the quantum (MC) and deterministic [quantum-corrected Landau-Lifshitz (cLL)] descriptions of radiation reaction. It also provided us with an additional description, henceforth referred to as the Fokker-Planck (FP) description, of radiation reaction to complement the standard MC and cLL descriptions. We then computed the equation of evolution of the successive (energy) moments of the electron distribution function for all three descriptions. Doing so, we were able to show that the average energy was well reproduced by all three models. The FP or MC were needed to reproduce correctly the energy spread (second order moments) while the third order moment was predicted correctly only by the MC method. Using this, we investigated the domain of validity of the various descriptions to correctly model the evolution of the electron distribution function. In the next section, we will focus on the ability of the various models to correctly describe the spectral properties of the radiation associated to RR.

7.2.3 *High Energy Photon Spectrum Accounting for Radiation Reaction*

The Master Eqs. (5.28a) and (5.28b) capture the physical picture of RR being the cumulative effect of the successive and incoherent emissions of high-energy photons by the radiating electrons. Hence, the instantaneous photon energy spectrum can be obtained at any given time t by multiplying Eq. (5.28b) by γ_γ, leading:

$$\frac{dP_{\mathrm{rad}}^{\mathrm{RR}}}{d\gamma_\gamma}(t, \mathbf{x}, \gamma_\gamma, \mathbf{\Omega}) = \int_1^{+\infty} d\gamma \, \frac{dP_{\mathrm{inst}}}{d\gamma_\gamma} \, f_e(t, \mathbf{x}, \gamma, \mathbf{\Omega}) \,, \tag{7.11}$$

where $dP_{\mathrm{inst}}/d\gamma_\gamma$ is the (single electron) instantaneous power spectrum given by Eq. (7.5). The remarkable thing here is that, as the effect of RR is encompassed into the electron distribution function f_e, Eq. (7.11) returns the spectrum of high-energy photons accounting for the effect of RR. It can be computed at any given time t provided that the electron distribution function is known at this time. The remarkable thing here is that the final radiated spectrum (with RR) depends only on the instantaneous electron distribution function. In other words, all the effects of the RR will materialise in f_e. This allows us to readily apply all of the predictions on the electron distribution function discussed in Ref. [25] to assess the effects of RR on the high-energy photon spectrum. In the next Sect. 7.3, we will in particular demonstrate that both the MC and FP descriptions for the evolution of the electron populations, which were shown to correctly describe the evolution of the first two energy moments of the electron distribution function (namely the electron mean energy and energy dispersion) provide similar predictions for the emitted photon spectra, while the cLL description fails in predicting the correct photon spectrum at large χ.

Let us finally note that, interpreting f_e not as a distribution function for a set (population) of electrons but as the probability density function for a single electron to have an energy between γ and $\gamma + d\gamma$ at a time t, Eq. (7.11) can also be employed to compute the radiation spectrum of a single electron taking into account RR.

7.3 Numerical Simulations

In what follows we first detail the numerical method developed in the PIC code SMILEI to simulate the evolution of a radiating electron population in a given external electromagnetic accounting for RR, focusing in particular on how we extract the emitted radiation properties. We then present the results of simulations considering both the head-on collision of an ultra-relativistic electron beam with an electromagnetic plane-wave, and the evolution of a hot electron population in a constant and homogeneous magnetic field.

7.3.1 Method

To investigate the properties of high-energy photons emitted by radiating ultra-relativistic electrons in various configurations, two series of simulations (one for an electron bunch with large mean energy and small energy dispersion and the other considering a hot electron population) have been performed with the PIC code SMILEI [30]. This open-source code has been upgraded to allow for the treatment of radiation reaction either by considering the quantum-corrected Landau-Lifshitz (cLL) force acting on an electron population, the Fokker-Planck (FP) pusher developed in Ref. [25] which complement the deterministic cLL description by adding a stochastic diffusion term, or the full Monte-Carlo (MC) procedure allowing to generate high-energy (macro-)photons. The implementation of these three descriptions closely follows what is presented in Ref. [25] (see also Refs. [9, 39] for the cLL and MC descriptions). It has been improved to allow for a better handling of the vectorization capabilities of new high-performance super-computers, but these improvements are beyond the scope of this work and will be presented elsewhere.

From these simulations, we have been able to extract the spectral properties (energy and angular distributions) of the emitted radiation. This is particularly simple when using the MC procedure as (macro-)photons are naturally created at runtime.

The spectral properties of the radiated light can then be easily reconstructed (at each time-steps or integrated over the full simulation time) by *depositing* the (macro-)photons energy over a regular grid in energy (and/or angle). When considering the two other (cLL or FP) descriptions, the spectral properties of the emitted radiation where computed, at runtime, considering that all electrons emit, at each timestep and in their direction of propagation $\mathbf{\Omega}$, the full spectrum given by Eq. (7.5). Let us recall here that the descriptions of radiation reaction used in this work rely on the LCFA.

This procedure allowed us to access the spectral properties of the emitted radiation, accounting for radiation reaction, in various configurations. In what follows, we detail the results obtained considering (i) an ultra-relativistic electron bunch head-on collision with an electromagnetic plane-wave (Sect. 7.3.3) and (ii) a hot electron population (broad zero-drift Maxwell-Jüttner distribution) radiating in a constant magnetic field (Sect. 7.3.4). In all these simulations, the electron feedback on the external electromagnetic field was turned off (no-current deposition) to focus on the particle dynamics in the external field without accounting for Coulomb repulsion. The timestep and number of macro-particles were chosen to ensure convergence of the simulations and will be detailed for each case.

7.3.2 Results

In our previous work [25], it was shown that not only the average quantum parameter χ but also the initial energy spread $\hat{\sigma}_\gamma \equiv \langle(\gamma - \gamma_0)^2\rangle^{1/2}/\gamma_0$ (with γ_0 the electron distribution initial mean Lorentz factor) is important in determining the overall dynamics of a radiating electron population. Thus, in what follows, we consider two complementary set-ups in the limits $\hat{\sigma}_\gamma \ll 1$ and $\hat{\sigma}_\gamma \simeq 1$, respectively. First (Sect. 7.3.3), an ultra-relativistic electron beam is considered, following a Maxwell-Jüttner distribution with drift velocity $\mathbf{v}_d = c\sqrt{\gamma_0^2 - 1}/\gamma_0\hat{\mathbf{x}}$ with $\gamma_0 = 1800$ and (proper) temperature $T_0 = 2.5 \times 10^{-3}\ mc^2$. Second (Sect. 7.3.4), a hot electron population with zero-drift Maxwell distribution and temperature $T_0 = 600\,mc^2$ [corresponding to an initial average electron Lorentz factor $\gamma_0 = \langle\gamma\rangle(t = 0) = 1800$] is considered. Note that, in SMILEI, the correct loading of the relativistic Maxwell-Jüttner distribution is ensured following the method proposed in Ref. [40].

Note that both sets of simulations were conducted in 1D3V (one dimension in space and three dimension in momentum). The simulation box size were chosen so that boundary condition would not pay any role. The spatial (Δx) and temporal ($c\Delta t$) resolutions were chosen to ensure convergence. For the first set-up (head-on collision), we used $\Delta x = \lambda_0/128$ and $c\Delta t = 0.95\,\Delta x$, with λ_0 the wavelength of the UHI electromagnetic wave. For the second set-up (constant magnetic field), we used $\Delta t = 2\pi/\omega_c/128$ and $\Delta x = c\,\Delta t/0.95$, with ω_c the characteristic cyclotron frequency (see below for details).

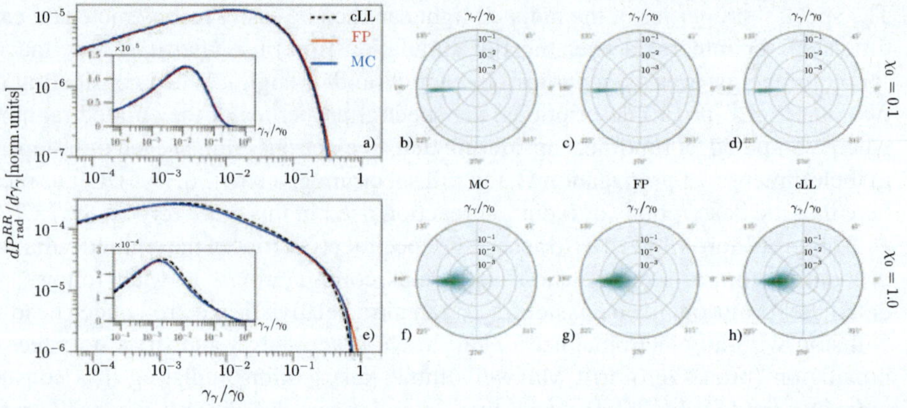

Fig. 7.1 Time integrated high-energy photon spectra considering a high-energy electron bunch head-on collision with a large amplitude electromagnetic plane wave. Top panels (a–d) correspond to an initial electron quantum parameter $\chi_0 = 0.1$, bottom panels (e–h) to $\chi = 1.0$. Black dashed lines correspond to the deterministic, quantum-corrected Landau-Lifshitz (cLL) simulations, red lines to the stochastic Fokker-Planck (FP) simulations and blue lines to the Monte-Carlo (MC) simulations. Left panels (a, e) report the energy spectra integrated over the full simulation time (the y-axis is reported in linear scale in the inserts). Panels (b, f) show the high-energy photon energy-angle-distributions computed using the MC method. Panels (c, g) show the high-energy photon energy-angle-distributions computed using the FP method. Panels (d, h) show the high-energy photon energy-angle-distributions computed using the cLL method. The angle θ measures the direction of emission of the photon, in the polarization plane, considering $\theta = 0$ as the direction of propagation of the laser pulse (conversely, $\theta = 180°$ is the initial direction of propagation of the electron beam) [41]

7.3.3 Electron Beam Head-On Collision with an UHI Plane-Wave

We start by simulating the interaction of an ultra-relativistic electron beam colliding head-on with a large amplitude, linearly polarized electromagnetic plane-wave (henceforth referred to as the laser), a configuration of particular interest for current and soon-to-come experiments [20–22] on extreme light laser facilities. The ultra-relativistic electron beam is initialized as following a Maxwell-Jüttner distribution with temperature $T_0 = 2.5 \times 10^{-3}\ mc^2$ and drift velocity $\mathbf{v}_d$ corresponding to the Lorentz factor $\gamma_0 = 1800$. This is similar to one of the configuration investigated in Ref. [25] and corresponds to an initial energy spread $\hat{\sigma}_\gamma \simeq 5 \times 10^{-2}$.

The amplitude of this plane-wave is chosen so that $\chi_0 = 2a_0\gamma_0\, \tau_e\omega_0/\alpha = 10^{-1}$ and 1 (corresponding to normalized vector potential $a_0 = 11.4$ and 114, respectively, and considering $\lambda_0 = 1\ \mu$m the laser wavelength). The end of the simulation is taken when the energy decrease of the electron population has considerably slowed down (i.e. we approach a regime in which radiation losses are not important). We used $t_{\rm end} \simeq 190/\omega_0$ for $\chi_0 = 0.1$ and $t_{\rm end} \simeq 60/\omega_0$ for $\chi_0 = 1$, with $\omega_0 = 2\pi c/\lambda_0$ the laser angular frequency.

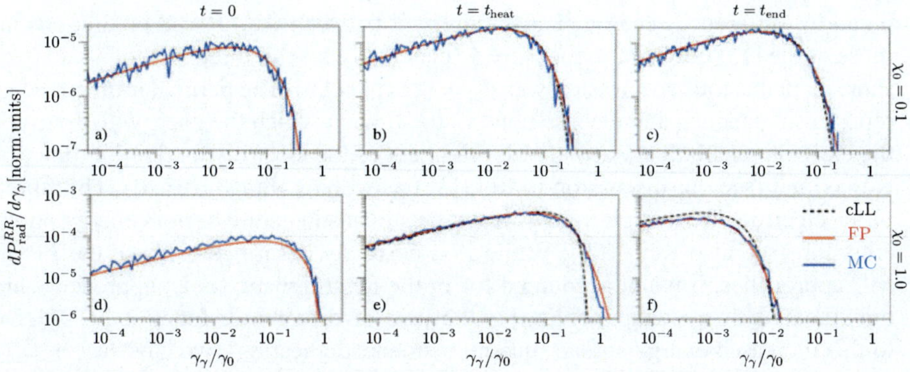

Fig. 7.2 Instantaneous high-energy photon spectra considering a high-energy electron bunch head-on collision with a large amplitude electromagnetic plane wave. Top panels (a–c) correspond to an initial electron quantum parameter $\chi_0 = 0.1$, bottom panels (d–f) to $\chi = 1.0$. Color-coding is the same than for Fig. 7.1. Spectra are reported at the early time of interaction $t = 0$ (a, d), at the time of maximum energy spread of the electron distribution function [$t_{\rm heat} \simeq 100\,\omega_0^{-1}$ for $\chi_0 = 0.1$ (panel b) and $t_{\rm heat} \simeq 4.4\,\omega_0^{-1}$ for $\chi_0 = 1.0$ (panel e)], and at the end of the simulation (c, f) [41]

For both simulations, the electron distribution function evolves exactly as reported in Fig. 11 of Ref. [25]. The simulation results concerning the radiation spectra are summarized in Figs. 7.1 and 7.2 considering the time-integrated or instantaneous spectra, respectively. Let us start by noticing that all three descriptions (cLL, FP and MC) predict similar time-integrated energy and angular distributions (Fig. 7.1). Note also that, as expected for this interaction configuration, the photon emission is beamed in the direction of the electron beam ($\theta \sim -180°$, with θ the angle of emission in the laser polarization plane). In particular, for the case $\chi_0 = 0.1$, all three descriptions provide the very same prediction for the time-integrated radiation energy spectrum (Fig. 7.1a). Discrepancies can be seen when $\chi_0 = 1$ between the FP/MC models on the one hand and the (deterministic) cLL description on the other hand. In particular, we note a hardening of the photon spectrum (increase of the radiated power at high energies) for the FP/MC predictions, while the (deterministic) cLL one overestimates the power spectrum at intermediate photon energies. Both stochastic descriptions (FP and MC) are however in very good agreement for both initial values of χ_0. A small discrepancy for the highest energy photons can still be observed in between these two methods which follows from the spurious up-scattering of electrons appearing at high-χ in the FP model (see Ref. [25]). Still, the discrepancies in between the two stochastic methods are much more tenuous than those observed in the electron distribution functions reported in Ref. [25].

In order to study more precisely the radiation spectrum, we now turn to the instantaneous spectra reported at different times in Fig. 7.2. At $t = 0$, Fig. 7.1 (panels a and d), the three models give the exact same spectrum. This is due to the fact that the total spectrum only depends on the electron distribution function [see Eq. (7.11)], which is the same initially in the three simulations. At later times, Fig. 7.1 (panels b and e), differences can be clearly observed between the FP/MC predictions and those of the cLL description for $\chi_0 = 1$ (panel e), while all three methods lead to the same instantaneous spectra at lower $\chi_0 = 0.1$ (panel b). The particular time $t = t_{\text{heat}}$ reported in panels b and e correspond to the time at which the energy dispersion of the electron population (second order moment of the electron distribution function) is maximal (see our discussion in Ref. [25] as well as Refs. [16, 26]). This spread of the electron distribution follows from the stochastic nature of high energy photon emission (see also Ref. [37]). While it is correctly described by both the FP and MC approaches, it is not accounted for in the deterministic (cLL) approach which only predicts a narrowing (cooling) of the electron distribution function [7, 25]. Due to this increased energy spread (quantum stochastic heating), we have at $t = t_{\text{heat}}$ a non-negligible number of electron with $\gamma > \langle\gamma\rangle$ in the FP/MC models with respect to the deterministic cLL model allowing for the emission of very high energy photons (hence the observed hardening of the radiation). We note for $\chi_0 = 1$ a slight overestimate with the FP method of the power emitted at large photon energy, which as discussed above, follows from the FP-inherent spurious electron up-scattering. In addition, the number of particles with an energy $\gamma \sim \langle\gamma\rangle$ is larger in the cLL model than in the FP/MC models and we observe a spectrum which is more intense at intermediate energy in the cLL descriptions than in the FP/MC ones. Similarly, we find more electrons at lower energy $\gamma < \langle\gamma\rangle$ in the FP/MC models than in the cLL model, and the energy yield at these low energies is larger considering the stochastic models (FP & MC) than considering the deterministic one (cLL). Finally, at $t = t_{\text{end}}$, we make the same observations as for $t = t_{\text{heat}}$, with the additional particularity that the MC and FP models are now in perfect agreement, which is coherent with the analysis carried in Ref. [25] that the FP and MC models, while yielding different predictions at high χ and early times, are in good agreement at longer times.

To conclude with this first series of simulations, we would like to point out that the discrepancies in the emitted radiation spectra predicted by all three methods are much more tenuous than those observed in the electron distributions as discussed in Ref. [25]. This follows from the fact that a single energetic particle along its trajectory will emit photons with a very wide range of energies resulting in the smoothing of the overall photon spectrum, in particular when considering time-integrated spectra. Furthermore, while some large discrepancies on the highest (>2) energy moments of the electron distribution were reported in between the FP and MC predictions for large $\chi_0 \simeq 1$, both FP and MC methods are here demonstrated to provide interestingly similar predictions on the emitted radiation properties. This indicates that the emitted radiation properties depend mainly on the first two energy moments of the electron distribution function.

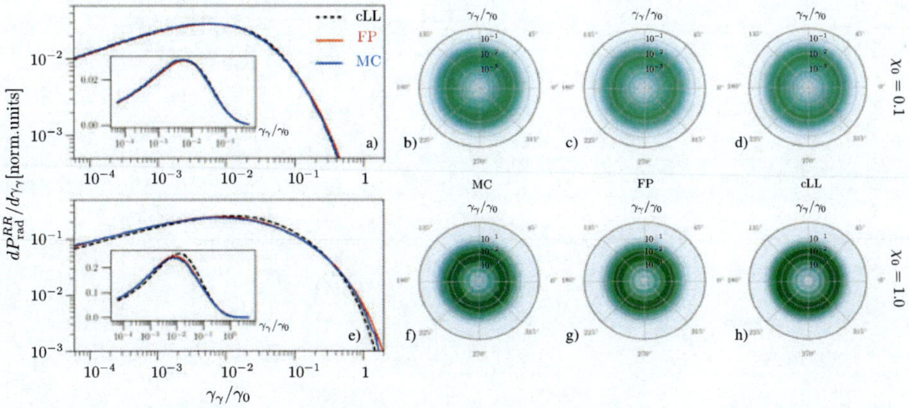

Fig. 7.3 Time integrated high-energy photon spectra considering a hot (Maxwell-Jüttner) electron population radiating in a constant magnetic field. Top panels (a–d) correspond to an initial electron quantum parameter $\chi_0 = 0.1$, bottom panels (e–h) to $\chi = 1.0$. Color-coding is the same than for Fig. 7.1. Left panels (a, e) report the energy spectra integrated over the full simulation time (the y-axis is reported in linear scale in the inserts). Panels (b, f) show the high-energy photon energy-angle-distributions computed using the MC method. Panels (c, g) show the high-energy photon energy-angle-distributions computed using the FP method. Panels (d, h) show the high-energy photon energy-angle-distributions computed using the cLL method. Here, the angle θ denotes the direction of photon emission in the plane perpendicular to the magnetic field lines [41]

7.3.4 *Hot (Maxwell-Jüttner) Electron Population Radiating in a Constant Magnetic Field*

We now simulate the evolution of a hot Maxwell-Jüttner electron population [with zero drift and temperature $\theta = T/(mc^2) = 600$] in a constant-uniform magnetic field with magnitude corresponding to initial average electron quantum parameters $\chi_0 = \langle\chi\rangle(t = 0) = 10^{-1}$ and 1 (correspondingly, 225 kT and 2.25 MT). This situation now refers to an initially broad energy distribution with an initial energy spread $\hat{\sigma}_\gamma \simeq 0.57$, a situation that may be encountered in various environments such as following the strong heating of a dense target by a laser pulse or in some astrophysical scenarii involving hot electron-positron plasmas evolving in high magnetic fields.

The duration of the simulations was here taken arbitrarily as $t_{\text{end}} \simeq 235/\omega_c$ for $\chi_0 = 0.1$, and $t_{\text{end}} \simeq 16/\omega_c$ for $\chi_0 = 1$, with $\omega_c = eB/(m\gamma_0)$ the cyclotron frequency for an electron with the average Lorentz factor $\gamma_0 = \langle\gamma\rangle(t = 0) = 1800$. As shown in Ref. [25], for this particular case of a broad Maxwell-Jüttner distribution, all three descriptions predict a cooling (narrowing) of the electron distribution function and the reported simulations were long enough for the electron populations to significantly cool down. The simulation results are presented in Figs. 7.3 and 7.4. For this particular simulations, and similarly to what was reported in the previous section, all three models are in perfect agreement for $\chi_0 = 0.1$ and only slight deviations are observed at $\chi_0 = 1$ in the time-integrated spectra (Fig. 7.3, note however the isotropy of the

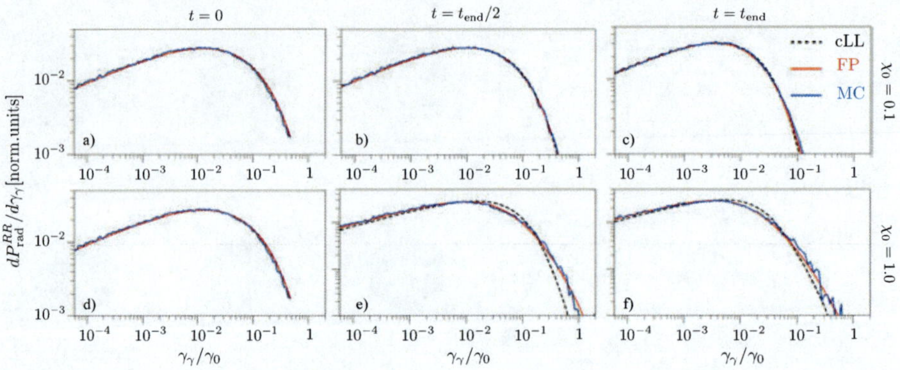

Fig. 7.4 Instantaneous high-energy photon spectra considering a hot (Maxwell-Jüttner) electron population radiating in a constant magnetic field. Top panels (a–c) correspond to an initial electron quantum parameter $\chi_0 = 0.1$, bottom panels (d–f) to $\chi = 1.0$. Color-coding is the same than for Fig. 7.1. Spectra are reported at the early time of interaction $t = 0$ (a, d), at half of the simulation (b, e), and at the end of the simulation (c, f) [41]

angular distribution in the plane transverse to the magnetic fields which follows from the symmetry of the system, $\mathbf{v}_d = 0$). The excellent agreement at all times is confirmed in Fig. 7.4 for $\chi_0 = 0.1$ while discrepancies at larger initial quantum parameter $\chi_0 = 1$ are found to follow the same line as observed in the previous Sect. 7.3.3. Again, in light of our previous study on the successive moments of the electron distribution function, and in particular in the ability of only the FP and MC descriptions to correctly capture the temporal evolution of the electron population first two moments, the excellent agreement between FP and MC simulations observed here once more indicate that the emitted radiation spectrum is mainly sensitive to this first two moments.

7.4 Conclusions

In this chapter we showed that the radiated spectrum in laser-electron beam interaction or in the evolution of a hot plasma in a constant uniform magnetic field accounting for RR effects can be easily reconstructed, in the LCFA, from the instantaneous electron distribution function. We can then apply predictions to the electron distribution function [25] to the radiated spectrum. In particular, a reduced description of the distribution function evolution involving two energy moments (when the Fokker Planck modelling is appropriate) or three energy moments (when the Monte-Carlo modelling is necessary) of f_e can be used in principle. In practice, the excellent agreement of the FP model with the MC procedure, supported by PIC simulations, up to $\chi_0 = 1$ shows that only the first two moments of the electron distribution are actually sufficient to describe the qualitative features of the photon spectrum. Indeed, a remarkable result is that, while the electron energy distributions from the three models showed large differences as the quantum parameter (χ) was getting close to one, differences in the radiated spectrum from the different models are smoothed out even for $\chi \sim 1$ so that, as far as radiation emission only is considered, the range of parameters for which the FP description is correct becomes very broad. Only a hardening of the photon spectrum is observed consistently for the FP and MC that does not exist in the classical (quantum-corrected, cLL) description. The fact that this hardening is correctly modeled by both the FP and MC procedures indicates that it follows from the broadening of the electron energy distribution (correctly modeled by both FP and MC procedures, and by the cLL one), and is weakly affected by the quenching process (related to the third energy moment of the electron distribution and not correctly handled by the FP description, as shown in Ref. [25]). Yet, this hardening vanishes at long enough times and is further reduced considering time-integrated measurements. It thus appears clear that, in order to discriminate experimentally between the different models and evidence more easily pure quantum effects, a direct observation of the electron (energy) distribution function will prove more advantageous than a study of the photon spectrum.

With respect to previous literature, we would like to point out that it was already suggested by Thomas *et al.* [42] (see also Ref. [43]) that, in the case of a narrow electron beam interacting with an intense laser pulse, the measure of the particle energy spectrum (and in particular its shift toward low energy) would give a more reliable observable than the radiation spectrum itself to evidence the onset of quantum effects. Yet, in this previous work, although values of $\chi \simeq 0.1$ or 1 were used, the quantum effects studied by Thomas *et al.* were limited to the quantum reduction of the radiated power as described by the $g(\chi)$ term and accounted for in the cLL description of the present paper. As such, this work did not assess additional quantum effects, such as the stochastic nature of high-energy photon emission that is at the center of our study. By correctly including the stochastic nature of the quantum description by use of the MC procedure, Blackburn *et al.* [13] suggested that, for a certain range of parameters, the hardening of the photon spectrum could be measured and may give a more reliable diagnostics of quantum effects on radiation reaction. Our results allow to generalize Blackburn *et al.* results by showing that the hardening of the photon spectrum can only be observed for short enough times, of the order of or smaller than the characteristic time $t_{\rm heat}$ over which quantum stochasticity leads to a broadening of the electron distribution. As a result, observing this effect in future experiments would require either ultra-short laser pulses or (technically challenging) time-resolved photon spectrum measurements. Clearly, measuring the electron energy distribution will provide a more complete diagnostics of quantum effects on radiation reaction.

References

1. Landau LD, Lifshitz EM (1947) The classical theory of fields. Butterworth-Heinemann, Oxford
2. Spohn H (2004) Dynamics of charged particles and their radiation field. Cambridge University Press, Cambridge
3. Di Piazza A, Müller C, Hatsagortsyan K, Keitel CH (2012) Extremely high-intensity laser interactions with fundamental quantum systems. Rev Mod Phys 84:1177
4. Jaroschek CH, Hoshino M (2009) Radiation-dominated relativistic current sheets. Phys Rev Lett 103:075002
5. Cerutti B, Werner GR, Uzdensky DA, Begelman MC (2014) Gamma-ray flares in the Crab Nebula: a case of relativistic reconnection? Phys Plasmas 21(5):056501
6. Cerutti B, Philippov AA, Spitkovsky A (2016) Modelling high-energy pulsar light curves from first principles. Mon. Not. R. Astron. Soc. 457(3):2401
7. Tamburini M, Pegoraro F, Piazza AD, Keitel CH, Macchi A (2010) Radiation reaction effects on radiation pressure acceleration. New J Phys 12(12):123005
8. Sokolov IV, Naumova NM, Nees JA, Mourou GA (2010) Pair creation in QED-strong pulsed laser fields interacting with electron beams. Phys Rev Lett 105:195005
9. Duclous R, Kirk JG, Bell AR (2011) Monte Carlo calculations of pair production in high-intensity laser-plasma interactions. Plasma Phys Controlled Fusion 53(1):015009
10. Nerush EN, Kostyukov IY, Fedotov AM, Narozhny NB, Elkina NV, Ruhl H (2011b) Laser field absorption in self-generated electron-positron pair plasma. Phys Rev Lett 106:035001
11. Ridgers CP, Brady CS, Duclous R, Kirk JG, Bennett K, Arber TD, Robinson APL, Bell AR (2012) Dense electron-positron plasmas and ultraintense γ rays from laser-irradiated solids. Phys Rev Lett 108:165006
12. Capdessus R, d'Humières E, Tikhonchuk VT (2013) Influence of ion mass on laser-energy absorption and synchrotron radiation at ultrahigh laser intensities. Phys Rev Lett 110:215003
13. Blackburn TG, Ridgers CP, Kirk JG, Bell AR (2014) Quantum radiation reaction in laser-electron-beam collisions. Phys Rev Lett 112:015001
14. Gonoskov A et al (2015) Extended particle-in-cell schemes for physics in ultrastrong laser fields: review and developments. Phys Rev E 92:023305
15. Lobet M, Ruyer C, Debayle A, d'Humières E, Grech M, Lemoine M, Gremillet L (2015) Ultrafast synchrotron-enhanced thermalization of laser-driven colliding pair plasmas. Phys Rev Lett 115:215003
16. Vranic M, Grismayer T, Fonseca RA, Silva LO (2016) Quantum radiation reaction in head-on laser-electron beam interaction. New J Phys 18(7):073035
17. Martins JL, Vranic M, Grismayer T, Vieira J, Fonseca RA, Silva LO (2016) Modelling radiation emission in the transition from the classical to the quantum regime. Plasma Phys Controlled Fusion 58(1):014035
18. Lobet M, Davoine X, d'Humières E, Gremillet L (2017) Generation of high-energy electron-positron pairs in the collision of a laser-accelerated electron beam with a multipetawatt laser. Phys Rev Accel Beams 20:043401
19. Grismayer T, Vranic M, Martins JL, Fonseca RA, Silva LO (2017) Seeded QED cascades in counterpropagating laser pulses. Phys Rev E 95:023210
20. Cros B, Paradkar B, Davoine X, Chancé A, Desforges F, Dobosz-Dufrénoy S, Delerue N, Ju J, Audet T, Maynard G, Lobet M, Gremillet L, Mora P, Schwindling J, Delferrière O, Bruni C, Rimbault C, Vinatier T, Piazza AD, Grech M, Riconda C, Marquès J, Beck A, Specka A, Martin P, Monot P, Normand D, Mathieu F, Audebert P, Amiranoff F (2014) Laser plasma acceleration of electrons with multi-PW laser beams in the frame of CILEX. Nucl Instrum Methods Phys Res Sect A 740:27–33. Proceedings of the first European Advanced Accelerator Concepts Workshop 2013
21. Cole JM, et al (2017) Experimental observation of radiation reaction in the collision of a high-intensity laser pulse with a laser-wakefield accelerated electron beam. arXiv:1707.06821
22. Poder K, et al (2017) Evidence of strong radiation reaction in the field of an ultra-intense laser. arXiv:1709.01861

23. Extreme light infrastructure (ELI). https://eli-laser.eu/
24. Neitz N, Piazza AD (2013) Stochasticity effects in quantum radiation reaction. Phys Rev Lett 111(054802):054802
25. Niel F, Riconda C, Amiranoff F, Duclous R, Grech M (2018) From quantum to classical modeling of radiation reaction: a focus on stochasticity effects. Phys Rev E 97:043209
26. Ridgers CP (2017) Signatures of quantum effect on radiation reaction in laser-electron-beam collisions. arXiv:1708.04511
27. Harvey CN, Gonoskov A, Ilderton A, Marklund M (2017) Quantum quenching of radiation losses in short laser pulses. Phys Rev Lett 118:105004
28. Di Piazza A, Hatsagortsyan KZ, Keitel CH (2010) Quantum radiation reaction effects in multiphoton compton scattering. Phys Rev Lett 105:220403
29. Neitz N, Di Piazza A (2014) Electron-beam dynamics in a strong laser field including quantum radiation reaction. Phys Rev A 90:022102
30. Derouillat J, Beck A, Pérez F, Vinci T, Chiaramello M, Grassi A, Flé M, Bouchard G, Plotnikov I, Aunai N, Dargent J, Riconda C, Grech M (2018) SMILEI: a collaborative, open-source, multipurpose particle-in-cell code for plasma simulation. Comput Phys Commun 222(Supplement C):351–373
31. Di Piazza A (2018) Implementing nonlinear compton scattering beyond the local-constant-field approximation. Phys Rev A 98(012134):012134
32. Nikishov AI, Ritus VI (1964) Quantum processes in the field of a plane electromagnetic wave and in a constant field. Sov Phys J Exp Theor Phys 19(529):529
33. Jackson JD (1999) Classical electrodynamics, 3rd edn. Wiley, New York
34. Erber T (1966) High-energy electromagnetic conversion processes in intense magnetic fields. Rev Mod Phys 38:626–659
35. Ridgers C, Kirk J, Duclous R, Blackburn T, Brady C, Bennett K, Arber T, Bell A (2014) Modelling gamma-ray photon emission and pair production in high-intensity laser-matter interactions. J Comput Phys 260:273–285
36. Elkina NV, Fedotov AM, Kostyukov IY, Legkov MV, Narozhny NB, Nerush EN, Ruhl H (2011a) QED cascades induced by circularly polarized laser fields. Phys Rev ST Accel Beams 14(054401):054401
37. Neitz N, Piazza AD (2013) Stochasticity effects in quantum radiation reaction. Phys Rev Lett 111(054802):054802
38. Lapeyre B, Pardoux E, Sentis R (1998) Méthode de Monte-Carlo pour les équations de transport et de diffusion. Springer, Heidelberg
39. Lobet M, d'Humières E, Grech M, Ruyer C, Davoine X, Gremillet L (2016) Modeling of radiative and quantum electrodynamics effects in PIC simulations of ultra-relativistic laser-plasma interaction. J Phys Conf Ser 688(1):012058
40. Zenitani S (2015) Loading relativistic Maxwell distributions in particle simulations. Phys Plasmas 22(4):042116
41. Niel F, Riconda C, Amiranoff F, Lobet M, Derouillat J, Pérez F, Vinci T, Grech M (2018b) From quantum to classical modeling of radiation reaction: a focus on the radiation spectrum. Plasma Phys Controlled Fusion 60(9):094002
42. Thomas AGR, Ridgers CP, Bulanov SS, Griffin BJ, Mangles SPD (2012) Strong radiation-damping effects in a gamma-ray source generated by the interaction of a high-intensity laser with a wakefield-accelerated electron beam. Phys Rev X 2:041004
43. Ilderton A, Torgrimsson G (2013a) Radiation reaction from QED: lightfront perturbation theory in a plane wave background. Phys Rev D 88:025021

Chapter 8
QED Cascades in Counterpropagating Plane-Waves

8.1 Outline

One of the first effects that can be studied with high-intensity lasers is radiation reaction [1, 2]. Several configurations are possible to study this effect, among which we find: the head-on collision between a high-intensity laser and a high energy electron beam, or the interaction of a high-intensity laser with a hot Maxwell-Jüttner distribution. These two configurations were studied in details, first focusing on the electron distribution function in Chaps. 5 and 6, second looking at the photon spectrum in Chap. 7. At higher intensities, another effect of interest is the creation of electron-positron pairs via nonlinear Breit-Wheeler. This effect can be studied using the head-on collision of a high-intensity laser with a high energy electron (resp. photon) beam. In this case, the incident high-energy electrons (resp. photons) emit high energy photons by nonlinear Compton scattering (resp. decompose into electron-positron pairs via nonlinear Breit-Wheeler) which subsequently decompose into electron-positron pairs (resp. emit high-energy photons) etc. This process stops when the new high-energy photons have an energy smaller than $2mc^2$ and are thus not able to subsequently decompose into an electron-positron pair. We talk about *showers* [3]. Another configuration of interest for pair creation is that of the interaction of two counter-propagating plane-waves with a seed electron. In this situation, the same occurs as what described previously for *showers*, with the difference that the external field is now able to restore the particle's energy [4–8]. The chain reaction is therefore not limited by the energy of the initial particle but by the time of stay of the particles in the external field (or under more extreme conditions by the energy of the external field itself) [6]. We talk about *cascades*. Cascades have been the subject of extensive numerical [7–9] as well as analytical studies [10, 11]. Here we wish to investigate the properties of the electron's distribution in the interaction with two counter-propagating plane-waves. This is a first step towards a better understanding of QED cascades.

This Chapter is structured as follows. In Sect. 8.2, we review the motion of an electron in the magnetic nodes of two counter-propagating plane-waves without radiation-reaction. In Sect. 8.3, we then adapt the results of Chap. 5 about the

F. Niel, *Classical and Quantum Description of Plasma and Radiation in Strong Fields*, Springer Theses, https://doi.org/10.1007/978-3-030-73547-0_8

moments of the electron distribution function to the present case where the so-called Vlasov terms are no longer negligible. From the asymptotic moments, we deduce the analytical expression for the asymptotic stationary electron distribution function. In Sect. 8.4, we then investigate the case where $\langle\chi\rangle > 1$ and where we need to consider pair production. Finally, in Sect. 8.5, we summarize our findings.

8.2 Motion of a Single Electron in Two Counter-Propagating Plane-Waves

8.2.1 Without RR

Let us first consider the configuration of two circularly polarized plane-waves. The obtained field is a stationary wave given by

$$\tilde{\mathbf{E}} = a_0\Big(0, \cos(kx)\cos(\omega t), -\cos(kx)\sin(\omega t)\Big), \tag{8.1a}$$

$$\tilde{\mathbf{B}} = a_0\Big(0, -\sin(kx)\cos(\omega t), -\sin(kx)\sin(\omega t)\Big), \tag{8.1b}$$

where $\tilde{\mathbf{E}} = e\mathbf{E}/(mc\omega)$ and $\tilde{\mathbf{B}} = e\mathbf{B}/(m\omega)$ are the electric and magnetic fields normalized in Compton units. As shown by [12–15], for $a_0 > 1$, the motion in such a field is in general chaotic. Here, we avoid this complication by placing ourselves in the anti-nodes (magnetic nodes in which $B = 0$) where the field is simply that of a uniformly rotating electric field described by

$$\tilde{\mathbf{E}}(t) = a_0\,[0, \cos(\omega t), \sin(\omega t)]\;. \tag{8.2}$$

This model is of outmost importance for the study of QED cascades as it provides good estimations for the full stationary wave situation [6–8, 10]. In this case, the equation of motion without RR simply reads

$$\dot{\mathbf{p}}(t) = -\tilde{\mathbf{E}}(t)\,, \tag{8.3}$$

where $\boldsymbol{p}$ is normalized in mc and t in ω ($\tilde{t} = \omega t$). Its solution is obtained by direct integration as

$$p_y(t) = p_{y_0} - a_0\left[\sin(\tilde{t}) - \sin(\tilde{t}_0)\right], \tag{8.4a}$$

$$p_z(t) = p_{z_0} + a_0\left[\cos(\tilde{t}) - \cos(\tilde{t}_0)\right], \tag{8.4b}$$

where $p_{y_0} = p_y(t_0)$ and $p_{z_0} = p_z(t_0)$ and p_y and p_z are normalized in mc. From this solution and the expression for the field Eq. (8.2), we deduce the energy of an electron initially at rest (at $\tilde{t}_0 = 0$) as

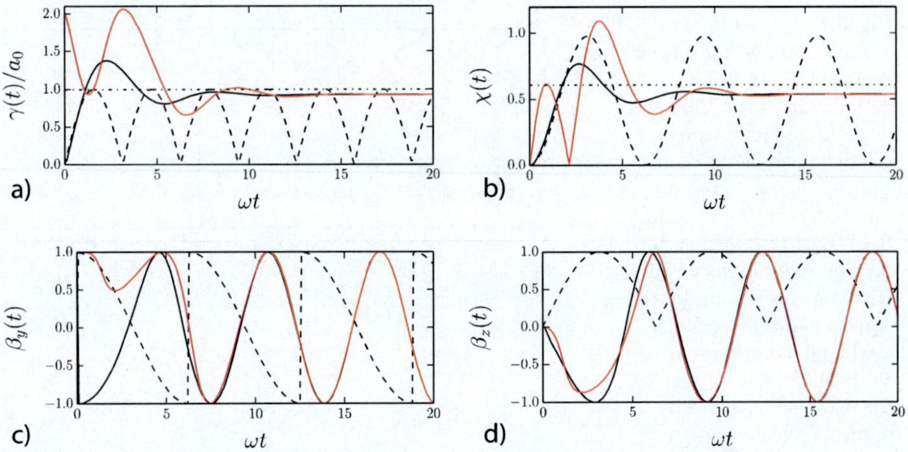

Fig. 8.1 Temporal evolution of **a** the Lorentz factor normalized to a_0 **b** the electron quantum parameter **c** the normalized speed along the y axis and **d** the normalized speed along the z axis. For an electron initially at rest and without RR (black dashed line), an electron initially at rest with quantum corrected classical RR (black line) and an electron with initial momentum $p_{y_0} = 1000$ and $p_{z_0} = 0$ with quantum corrected classical RR (red line). In all cases $a_0 = 500$

$$\gamma(t) = \sqrt{1 + 4a_0^2 \sin^2\left(\frac{\tilde{t}}{2}\right)}. \tag{8.5}$$

This leads to the electron quantum parameter

$$\chi(t) = \frac{\hbar\omega_0}{mc^2} a_0 \sqrt{1 + 4a_0^2 \sin^4\left(\frac{\tilde{t}}{2}\right)}, \tag{8.6}$$

and the normalized velocities $\beta_x = v_x/c$ and $\beta_y = v_y/c$

$$\beta_y = \frac{a_0 \sin(\tilde{t})}{\sqrt{1 + 4a_0^2 \sin^2\left(\frac{\tilde{t}}{2}\right)}}, \tag{8.7a}$$

$$\beta_z = \frac{a_0[1 - \cos(\tilde{t})]}{\sqrt{1 + 4a_0^2 \sin^2\left(\frac{\tilde{t}}{2}\right)}}. \tag{8.7b}$$

These results are plotted in Fig. 8.1 in black dashed line for an electron initially at rest. We see that the electron oscillates in energy, speed, and quantum parameter and drifts in space (Fig. 8.2 in black dashed line for an electron initially at rest and red dashed line for an electron with an initial momentum $p_y = 1000mc$).

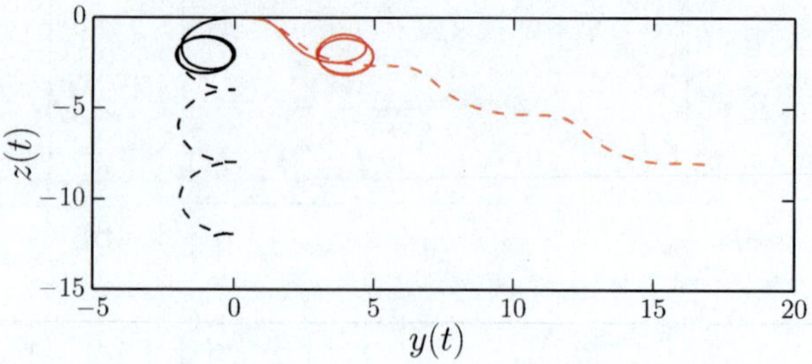

Fig. 8.2 Trajectory of a particle initially at rest (black line) and with initial momentum $p_{y_0} = 1000$ and $p_{z_0} = 0$ (red line) with (straight line) and without (dashed line) RR in a uniformly rotating electric field (Eq. (8.2)). After a few periods, the trajectory with RR stabilizes to a circle with a given radius independent of the initial momentum. Without RR, the distance to the initial position of the particle increases with time without stabilizing

8.2.2 *With RR*

When radiation reaction is added, the equation of motion reads

$$mc^2 \frac{d\gamma}{dt} = -ec\, \boldsymbol{\beta}\cdot\mathbf{E} - mc^2\, S(\chi)\,. \tag{8.8}$$

The nonlinearity of the damping term and the fact that the source term $\boldsymbol{\beta} \cdot \mathbf{E}$ is not independent of the current value of γ makes this equation hard to solve analytically. We therefore solve it numerically in the case of a) an electron initially at rest and b) an electron with an initial momentum $p_{y_0} = 1000$ and $p_{z_0} = 0$ for $a_0 = 500$. The results are plotted in Fig. 8.1. We see that after a few periods, the energy and quantum parameter stabilize [14, 16] to approximately a_0 for the normalized energy and a_0^2/a_S for the quantum parameter (Fig. 8.1a and b), where $a_S = eE_S/(mc\omega)$ is the Schwinger field in Compton units. Moreover, these asymptotic values don't depend on the initial energy of the particle. In the same way, independently of the initial momentum of the particle, the normalized speeds along y and z, β_y and β_z stabilize (with RR) after a few periods and oscillate at ω between -1 and $+1$ (Fig. 8.1c and d).

After a few periods, the motion of the electron is circular with a phase-shift φ with respect to the external field (see Fig. 8.2). Assuming that

$$\boldsymbol{\beta} = |\boldsymbol{\beta}|[0, \cos(t+\varphi), \sin(t+\varphi)]\,, \tag{8.9}$$

and using the equation of evolution of the normalized speed $\boldsymbol{\beta}$ (Eq. (5.41))

$$\frac{d\boldsymbol{\beta}}{dt} = -\frac{e}{p}(\mathbb{1} - \boldsymbol{\beta}\otimes\boldsymbol{\beta})\cdot\mathbf{E}\,, \tag{8.10}$$

we get

$$\varphi = \arcsin \tilde{\gamma} \tag{8.11a}$$

$$\boldsymbol{\beta} \cdot \tilde{\boldsymbol{E}} = a_0\sqrt{1-\tilde{\gamma}^2} \tag{8.11b}$$

$$\chi = \tilde{\gamma}^2 \frac{a_0^2}{a_S} = \tilde{\gamma}\frac{a_0^2}{a_S}|\sin\varphi| \tag{8.11c}$$

where we assumed $|\boldsymbol{\beta}| = 1$ since we are interested only in ultra-relativistic electrons. Inserting these expressions in Eq. (8.8) with the stationarity condition yields

$$a_0\sqrt{1-\tilde{\gamma}^2} - A\tilde{S}(\tilde{\gamma}^2 a_0^2/a_S) = 0\,. \tag{8.12}$$

In the strong field limit, the work of the field $\boldsymbol{\beta} \cdot \tilde{\boldsymbol{E}} \simeq a_0$. Equation (8.12) therefore simplifies to $a_0 = 0.56A\chi^{2/3}$, where we used the asymptotic expression of $S(\chi)$ for $\chi \gg 1$ (Eq. (C.2)). Using the expression of χ given by Eq. (8.11c), this solves to

$$\gamma_{\mathrm{as}} = \frac{\sqrt{a_S}}{(0.56A)^{3/4}} a_0^{3/4}\,, \tag{8.13}$$

which is exactly the dependance given by [6]. Using Eq. (8.11b) and (8.11c), we deduce

$$(\boldsymbol{\beta} \cdot \tilde{\boldsymbol{E}})_{\mathrm{as}} = \sqrt{a_0 - \sqrt{a_0 a_S}/(0.56A)^{3/2}}\,, \tag{8.14}$$

$$\chi_{\mathrm{as}} = \frac{a_0^{3/2}}{(0.56A)^{3/4}}\,. \tag{8.15}$$

These values are plotted in black dashed lines on Fig. 8.3 where they are compared to the numerical simulations in red. Moreover, we plot the time it takes for the electron to reach the stationary state T_{stat} for an electron initially at rest (orange crosses) and with an initial momentum $p_{y_0} = 1000$ (blue dots). We see that for $a_0 > 1000$, the stationary state is reached in less than half a laser period.

8.3 Evolution of the Electron Distribution Function in a Uniformly Rotating Electric Field Without Pair Production

In the same way as in Chap. 5, we use the equation of evolution for the electron distribution function $f_e(t, \boldsymbol{r}, \gamma, \boldsymbol{\Omega})$ and the corresponding photon distribution function $f_\gamma(t, \boldsymbol{r}, \gamma, \boldsymbol{\Omega})$, accounting for the effect of high energy photon emission. These read

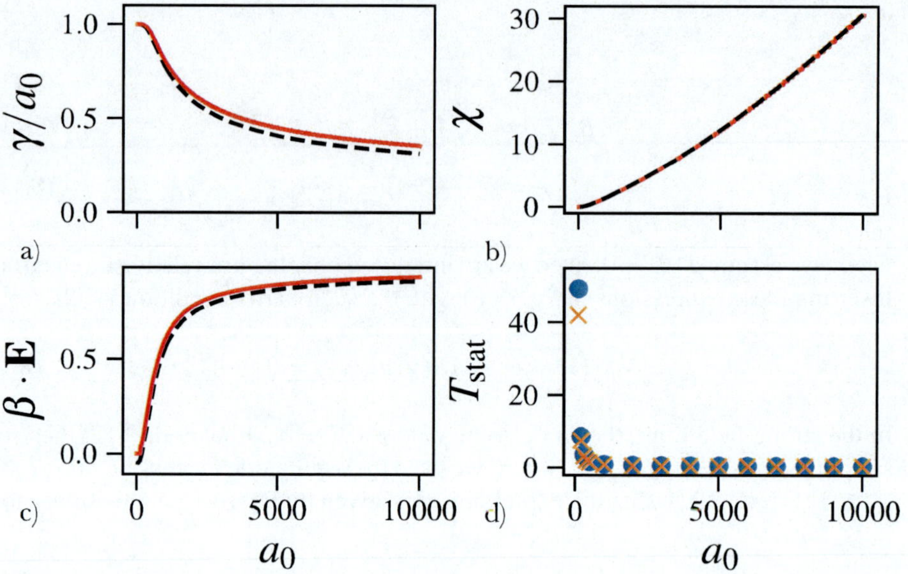

Fig. 8.3 Asymptotic **a** normalized gamma factor **b** electron quantum parameter **c** work of the external field, in the interaction of an electron initially at rest in a uniformly rotating electric field. The black dashed lines represent Eqs. (8.13) to (8.15) while the red lines represent the results of the numerical simulations. The time it takes to reach the stationary state T_{stat} is represented in **d** in orange crosses for an electron initially at rest and for an electron with an initial momentum $p_{y_0} = 1000$ in blue dots

$$\frac{df_e}{dt} = \mathcal{C}_{\text{nCs}}[f_e]\,, \tag{8.16a}$$

$$\frac{df_\gamma}{dt} = \mathcal{C}_\gamma[f_e]\,, \tag{8.16b}$$

where

$$\mathcal{C}_{\text{nCs}}[f_e] = \int_0^{+\infty} d\gamma_\gamma\, w_\chi^{\text{nCs}}(\gamma+\gamma_\gamma) f_e(\gamma+\gamma_\gamma, \gamma_\gamma) - f_e(\gamma)\int_0^{+\infty} d\gamma_\gamma\, w_\chi^{\text{nCs}}(\gamma, \gamma_\gamma)\,, \tag{8.17a}$$

$$\mathcal{C}_\gamma[f_\gamma] = \int_1^{+\infty} d\gamma\, w_\chi^{\text{nCs}}(\gamma, \gamma_\gamma) f_e(\gamma)\,, \tag{8.17b}$$

$$\frac{df_e}{dt} = \partial_t f_e + \nabla\cdot[c\beta\boldsymbol{\Omega} f_e] - \frac{1}{mc^2}\partial_\gamma[ec\beta(\boldsymbol{\Omega}\cdot\mathbf{E})f_e] - \frac{e}{p}\nabla_{\boldsymbol{\Omega}}\cdot[(\mathbb{1}-\boldsymbol{\Omega}\otimes\boldsymbol{\Omega})\cdot(\mathbf{E}+\beta\boldsymbol{\Omega}\times\mathbf{H})f_e]\,.$$

8.3.1 Equations of Evolution of the Energy Moments of the Electron and Photon Distribution Functions in the Presence of a Source

We then use the same procedure as in Chap. 5 and compute the successive energy moments. Defining $F_e(\gamma, t) = \int d^3r\, d^2\Omega\; f_e(\boldsymbol{r}, \boldsymbol{\Omega}, \gamma, t)$ and $N_e = \int d\gamma\; F_e(\gamma, t)$ (which does not depend on time since we don't consider pair production for the moment), we compute the successive energy moments as

$$\langle\gamma\rangle = \frac{\int d\gamma\, \gamma\, F_e(\gamma, t)}{N_e}, \tag{8.18}$$

and for $n > 1$,

$$\mu_n = \frac{\int d\gamma\, (\gamma - \langle\gamma\rangle)^n F_e(\gamma, t)}{N_e}. \tag{8.19}$$

This leads, up to the 3rd order moment, and keeping the source terms $-ne\langle(\gamma - \langle\gamma\rangle)^{n-1}\boldsymbol{\beta}\cdot\boldsymbol{E}\rangle/(mc)$

$$\frac{d\langle\gamma\rangle}{dt} = -\langle\boldsymbol{\beta}\cdot\tilde{\mathbf{E}}\rangle - \langle S(\chi)\rangle, \tag{8.20a}$$

$$\begin{aligned}\frac{d\sigma_e^2}{dt} &= -2\langle(\gamma - \langle\gamma\rangle)\boldsymbol{\beta}\cdot\tilde{\mathbf{E}}\rangle + \langle R(\chi,\gamma)\rangle \\ &\quad - 2\langle(\gamma - \langle\gamma\rangle)S(\chi)\rangle,\end{aligned} \tag{8.20b}$$

$$\begin{aligned}\frac{d\mu_3}{dt} &= -3\langle(\gamma - \langle\gamma\rangle)^2\boldsymbol{\beta}\cdot\tilde{\mathbf{E}}\rangle - 3\langle(\gamma - \langle\gamma\rangle)^2 S(\chi)\rangle \\ &\quad + 3\langle(\gamma - \langle\gamma\rangle)R(\chi,\gamma)\rangle - \langle A_3(\chi,\gamma)\rangle,\end{aligned} \tag{8.20c}$$

where $\tilde{\boldsymbol{E}} = e\boldsymbol{E}/(mc\omega)$, $\sigma_e^2 = \langle(\gamma - \langle\gamma\rangle)^2\rangle$, $S(\chi) = A_1(\chi)$ and $R(\chi,\gamma) = A_2(\chi,\gamma)$.

We now compute the equations of evolution of the successive moments of the photon distribution. Contrarily to the electrons, the number of photon continuously increases. We therefore consider in addition the equation of evolution of the number of photons. The system of equations of evolution of the successive energy moments of the photon distribution then reads

$$\frac{dN_\gamma}{dt} = \langle A_0(\chi, \gamma) \rangle, \tag{8.21}$$

$$\frac{d\langle \gamma_\gamma \rangle}{dt} = \left[\langle S(\chi) \rangle - \langle \gamma_\gamma \rangle \langle A_0(\chi, \gamma) \rangle \right] \frac{N_e}{N_\gamma}, \tag{8.22}$$

$$\frac{d\sigma_\gamma^2}{dt} = \Big[\langle R(\chi, \gamma) \rangle - 2 \langle (\gamma - \langle \gamma \rangle) S(\chi) \rangle \tag{8.23}$$

$$+ \langle \gamma_\gamma \rangle^2 \langle A_0(\chi, \gamma) \rangle - \sigma_\gamma^2 \langle A_0(\chi, \gamma) \rangle \Big] \frac{N_e}{N_\gamma}.$$

8.3.2 Asymptotic Energy Spread

Using the same expansion as in [17], Eq. (8.20b) simplifies to

$$\frac{d\sigma_e^2}{dt} \simeq -2 \langle (\gamma - \langle \gamma \rangle) \boldsymbol{\beta} \cdot \tilde{\mathbf{E}}) \rangle + A \langle \gamma \rangle h \left(\langle \chi \rangle \right) \tag{8.24}$$

$$- \operatorname{cov}(\gamma, \chi) A \left[2\tilde{S}' \left(\langle \chi \rangle \right) - h' \left(\langle \chi \rangle \right) \right],$$

where $\operatorname{cov}(\gamma, \chi) = \langle (\gamma - \langle \gamma \rangle)(\chi - \langle \chi \rangle) \rangle$ and $A = 2\alpha^2/(3\tau_e)$. From the simulations, we see that $|2\langle (\gamma - \langle \gamma \rangle) \boldsymbol{\beta} \cdot \tilde{\mathbf{E}}) \rangle| \ll |A \langle \gamma \rangle h \left(\langle \chi \rangle \right) |$ and that the relation between the covariance of γ and χ is linked to the electron energy spread σ_e as $\operatorname{cov}(\gamma, \chi) \simeq 1.55 \langle \chi \rangle \sigma_e^2 / \langle \gamma \rangle$. Moreover, in the same way as for a single particle with RR, the electron distribution function reaches a stationary state after a certain time (see Fig. 8.5). Using the stationarity condition $d\sigma_e^2/dt = 0$, we therefore get

$$\hat{\sigma}_e^\infty = \sqrt{\frac{h \left(\langle \chi \rangle \right)}{1.55 \langle \chi \rangle \left[2S' \left(\langle \chi \rangle \right) - h' \left(\langle \chi \rangle \right) \right]}}. \tag{8.25}$$

The result is confronted to PIC simulations for a_0 going from 200 to 50000 and is plotted in Fig. 8.4. Note that, up to a numerical factor, the value of the asymptotic electron energy spread $\hat{\sigma}_e^\infty$ is the same as the maximum energy spread found for the head-on collision in Chap. 5. Moreover, we observe that for a_0 large, the asymptotic energy spread reaches a maximum value of $\hat{\sigma}_e^\infty \simeq 0.62$.

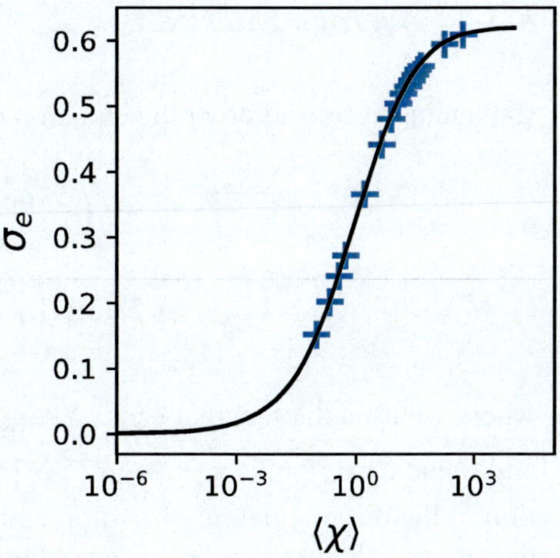

Fig. 8.4 Asymptotic electron energy spread in the interaction of an electron distribution with a uniformly rotating electric field as a function of the average quantum parameter $\langle\chi\rangle$. The black line represents the analytical equation (8.25) and the blue crosses the results of PIC simulations

8.3.3 Asymptotic Third Order Moment

Expanding Eq. (8.20c) in the same way as previously yields

$$
\begin{aligned}
\frac{d\mu_3}{dt} \simeq & -\frac{\mu_3}{\langle\gamma\rangle} A\Big[\langle\chi\rangle a_3'(\langle\chi\rangle) - 3\langle\chi\rangle h'(\langle\chi\rangle) + 3\langle\chi\rangle \tilde{S}'(\langle\chi\rangle)\Big] \\
& + \sigma_e^2 A\Big[- a_3(\langle\chi\rangle) - 2\langle\chi\rangle a_3'(\langle\chi\rangle) + 3h(\langle\chi\rangle) \\
& + 3\langle\chi\rangle h'(\langle\chi\rangle)\Big] - \langle\gamma\rangle^2 A\, a_3(\langle\chi\rangle) \\
& - \frac{3e}{mc}\langle(\gamma - \langle\gamma\rangle)^2 \boldsymbol{\beta}\cdot\mathbf{E})\rangle + \frac{3e}{mc}\sigma_e^2\langle\boldsymbol{\beta}\cdot\mathbf{E}\rangle ,
\end{aligned}
\tag{8.26}
$$

for which the stationarity condition yields

$$
\begin{aligned}
\hat{\mu}_3^{\infty} = & \, \hat{\sigma}_e^{\infty^2} \frac{f_2(\langle\chi\rangle)}{f_1(\langle\chi\rangle)} - \frac{a_3(\langle\chi\rangle)}{f_1(\langle\chi\rangle)} \\
& - \frac{3e}{Amc}\left[\frac{\langle(\gamma - \langle\gamma\rangle)^2 \boldsymbol{\beta}\cdot\boldsymbol{E}\rangle}{\langle\gamma\rangle^2} - \hat{\sigma}_e^{\infty^2}\langle\boldsymbol{\beta}\cdot\boldsymbol{E}\rangle\right],
\end{aligned}
\tag{8.27}
$$

with

$$
f_1(\chi) = \chi a_3'(\chi) - 3\chi h'(\chi) + 3\chi \tilde{S}'(\chi)\,, \tag{8.28a}
$$

$$
f_2(\chi) = -a_3(\chi) - 2\chi a_3'(\chi) + 3h(\chi) + 3\chi h'(\chi)\,. \tag{8.28b}
$$

8.3.4 Average Energy

Expanding at second order the equation of evolution for the average energy yields

$$a_0\sqrt{1-\tilde{\gamma}^2} - A\,\tilde{S}\left(\tilde{\gamma}^2\frac{a_0^2}{a_S}\right) - 2.83A\left(\tilde{\gamma}^2\frac{a_0^2}{a_S}\right)\hat{\sigma}_e^{\infty^2}\,\tilde{S}''\left(\tilde{\gamma}^2\frac{a_0^2}{a_S}\right) = 0\,, \tag{8.29}$$

where we used the fact that $\sigma_\chi^2 \simeq 2.83\langle\chi\rangle^2\hat{\sigma}_e^2$. Compared to Eq. (8.12), we have an additional term $\left[2.83A\left(\tilde{\gamma}^2\frac{a_0^2}{a_S}\right)\hat{\sigma}_e^{\infty^2}\,\tilde{S}''\left(\tilde{\gamma}^2\frac{a_0^2}{a_S}\right)\right]$. This term represents the contribution of the discrete nature of photon emission to the average asymptotic energy. In the strong field limit ($a_0 \gg 1$), we obtain $a_0 - 0.43A\chi^{2/3} = 0$, which yields

$$\chi_{\mathrm{as}} = \frac{a_0^{3/2}}{(0.43A)^{3/4}}\,, \tag{8.30}$$

where we used the asymptotic expansion of $S(\chi)$, $S''(\chi)$ and the asymptotic value of $\hat{\sigma}_e$. The discrete nature of photon emission increases the asymptotic stationary energy of the electron.

8.3.5 Evolution of an Initial Cold Electron Distribution Function

We simulate the interaction of a set of 10 000 particles initially at rest with a uniformly rotating electric field of the form Eq. (8.2) with $a_0 = 45$, 145 and 500 (corresponding roughly to $\chi = 10^{-2}$, 10^{-1} and 1). The end of the simulation is taken when the electron distribution function has converged to its asymptotic form, that is to say $t_{\mathrm{end}} = 2000\omega^{-1}$, $200\omega^{-1}$ and $20\omega^{-1}$ for $a_0 = 45$, 145 and 500 (resp.). The results of the simulations are summarized on Fig. 8.5.

The first row a) corresponds to $\chi_0 = 10^{-2}$, the second b) to $\chi_0 = 10^{-1}$ and the third c) to $\chi_0 = 1$. The first three columns correspond to the evolution of the distribution function $f_e(t,\gamma)$ respectively in the case of the Monte-Carlo simulation (MC), the stochastic (Fokker-Planck) pusher (FP) and the deterministic (cLL) radiation reaction pusher [including the quantum correction $g(\chi)$]. The fourth column corresponds to the (normalized) difference between the average energy extracted from the Monte-Carlo simulations and the average energy obtained from the stochastic pusher (red line), and that obtained using the deterministic pusher (black line). Both are normalized to the normalized field strength a_0: $\Delta\gamma_\alpha/\gamma_0 = (\langle\gamma\rangle_{\mathrm{MC}} - \langle\gamma\rangle_\alpha)/\gamma_0$, with $\alpha =$ cLL or FP. Finally the last two rows correspond to the standard deviation normalized to the average energy $\hat{\sigma}_\gamma = \sigma_\gamma/\langle\gamma\rangle$ and to the normalized moment of

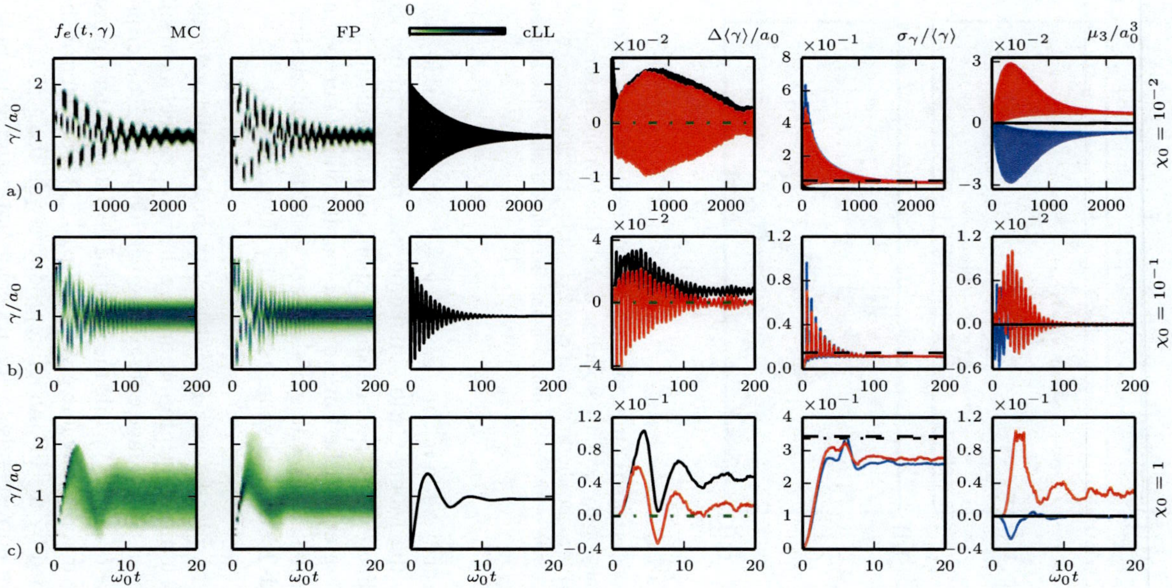

Fig. 8.5 Simulations of an electron bunch initial at rest with no temperature in a uniformly rotating electric field Eq. (8.2) with a_0 such that **a** $\chi_0 = 10^{-2}$, **b** $\chi_0 = 10^{-1}$ and **c** $\chi_0 = 1$ if $\gamma = a_0$. The first three panels of each row shows the electron distribution functions from the Monte-Carlo simulations (MC, first panels), stochastic (Fokker-Planck) simulations (FP, second panels) and quantum-corrected deterministic simulations (cLL, third panels). The fourth panels show the difference in the prediction of the mean electron energy in between the MC simulation and the deterministic (black line) and FP (red line) simulations. The two last panels (in each row) correspond to the moments of order 2 (energy variance) and 3 for the MC (blue line), FP (red line) and deterministic (black line) simulations

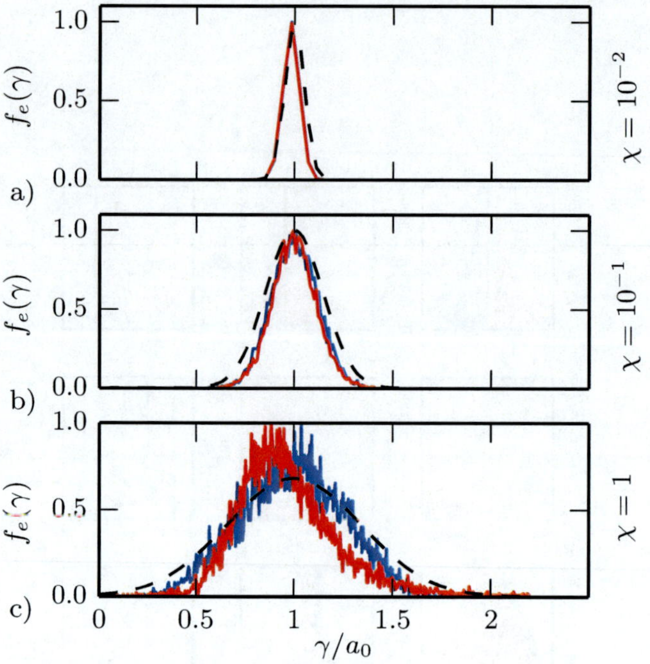

Fig. 8.6 Asymptotic electron distribution function in the interaction of a uniformly rotating electric field with an initially cold electron distribution at rest. The first line **a** corresponds to $a_0 = 45$, the second **b** to $a_0 = 145$ and the third one **c** to $a_0 = 454$. The blue line corresponds to the MC simulation, the red one to the FP simulation and the black dashed line to the analytical solution Eq. (8.31)

order three normalized to a_0^3: $\mu_3/a_0^3 = \langle(\gamma - \langle\gamma\rangle)^3\rangle/a_0^3$ (in all plots, the blue line corresponds to the Monte-Carlo simulation, the red line to the stochastic pusher and the black line to the deterministic pusher [with the quantum correction $g(\chi)$]. In addition, the green dashed line on the fourth column corresponds to the value $\Delta\gamma = 0$ (to facilitate the reading of the plot), while the black dashed line on the fifth row corresponds theoretical value as given by Eq. (8.25).

Let us first consider the case $\chi_0 = 10^{-2}$. As shown in the previous Section, in this case the Vlasov terms are negligible and we therefore expect the results of Chap. 5 to hold. In particular, the average energy as given by the three descriptions is approximately the same (Fig. 8.5d) while the standard deviation, after a phase of oscillation, stabilizes to the theoretical value given by Eq. (8.25). The electron distribution function, as described by the FP model, reproduces well the MC simulation, while the cLL description corresponds to the trajectory of a single particle (since we start from a cold distribution) and reproduces only the average of the FP/MC distribution functions (see Sect. 5.5.4).

We now examine the case $\chi_0 = 10^{-1}$. The FP and MC descriptions still agree well. In particular, the difference in the mean energies between the two models remains

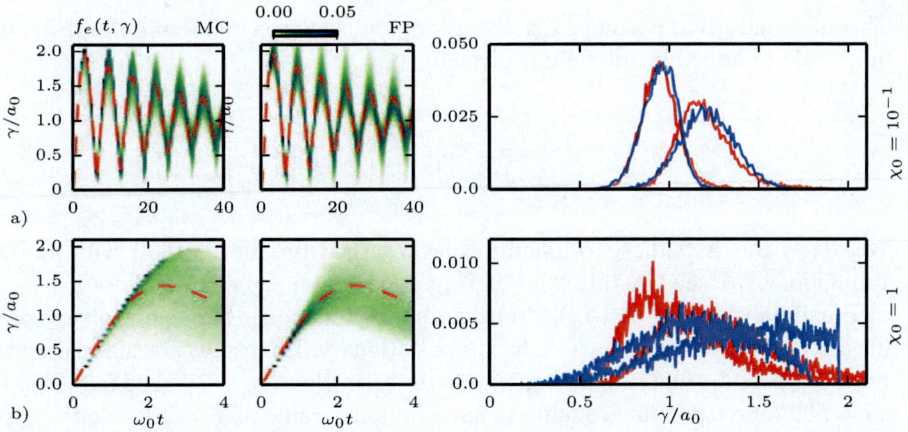

Fig. 8.7 Simulations of a cold electron beam initially at rest in a uniformly rotating electric field for **a** $a_0 = 145$, **b** $a_0 = 454$. This figure focuses on the early times of interaction during which the mean energy and energy dispersion increase. The first column corresponds to MC simulations, the second to the FP ones. The red dashed line corresponds to the mean energy. The last row shows snap-shots of the electron distribution functions at different times $t = 20\omega^{-1}$ and $t = 40\omega^{-1}$ for $a_0 = 145$ and $t = 2\omega^{-1}$ and $t = 4\omega^{-1}$ for $a_0 = 454$. The red lines correspond to FP simulations, the blue one to MC simulations

small, and smaller than the difference between the MC and cLL models. The variance as given by the FP and MC models agrees well and agrees asymptotically with the theoretical value given by Eq. (8.25). Finally, the third order moment oscillates at initial times both in the FP and MC descriptions and their value disagree slightly, while their asymptotic values agree well are is approximately equal to zero (the asymptotic distribution function both in the MC and FP descriptions are approximately symmetric around the average value.).

Finally, we examine the case $\chi_0 = 1$. In this case, the average of the MC is well reproduced by the FP, while the cLL exhibits a discrepancy even at asymptotic times. The variance is still well reproduced by the FP while it exhibits a slight difference with the theoretical value $\hat{\sigma}_e^\infty$. The third order moment in this case is not well reproduced by the FP description (as in the case of the collision of an electron bunch with a counter propagating plane-wave). In the case of the MC, the third order moment goes to zero after an oscillatory phase of a few ω^{-1} indicating a symmetric asymptotic distribution function (contrarily to the FP which exhibits a residual asymptotic μ_3 and therefore the asymptotic distribution function will not be symmetric).

The good agreement between both quantum models (MC and FP) with respect to the temporal evolution of the distribution function can be clearly seen in Fig. 8.6a and 8.6b, where we superimposed the electron distribution functions obtained from the stochastic (FP) pusher (red line) and Monte Carlo approach (blue line) at asymptotic time $t = t_{\text{end}}$. According to the numerical simulations (Fig. 8.5), the asymptotic average energy is approximately equal to a_0, the asymptotic standard deviation is given by Eq. (5.81) and the asymptotic third order moment is approximately zero.

This motivates us to postulate that the asymptotic electron distribution function in a uniformly rotating electric field is given by

$$f_e(\gamma) = \frac{1}{\sqrt{2\pi\hat{\sigma}_\gamma^{\text{thr}}}} \exp\left[-\frac{(\gamma/a_0 - 1)^2}{2(\hat{\sigma}_\gamma^{\text{thr}})^2}\right]. \tag{8.31}$$

We verify this hypothesis (plotted in black dashed line on Fig. 8.6) with the MC simulations. We see that this solution is in good agreement even for $\chi \sim 1$.

Finally, let us examine the electron distribution function in the heating phase before the electron distribution function becomes stationary. The results are summarized on Fig. 8.7 where the first row corresponds to the case where $a_0 = 145$ and the second to $a_0 = 500$ (corresponding roughly to an asymptotic value of $\chi = 10^{-1}$ and 1 resp.). The first column corresponds to the electron distribution function in the MC case and the second one to the FP case. In both cases, the red dashed line corresponds to the average energy. The third row corresponds to the electron distribution function at times $t = 10\omega^{-1}$ and $20\omega^{-1}$ for $\chi_0 = 10^{-1}$ and $t = 2.5\omega^{-1}$ and $5\omega^{-1}$ for $\chi_0 = 1$. For $\chi_0 = 10^{-1}$, the FP agrees well with the MC result. In both cases the average energy as well as the variance oscillate with time. When $\chi_0 = 1$, the FP and MC electron distribution functions display a discrepancy in the heating phase. Similarly with the case of the head-on collision of Chap. 5, the FP algorithm displays upscatterings in the heating phase. Moreover, the third order moment is not well reproduced either. Finally, we consider the interaction of an initially at rest set of electrons interacting with different rotating electric fields with $a_0 = 1250$, 2500 and 5000 (corresponding to asymptotic values of electron quantum parameters $\chi_0 \simeq 3$, 15 and 60). The results are presented on Fig. 8.8. The first column represents the electron distribution function as simulated by the MC algorithm, the second, the average electron energy normalized to a_0, the third the variance in energy σ_γ^2 normalized to the average energy squared and the fourth, the third order moment μ_3 normalized to the cube of the average energy. The first line a) represents the case where $a_0 = 1250$, the second b) to $a_0 = 2500$ and the third to $a_0 = 5000$. First, we notice that the electron distribution function reaches its stationary state faster and faster with increasing a_0 and that the "overshoot" (that is to say, the propensity to overcome the asymptotic value in the heating phase) also decreases with increasing a_0. This is seen more easily on the average energy and on the energy spread (second and third column of Fig. 8.8).

8.3.6 *Interaction of a Hot Maxwell-Jüttner Distribution Function with a Uniformly Rotating Electric Field*

As was done in Chap. 6, we now consider the evolution of an electron population with an initially broad energy distribution, radiating in a uniformly rotating electric field. The electron energy distribution at the beginning of the simulation follows a (zero-drift) 2D Maxwell-Jüttner distribution:

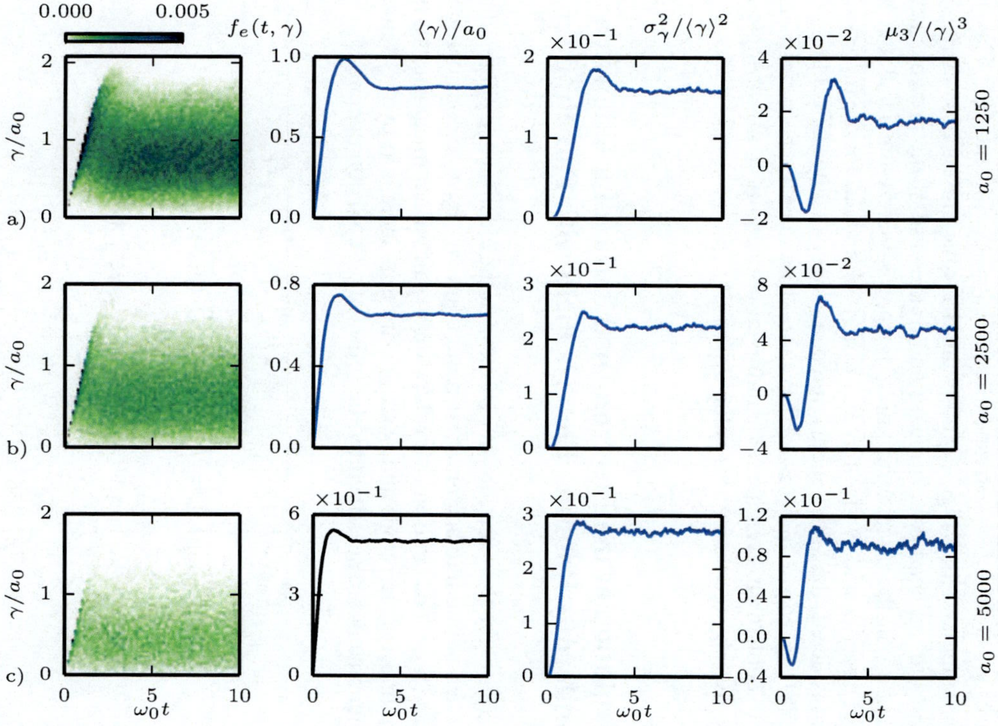

Fig. 8.8 Simulations of an electron bunch initial at rest with no temperature in a uniformly rotating electric field Eq. (8.2) with a_0 such that **a** $\chi_0 = 10^{-2}$, **b** $\chi_0 = 10^{-1}$ and **c** $\chi_0 = 1$ if $\gamma = a_0$. The first three panels of each row shows the electron distribution functions from the Monte-Carlo simulations (MC, first panels), stochastic (Fokker-Planck) simulations (FP, second panels) and quantum-corrected deterministic simulations (cLL, third panels). The fourth panels show the difference in the prediction of the mean electron energy in between the MC simulation and the deterministic (black line) and FP (red line) simulations. The two last panels (in each row) correspond to the moments of order 2 (energy variance) and 3 for the MC (blue line), FP (red line) and deterministic (black line) simulations

$$f_e(t=0,\gamma) = \frac{\mu^2}{\mu+1}\gamma \exp\left[-\mu(\gamma-1)\right], \tag{8.32}$$

where $\mu = (mc^2)/T = 1/20$ is the inverse of the normalized temperature corresponding to an initial electron mean Lorentz factor $\gamma_0 = \langle\gamma\rangle \simeq 200$, and initial energy standard deviation $\sigma_\gamma \simeq 0.7\gamma_0$. Note also that such a broad distribution also presents a large initial asymmetry and $\hat{\mu}_3 \simeq 0.49$ at the beginning of the simulation. Three cases are considered, corresponding to $a_0 = 45$, 145 and 500. The simulation results are summarized in Fig. 8.9 following the same presentation as Figs. 8.5.

Similarly to the head-collision case and to the case of the previous section, the average energy in the three descriptions is approximately the same and below the theoretical upper bound as given by Eq. (5.75). Moreover, for the three values of a_0 that we considered, the initial energy spread is above the theoretical threshold given by Eq. (5.81), $\sigma_\gamma(t_0) \simeq 0.7\gamma_0 > \sigma_\gamma^{\text{thr}}$. We therefore expect the electron distribution function to cool down, even in the quantum case. In addition, the asymptotic energy spread is given by the theoretical value Eq. (5.81).

8.4 Evolution of the Electron Distribution Function in a Uniformly Rotating Electric Field with Pair Production

The Eqs. (8.33) are the equations of evolution of the electron and photon distribution change when taking into account pair production. We have to consider the fact that photons can now decay into an electron/positron pair via nonlinear Breit-Wheeler and that the number of electrons and positrons are thus increased by this same process. These phenomena are described by the following 2 collision operators

$$C^e_{\text{nBW}}[f_\gamma] = \int_0^{+\infty} d\gamma_\gamma\, w^{\text{nBW}}_{\chi_\gamma}(\gamma_\gamma,\gamma) f_\gamma(\gamma_\gamma), \tag{8.33a}$$

$$C^\gamma_{\text{nBW}}[f_\gamma] = \int_1^{+\infty} d\gamma\, w^{\text{nBW}}_{\chi_\gamma}(\gamma_\gamma,\gamma) f_\gamma(\gamma_\gamma), \tag{8.33b}$$

with $\xi' = \chi/\chi_\gamma = \gamma/\gamma_\gamma$ and $\nu' = 1/[\chi_\gamma(1-\xi')\xi']$ and modify Eqs. (8.16) in the following way

$$\frac{df_e}{dt} = C_{\text{nCs}}[f_e] + C^e_{\text{nBW}}[f_\gamma], \tag{8.34a}$$

$$\frac{df_\gamma}{dt} = 2C_\gamma[f_e] - C^\gamma_{\text{nBW}}[f_\gamma], \tag{8.34b}$$

Note that, here, we don't make any difference between electrons and positrons as they have the same equations of evolutions.

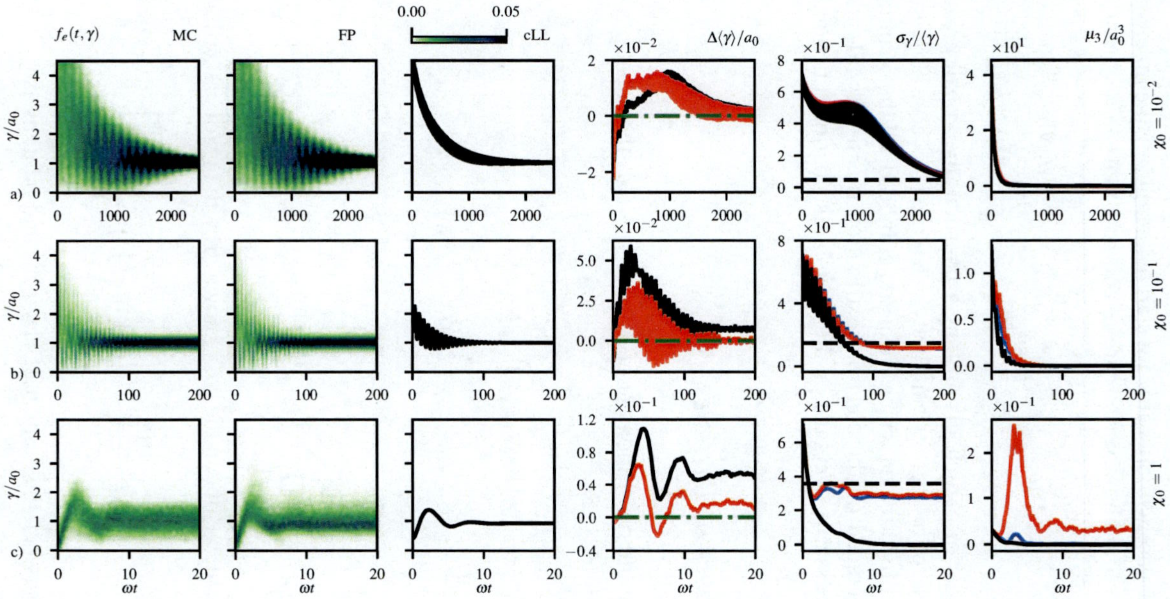

Fig. 8.9 Simulations of an electron bunch initial at rest with no temperature in a uniformly rotating electric field Eq. (8.2) with a_0 such that **a** $\chi_0 = 10^{-2}$, **b** $\chi_0 = 10^{-1}$ and **c** $\chi_0 = 1$ if $\gamma = a_0$. The first three panels of each row shows the electron distribution functions from the Monte-Carlo simulations (MC, first panels), stochastic (Fokker-Planck) simulations (FP, second panels) and quantum-corrected deterministic simulations (cLL, third panels). The fourth panels show the difference in the prediction of the mean electron energy in between the MC simulation and the deterministic (black line) and FP (red line) simulations. The two last panels (in each row) correspond to the moments of order 2 (energy variance) and 3 for the MC (blue line), FP (red line) and deterministic (black line) simulations

8.4.1 Equations of Evolution of the Energy Moments of the Electron and Photon Distribution Functions in the Presence of a Source and with Pair Production

Similarly to the case without pairs, we start by computing the moments of the Vlasov operator. One should however be careful and note that

$$d_t \mu_n = \frac{d_t\langle(\mu - \langle\mu\rangle)^n\rangle}{N} - \mu_n \frac{d_t N}{N}\,, \tag{8.35}$$

since the number of particles is no longer constant due to pair creation. Moreover, note that, in the following, for simplicity, we write $\langle . \rangle$ everywhere, independently of the distribution function on which we average. Indeed, all the quantities γ and χ are necessarily averaged on the *electron* distribution function, while the quantities γ_γ and χ_γ are necessarily averaged on the *photon* distribution function.

8.4.2 Number of Particles

Computing the 0th order moment of Eq. (8.34) yields the equations of evolution of the particle numbers

$$\frac{dN_e}{dt} = \langle B_0(\chi_\gamma, \gamma_\gamma)\rangle N_\gamma(t)\,, \tag{8.36a}$$

$$\frac{dN_\gamma}{dt} = 2\langle A_0(\chi, \gamma)\rangle N_e(t) - \langle B_0(\chi_\gamma, \gamma_\gamma)\rangle N_\gamma(t)\,, \tag{8.36b}$$

where[1]

$$B_n(\chi, \gamma) = \int_1^{+\infty} d\gamma\, \gamma^n\, w_{\chi_\gamma}^{\mathrm{nBW}}(\gamma_\gamma, \gamma) = \frac{2}{3}\frac{\alpha^2}{\tau_e}\, \gamma_\gamma^{n-1}\, b_n(\chi_\gamma)\,. \tag{8.37}$$

From this system, we obtain the well known growth rate of cascading as [6, 10, 18]

$$\Gamma = \frac{\langle W_{\mathrm{nBW}}\rangle}{2}\left[-1 + \sqrt{1 + \frac{8\langle W_{\mathrm{nCs}}\rangle}{\langle W_{\mathrm{nBW}}\rangle}}\right], \tag{8.38}$$

and the ratio between the number of electrons and photons as

$$\frac{N_e}{N_\gamma} = \frac{\langle W_{\mathrm{nBW}}\rangle}{4\langle W_{\mathrm{nCs}}\rangle}\left[1 + \sqrt{\frac{8\langle W_{\mathrm{nCs}}\rangle}{\langle W_{\mathrm{nBW}}\rangle}}\right]. \tag{8.39}$$

[1] See Appendix C.3 for more information about the functions B_n.

8.4.3 Average Energy

The first order moments yield the equations of evolution of the average energies

$$\frac{d\langle\gamma\rangle}{dt} = \langle\boldsymbol{\beta}\cdot\tilde{\mathbf{E}}\rangle - \langle S(\chi)\rangle + \Big[\langle B_1(\chi_\gamma)\rangle - \langle\gamma\rangle\langle B_0(\chi_\gamma,\gamma_\gamma)\rangle\Big]\frac{N_\gamma(t)}{N_e(t)}, \tag{8.40}$$

$$\frac{d\langle\gamma_\gamma\rangle}{dt} = \big[2\langle S(\chi)\rangle - 2\langle\gamma_\gamma\rangle\langle A_0(\chi,\gamma)\rangle\big]\frac{N_e}{N_\gamma} - \Big[\langle\gamma_\gamma B_0(\chi_\gamma,\gamma_\gamma)\rangle - \langle\gamma_\gamma\rangle\langle B_0(\chi_\gamma,\gamma_\gamma)\rangle\Big]. \tag{8.41}$$

8.4.4 Asymptotic Energy Spread

The evolution of the energy spread is given by the second order moment

$$\frac{d\sigma_e^2}{dt} = \langle R(\chi,\gamma)\rangle - 2\langle(\gamma-\langle\gamma\rangle)S(\chi)\rangle - 2\langle(\gamma-\langle\gamma\rangle)\boldsymbol{\beta}\cdot\tilde{\boldsymbol{E}}\rangle + \frac{N_\gamma}{N_e}\Big[\langle B_2(\chi_\gamma,\gamma_\gamma)\rangle - 2\langle\gamma\rangle\langle B_1(\chi_\gamma)\rangle + \langle\gamma\rangle^2\langle B_0(\chi_\gamma,\gamma_\gamma)\rangle - \sigma_e^2\langle B_0(\chi_\gamma,\gamma_\gamma)\rangle\Big], \tag{8.42a}$$

$$\frac{d\sigma_\gamma^2}{dt} = \Big[2\langle R(\chi,\gamma)\rangle - 4\langle\gamma_\gamma\rangle\langle S(\chi)\rangle + 2\langle\gamma_\gamma\rangle^2\langle A_0(\chi,\gamma)\rangle - \sigma_\gamma^2\langle A_0(\chi,\gamma)\rangle\Big]\frac{N_e}{N_\gamma} - \langle\gamma_\gamma^2 B_0(\chi_\gamma,\gamma_\gamma)\rangle + 2\langle\gamma_\gamma\rangle\langle\gamma_\gamma B_0(\chi_\gamma,\gamma_\gamma)\rangle - \langle\gamma_\gamma\rangle\langle B_0(\chi_\gamma,\gamma_\gamma)\rangle. \tag{8.42b}$$

8.5 Conclusion

As mentioned previously, the model developed in this work and in particular in Chap. 5 is valid for arbitrary field and electron configurations. We apply it here to the configuration of two counter-propagating plane waves with an electron distribution initially at rest. This configuration is of extreme importance for the study of the so-called electromagnetic cascades. Here for simplicity, and as often done in theoretical works, we place ourselves in the magnetic anti-nodes, where the field is simply a uniformly rotating electric field. In such a field, we need to evaluate the so-called "Vlasov terms" in addition to the RR terms already considered at length in the previous chapters. In particular, we see that for any value of a_0, the Vlasov term coming into play in the equation of evolution of the average energy is the same in the MC method as in the cLL one, while the Vlasov term coming from the equation of evolution of the variance is approximately equal to 0 for $\chi \lesssim 1$. Moreover, the electron distribution function reaches a stationary state after less than half a laser period (for $a_0 > 1000$) and we compute analytically its asymptotic average energy and energy spread, for any a_0. Finally, we extend the previous work to the case where nonlinear Breit-Wheeler pair production has to be taken into account and we derive the equations of evolutions of the successive energy moments of the electrons, positrons and photons in this situation.

References

1. Cole JM, et al (2017) Experimental observation of radiation reaction in the collision of a high-intensity laser pulse with a laser-wakefield accelerated electron beam. arXiv:1707.06821
2. Poder K, et al (2017) Evidence of strong radiation reaction in the field of an ultra-intense laser. arXiv:1709.01861
3. Fedotov AM (2016) Quantum regime of laser-matter interactions at extreme intensities. arXiv:1612.02038v1
4. Bell AR, Kirk JG (2008) Possibility of prolific pair production with high-power lasers. Phys Rev Lett 101:200403
5. Kirk JG, Bell AR, Arka I (2009) Pair production in counter-propagating laser beams. Plasma Phys Controlled Fusion 51(8):085008
6. Fedotov AM, Narozhny NB, Mourou G, Korn G (2010) Limitations on the attainable intensity of high power lasers. Phys Rev Lett 105:080402
7. Elkina NV, Fedotov AM, Kostyukov IY, Legkov MV, Narozhny NB, Nerush EN, Ruhl H (2011b) QED cascades induced by circularly polarized laser fields. Phys Rev ST Accel Beams 14:054401
8. Grismayer T, Vranic M, Martins JL, Fonseca RA, Silva LO (2017) Seeded QED cascades in counterpropagating laser pulses. Phys Rev E 95:023210
9. Tamburini M, Di Piazza A, Keitel CH (2017) Laser-pulse-shape control of seeded QED cascades. Sci Rep 7(1):5694
10. Nerush EN, Bashmakov VF, Kostyukov IY (2011) Analytical model for electromagnetic cascades in rotating electric field. Phys Plasmas 18(8):083107
11. Samsonov AS, Nerush EN, Kostyukov IY (2018) Asymptotic electron motion in the strongly-radiation-dominated regime. Phys Rev A 98:053858
12. Doveil F (1981) Stochastic plasma heating by a large-amplitude standing. Phys Rev Lett 46(8):532
13. Bauer D (1995) Relativistic ponderomotive force, uphill acceleration, and transition to chaos. Phys Rev Lett 75(25):4622
14. Esirkepov T (2015) Attractors and chaos of electron dynamics in electromagnetic standing waves. Phys Lett A 379:2044–2054
15. Esirkepov T (2017) Paradoxical stabilization of forced oscillations by strong nonlinear friction. Phys Lett A 381:2559–2564
16. Bulanov SV, Esirkepov TZ, Kando M, Koga JK, Bulanov SS (2011) Lorentz-Abraham-Dirac versus Landau-Lifshitz radiation friction force in the ultra relativistic electron interaction with electromagnetic wave (exact solutions). Phys Rev E 84:056605
17. Niel F, Riconda C, Amiranoff F, Duclous R, Grech M (2018a) From quantum to classical modeling of radiation reaction: a focus on stochasticity effects. Phys Rev E 97:043209
18. Bashmakov VF, Nerush EN, Kostyukov IYu, Fedotov AM, Narozhny NB (2013) Effect of laser polarization on QED cascading. arXiv: 1310(4077)

Part III
Nonlinear Breit-Wheeler Pair Production with Laguerre-Gauss Beams

Chapter 9
Orbital Angular Momentum of Light: A State of the Art

9.1 Introduction

Since the pioneering work by Poynting [1] and the first experimental verification by Beth [2], physicists know that electromagnetic fields can carry angular momentum. However, these works all considered *circularly polarized* light beams. In 1992, Allen et al. [3] showed that *vortex beams*, and in particular *Laguerre-Gaussian beams*, could carry *orbital angular momentum*. Such beams have a helical phase front. So far, these beams have found applications in communication and other low intensity domains. In this chapter, we investigate the effect of Laguerre-Gauss beams on SFQED processes such as the nonlinear Breit-Wheeler process.

9.2 Paraxial Optics

So far, we have mainly considered plane-wave fields as a simplified, analytically tractable model for laser fields. In this chapter, we will study more realistic laser field models such as Gaussian fields and their higher-order modes such as the Laguerre-Gauss modes. These modes are derived in the formalism of paraxial optics. Paraxial optics is an approximation of CED in which the fields can be written as a product of a rapidly oscillating monochromatic term and a slowly varying enveloppe whose variation in the longitudinal direction are small compared to their variation in the transverse direction (with respect to the direction of propagation). In this Section, we derive the paraxial equation from CED (Chap. 2). Moreover, this will allow us to define all the notations that will be useful in the rest of this part.

F. Niel, *Classical and Quantum Description of Plasma and Radiation in Strong Fields*, Springer Theses, https://doi.org/10.1007/978-3-030-73547-0_9

9.2.1 The Helmholtz Equation

As stated in Chap. 2, electromagnetic fields in vacuum satisfy the homogeneous Maxwell's equations

$$\partial F_{\text{ext}} = 0, \tag{9.1a}$$
$$\partial F^{*}_{\text{ext}} = 0, \tag{9.1b}$$

which in 3-vector notation read

$$\nabla \cdot \mathbf{E} = 0, \tag{9.2a}$$
$$\nabla \cdot \mathbf{B} = 0, \tag{9.2b}$$
$$\nabla \times \mathbf{E} = -\partial_t \mathbf{B}, \tag{9.2c}$$
$$c^2\, \nabla \times \mathbf{B} = \partial_t \mathbf{E}. \tag{9.2d}$$

They yield the well-known wave equations

$$\nabla^2 \mathbf{E} - \frac{1}{c^2}\partial_t^2 \mathbf{E} = \mathbf{0}, \tag{9.3a}$$
$$\nabla^2 \mathbf{B} - \frac{1}{c^2}\partial_t^2 \mathbf{B} = \mathbf{0}. \tag{9.3b}$$

Considering monochromatic beams with frequency ω, we introduce the complex notation

$$\mathbf{E} = \Re\left[\boldsymbol{\mathcal{E}}\, e^{-i\omega t}\right], \tag{9.4a}$$
$$\mathbf{B} = \Re\left[\boldsymbol{\mathcal{B}}\, e^{-i\omega t}\right]. \tag{9.4b}$$

Inserting these in Eqs. (9.3), we obtain the vectorial *Hemholtz equations*

$$\nabla^2 \boldsymbol{\mathcal{E}} + k^2 \boldsymbol{\mathcal{E}} = \mathbf{0}, \tag{9.5a}$$
$$\nabla^2 \boldsymbol{\mathcal{B}} + k^2 \boldsymbol{\mathcal{B}} = \mathbf{0}, \tag{9.5b}$$

where $k = \omega/c$. Moreover, we introduce the 4-vector potential $A^{\mu} = (V, \mathbf{A})$ with the corresponding complex notation

$$V = \Re\left[\mathcal{V}\, e^{-i\omega t}\right], \tag{9.6a}$$
$$\mathbf{A} = \Re\left[\boldsymbol{\mathcal{A}}\, e^{-i\omega t}\right]. \tag{9.6b}$$

They are linked to the electromagnetic field by

$$\mathbf{E} = -\nabla V - \partial_t \mathbf{A}, \tag{9.7a}$$
$$\mathbf{B} = \nabla \times \mathbf{A}. \tag{9.7b}$$

In the Lorenz gauge, $(\partial A) = \nabla \cdot \mathbf{A} + \partial_t V/c^2 = 0$ and we can rewrite

$$\boldsymbol{\mathcal{E}} = i\omega \left[\mathbf{A} + \frac{1}{k^2} \nabla\nabla \cdot \mathbf{A} \right], \tag{9.8a}$$
$$\boldsymbol{\mathcal{B}} = \nabla \times \mathbf{A}. \tag{9.8b}$$

Finally, rewriting Eqs. (9.2a) and (9.2b) in terms of the envelops $\boldsymbol{\mathcal{E}}$ and $\boldsymbol{\mathcal{B}}$ yields

$$\boldsymbol{k} \cdot \boldsymbol{\mathcal{E}} = 0, \tag{9.9a}$$
$$\boldsymbol{k} \cdot \boldsymbol{\mathcal{B}} = 0, \tag{9.9b}$$

which means that $\boldsymbol{\mathcal{E}}$ and $\boldsymbol{\mathcal{B}}$ are transverse. We can thus write

$$\boldsymbol{\mathcal{E}} = \boldsymbol{t}\, \mathcal{E}, \tag{9.10a}$$
$$\boldsymbol{\mathcal{B}} = \boldsymbol{t}\, \mathcal{B}, \tag{9.10b}$$

where $\boldsymbol{t}$ and $\boldsymbol{t}'$ are transverse vectors and obtain the *scalar* Helmholtz equations

$$\boxed{\nabla^2 \mathcal{E} + k^2 \mathcal{E} = 0} \tag{9.11a}$$
$$\boxed{\nabla^2 \mathcal{B} + k^2 \mathcal{B} = 0} \tag{9.11b}$$

9.2.2 The Paraxial Wave Equation

In order to simplify the Helmholtz equation and find solutions for the fields, we will use the *paraxial approximation*. To do so, we choose the direction of propagation to be in the z direction and write

$$\mathcal{E} = u_{\mathcal{E}}\, e^{ikz}, \tag{9.12a}$$
$$\mathcal{B} = u_{\mathcal{B}}\, e^{ikz}. \tag{9.12b}$$

Inserting these in the scalar Helmholtz equations (9.11a) yields

$$\nabla_\perp^2 u_{\mathcal{E},\mathcal{B}} + \partial_z^2 u_{\mathcal{E},\mathcal{B}} + 2ik\, \partial_z u_{\mathcal{E},\mathcal{B}} = 0. \tag{9.13}$$

The paraxial approximation consists in neglecting $\partial_z^2 u_{\mathcal{E},\mathcal{B}}$ compared to the other terms. This is achieved when the variation of the field in the longitudinal direction z is small compared to the variation in the transverse dimension (x, y) and when the typical length of variation of the field in the longitudinal direction is small compared to the wavelength $\lambda = 2\pi/k$, i.e.

$$\left|\frac{\partial u_{\mathcal{E},\mathcal{B}}}{\partial z}\right| \ll \left|k u_{\mathcal{E},\mathcal{B}}\right|, \tag{9.14a}$$

$$\left|\frac{\partial u_{\mathcal{E},\mathcal{B}}}{\partial z}\right| \ll \left|\nabla_\perp u_{\mathcal{E},\mathcal{B}}\right|. \tag{9.14b}$$

Keeping in mind Eqs. (9.2a) and (9.2b), we have $\nabla_\perp u_{\mathcal{E},\mathcal{B}} = 0$ which makes the condition (9.14b) impossible to be verified [4]. As pointed out by [5], the solution is to instead work with the vector potential **A** in the Lorenz gauge

$$(\partial A) = \nabla \cdot \mathbf{A} + \frac{1}{c^2}\partial_t V = 0, \tag{9.15}$$

in which case we have $\nabla \cdot \mathbf{A} \neq 0$. Following the same steps as for the electromagnetic field, the scalar and vector potential satisfy, in the Lorenz gauge, the wave equations (9.3) from which we deduce the vectorial and scalar Helmholtz equations (9.5) and (9.11a). Writing

$$\boldsymbol{\mathcal{A}} = \boldsymbol{t}_A\, u\, e^{ikz}, \tag{9.16}$$

where we choose the transverse vector as $\boldsymbol{t}_A = (\alpha\hat{\boldsymbol{x}} + \beta\hat{\boldsymbol{y}})$ where α and β are two complex numbers satisfying $|\alpha|^2 + |\beta|^2 = 1$. Applying the paraxial approximation (9.14) finally yields the *paraxial wave equation*

$$\boxed{\nabla_\perp^2 u + 2ik\,\partial_z u = 0} \tag{9.17}$$

The two constants α and β will be linked to the polarization of the field. More precisely, we define the parameter σ (that will be shown to be linked to the spin of photons) as

$$\sigma = i(\alpha\beta^* - \alpha^*\beta), \tag{9.18}$$

where the * denotes the complex conjugate[1]. A circularly polarized field will have $\sigma = \pm 1$ while linearly polarized fields have $\sigma = 0$. We can then go back to the complex enveloppe of the electromagnetic field thanks to

[1]The parameter σ is necessarily real.

$$\mathcal{E} = i\omega \left[\alpha u \, \hat{\mathbf{x}} + \beta u \, \hat{\mathbf{y}} + \frac{i}{k} \left(\alpha \, \partial_x u + \beta \, \partial_y u \right) \hat{\mathbf{z}} \right] e^{ikz}, \tag{9.19}$$

$$\mathcal{B} = ik \left[-\beta u \, \hat{\mathbf{x}} + \alpha u \, \hat{\mathbf{y}} - \frac{i}{k} \left(\beta \, \partial_x u - \alpha \, \partial_y u \right) \hat{\mathbf{z}} \right] e^{ikz},$$

in the paraxial approximation.

9.3 Linear and Angular Momentum of Light

9.3.1 Linear Momentum of Light

In this chapter we use the SI units. We will therefore remind the main results of Chap 2 in a condensed way and in SI units. The energy and linear momentum densities of the electromagnetic field are given by

$$W = \frac{1}{2} \left(\epsilon_0 \mathbf{E}^2 + \mu_0^{-1} \mathbf{B}^2 \right), \tag{9.20a}$$

$$\mathbf{p} = \epsilon_0 \, \mathbf{E} \times \mathbf{B}, \tag{9.20b}$$

and are linked together by the local continuity equation

$$\partial_t W + \nabla \cdot \mathbf{S} = 0, \tag{9.21}$$

where $\mathbf{S} = c^2 \, \mathbf{p} = \mu_0^{-1} \, \mathbf{E} \times \mathbf{B}$ is the so-called *Poynting vector*. Similarly, this linear momentum is itself conserved and we have

$$\partial_t \, S_i + \partial^j \, T_{ij} = 0, \tag{9.22}$$

where

$$T_{ij} = \frac{1}{2} \delta_{ij} \left(\epsilon_0 \mathbf{E}^2 + \mu_0^{-1} \mathbf{B}^2 \right) - \epsilon_0 E_i E_j - \mu_0 B_i B_j, \tag{9.23}$$

is the linear momentum flux density. In practice, we will instead be interested in their cycle-averaged momenta $\langle \mathbf{p} \rangle$ and $\langle \mathbf{j} \rangle$ which read

$$\langle \mathbf{p} \rangle = \frac{\epsilon_0}{2} \left(\mathcal{E}^* \times \mathcal{B} + \mathcal{E} \times \mathcal{B}^* \right). \tag{9.24}$$

The total average linear momenta is then simply given by

$$\langle \mathbf{p} \rangle = \int d^3x \, \langle \mathbf{p} \rangle. \tag{9.25}$$

9.3.2 Angular Momentum of Light

As shown in Chap. 2, there are several possible definitions for the angular momentum. In particular, let us focus here on the *canonical* angular momentum

$$\mathbf{j}_{\rm can} = \epsilon_0\,(\mathbf{E}\times\mathbf{A}) + \epsilon_0\,E^i\,(\mathbf{r}\times\nabla)A^i, \tag{9.26}$$

and on the *Belinfante* angular momentum

$$\mathbf{j}_{\rm Bel} = \epsilon_0\,\mathbf{r}\times\mathbf{p} = \epsilon_0\,\mathbf{r}\times(\mathbf{E}\times\mathbf{B})\,, \tag{9.27}$$

which really represents the *total* angular momentum, even though it looks like an orbital angular momentum. From these expression, one is tempted to identify in the expression of the canonical angular momentum the spin and orbital parts like

$$\boldsymbol{l}_{\rm can} = \epsilon_0\,(\mathbf{E}\times\mathbf{A}), \tag{9.28a}$$
$$\mathbf{S}_{\rm can} = \epsilon_0\,E^i\,(\mathbf{r}\times\nabla)A^i. \tag{9.28b}$$

However, each term is manifestly not gauge invariant and therefore not observable. For a more detailed discussion of these issues, the reader is referred to [6–15]. Actually, these issues appear for non-paraxial or non-monochromatic beams. Here, we shall focus on paraxial optics where it makes sense to talk separately about orbital and spin angular momentum.

In particular, in this case we can define the cycle-averaged angular momentum $\langle\mathbf{j}\rangle$ as

$$\langle\mathbf{j}\rangle = \mathbf{r}\times\langle\mathbf{p}\rangle, \tag{9.29}$$

and the total average angular momentum as

$$\langle\mathbf{J}\rangle = \int d^3x\,\langle\mathbf{j}\rangle. \tag{9.30}$$

In the particular case of the paraxial fields given by Eqs. (9.19), the density of linear and angular momenta are then given by

$$\langle\mathbf{p}\rangle = \frac{\epsilon_0}{2}\left[i\omega\left(u\nabla u^* - u^*\nabla u\right) - \omega\sigma\hat{\mathbf{z}}\times\nabla|u|^2 + 2k\omega\,|u|^2\,\hat{\mathbf{z}}\right], \tag{9.31a}$$
$$\langle\mathbf{j}\rangle = \mathbf{r}\times\langle\mathbf{p}\rangle. \tag{9.31b}$$

All light beams carry angular momentum since we can always choose the axis **r** so that the component of the linear momentum perpendicular to this axis is non-zero. However a more interesting situation arises when we choose **r** to be in the direction of propagation z of the beam. In this case, there are essentially two ways in which the angular momentum can be non-zero: either the light beam is elliptically polarized and the electric and magnetic field rotate around the propagation axis, or the wavefront of the light beam has a helical shape.

In the case of a gaussian beam, the wavefronts (close to focus) are planes perpendicular to the axis of propagation 9.2. Since the electric and magnetic fields are contained in the wavefront, the linear momentum (9.20b) is perpendicular to the wavefront and therefore, in this case, purely longitudinal (along the z axis on Fig. 9.2). However, if the wavefront has a helical shape, the linear momentum, still being orthogonal to the wavefront surface, will have an azimutal component. From Eq. (9.27) we therefore deduce that the angular momentum density in the z direction will be non-zero and given by

$$j_z = \rho\, p_\varphi, \tag{9.32}$$

where ρ is the distance to the axis of propagation. In this part, we will mainly be interested in this second case. The exact computation of the angular momentum of a beam with helical wavefront will be carried in Sect. 9.6.

9.4 Gaussian Beam

One of the simplest and most important solution of the paraxial wave equation is the so-called *Gaussian beam* given by [16]

$$\boxed{u_G(\rho, z) = A \frac{w_0}{z_R\, w(z)} \exp\left[-i\psi_0(z)\right] \exp\left[\frac{ik\rho^2}{2R(z)}\right] \exp\left[-\frac{\rho^2}{w^2(z)}\right]} \tag{9.33}$$

where

$$\rho = \sqrt{x^2 + y^2}, \tag{9.34a}$$

$$w(z) = w_0\sqrt{1 + (z/z_R)^2}, \tag{9.34b}$$

$$z_R = \frac{\pi w_0^2}{\lambda}, \tag{9.34c}$$

$$R(z) = \frac{\pi}{2z} w^2(z), \tag{9.34d}$$

$$\psi_0(z) = \arctan\left(\frac{z}{z_R}\right). \tag{9.34e}$$

The distance to the axis of the beam is given by ρ, $w(z)$ is the radius of the beam which we define as the distance at which the amplitude u_G falls to u_G/e, w_0 is the *beam waist* which is the smallest radius of the beam (reached in $z = 0$), z_R is the *Rayleigh length* which is the distance from the beam waist at which the radius of the beam is equal to $\sqrt{2}w_0$, ψ_0 is called the *Gouy phase* and $R(z)$ is the radius of curvature of the wavefront. All these quantities are summarized in Fig. 9.1.

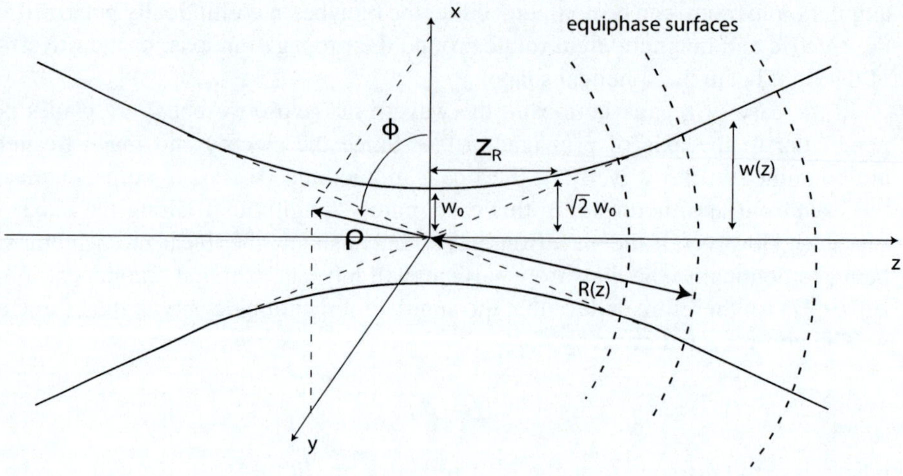

Fig. 9.1 Schematic representation of the different quantities involved in the expression of the Gaussian beam (9.33). The distance from the optical axis at which the amplitude falls from a factor $1/e$ is written $w(z)$. At focus, $w(0) = w_0$, which we call the waist. At a distance z_R (called the Rayleigh length) from focus, $w(z_R) = \sqrt{2}w_0$. The distance from the optical axis is parametrized by $\rho = \sqrt{x^2 + y^2}$ while the angle measured from the x axis is written $\phi = \text{artcan}(y/x)$

The wavefront of the Gaussian beam is obtained by looking at the points where the phase of Eq. (9.33) is constant. We obtain

$$z + \frac{\rho^2}{2R(z)} = \text{constant}, \tag{9.35}$$

where we neglected ψ_0. The wavefronts are therefore parabolas (see Fig. 9.2).

Fig. 9.2 Wavefront of the Gaussian beam Eq. (9.33). They are formed of parabolas as given by Eq. (9.35). Here the wavefronts are represented close to focus where they are formed of planes perpendicular to the propagation axis z

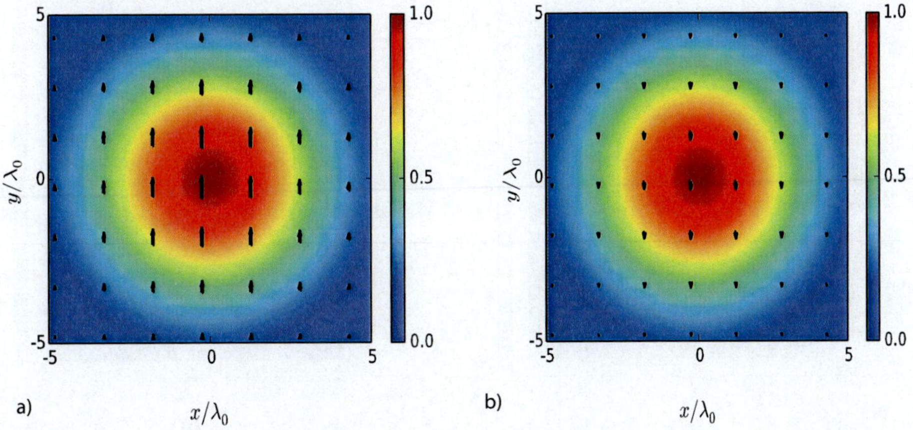

Fig. 9.3 Electric field of a linearly ($\sigma = 0$) polarized gaussian beam at focus ($z = 0$) at two different times (corresponding to (a) and (b)). The colormap represents the amplitude (9.33) of the field while the vector field represents the electric field as given by (9.19)

We have two interesting limiting cases : close to the focus of the beam, $z \ll z_R$ and $R(z) \gg 1$. The wavefront equation therefore reduces to $z =$ constant, that is to say, the wavefronts are just planes perpendicular to the z axis, similarly to the plane-wave case. On the contrary, far from the focus, $z \gg z_R$ and $R(z) \sim z$. The wavefronts are spherical, similarly to spherical waves. In both cases, close to the axis, the wavefronts are orthogonal to the direction of propagation z. The electric and magnetic fields being orthogonal to each other and to the direction of propagation, the linear momentum is purely longitudinal and therefore the longitudinal orbital angular momentum is equal to zero.

Using Eq. (9.19), we can then compute the electric and magnetic fields from the enveloppe (9.33). Taking $\alpha = 0$ and $\beta = 1$, we get $\sigma = 0$. The resulting field is a linearly polarized gaussian field represented in Fig. 9.3, where the colormap represents the amplitude of the enveloppe (9.33), while the vector map represents the electric field as given by (9.19) with u given by (9.33). The two figures (a) and (b) represent the same field at focus for two different times. We see that the field has a fixed direction while its amplitude oscillates.

Taking $\alpha = 1/\sqrt{2}$ and $\beta = i\sqrt{2}$, we get $\sigma = 1$ and the field is circularly polarized. It is represented at two different times in Fig. 9.4(a) and (b) similarly to the linearly polarized case. Here again, the colormap represents the amplitude of the enveloppe (9.33), while the vector map represents the electric field as given by (9.19) with u given by (9.33). Here the amplitude of the field is constant while its direction rotates.

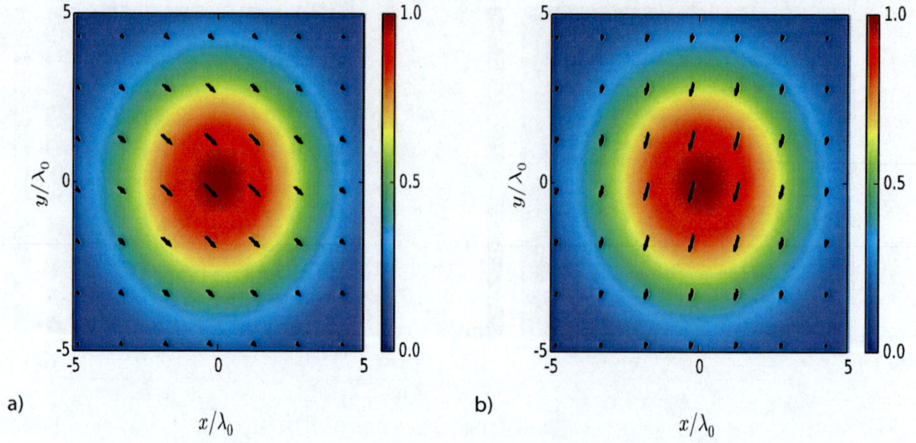

Fig. 9.4 Electric field of a circularly ($\sigma = 1$) polarized gaussian beam at focus ($z = 0$) at two different times (corresponding to (a) and (b)). The colormap represents the amplitude (9.33) of the field while the vector field represents the electric field as given by (9.19)

9.5 Higher-Order Beams

9.5.1 Laguerre-Gaussian Beams

We will now look at higher-order solutions with azimutal dependence of the form[2] $e^{il\varphi}$, where $\varphi = \arctan(y/x)$ is the azimutal angle and l an integer. The most general solution of the paraxial equation satisfying this Ansatz reads [16]

$$u(\rho, \phi, z) = \sum_{l,p} C_{p,l} \frac{w_0}{z_R\, w(z)} \left(\frac{\sqrt{2}\rho}{w(z)}\right)^{|l|} L_p^{|l|}\left(\frac{2\rho^2}{w(z)^2}\right) \times \exp\left[-\frac{\rho^2}{w(z)^2}\right] \exp\left[-i\psi_{pl}(z) + il\varphi + i\frac{k\rho^2}{2R(z)}\right], \tag{9.36}$$

where L_p^l are the generalized Laguerre polynomials, l is an integer, p a non-negative integer, $\psi_{pl}(z) = (2p + |l| + 1)\arctan\left(\frac{z}{z_R}\right)$ is the generalized Gouy phase and $C_{p,l}$ are normalization constants. Each term of this sum is called *Laguerre-Gaussian beam* and we will be interested in these beams in rest of this part

[2] The reason for this particular azimutal dependence will become apparent when we will study the orbital angular momentum of the electromagnetic field.

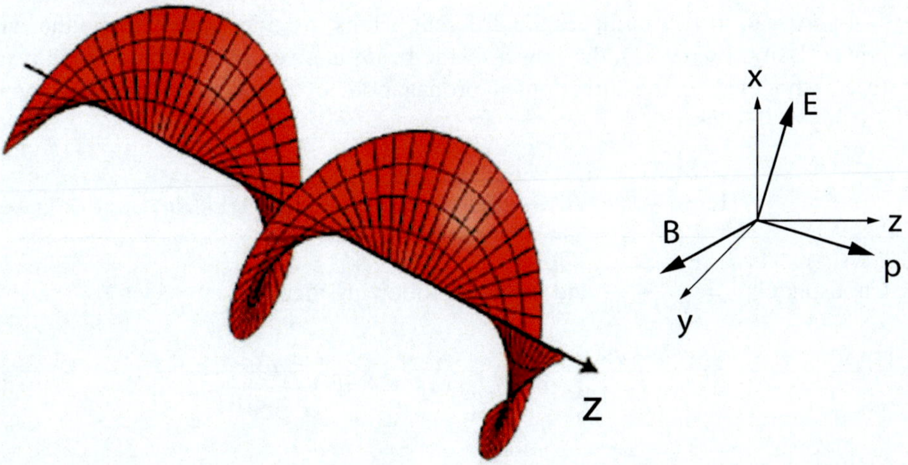

Fig. 9.5 Wavefront of the $LG_{(p,1)}$ beam Eq. (9.37). It is helically shaped making one complete revolution per wavelength. On the right, we represented the electric and magnetic fields as well as the linear momentum **p**. This latest one acquire an azimuthal component (because of the helical shape of the wavefront) at the origin of the angular momentum (9.32)

$$u_{lp}^{\mathrm{LG}}(\rho, \varphi, z) = C_{p,l} \frac{w_0}{w(z)} \left(\frac{\sqrt{2}\rho}{w(z)} \right)^{|l|} L_p^{|l|} \left(\frac{2\rho^2}{w(z)^2} \right) \times \exp\left[-\frac{\rho^2}{w(z)^2} \right] \exp\left[-i\psi_{pl}(z) + il\varphi + i\frac{k\rho^2}{2R(z)} \right] \quad (9.37)$$

Similarly to the case of the gaussian beam, we look at the wavefronts of the LG beams whose equations are given by

$$kz + l\varphi + \frac{k\rho^2}{2R(z)} = \text{constant}. \quad (9.38)$$

The behavior of the phase will therefore depend on the value of l. First of all, we note that for $l = 0$ we get, as expected, the wavefronts of the gaussian beam (9.35). For higher integer values of l, the phase fronts will look like intertwined spirals (see Fig. 9.5). Moreover, spirals will take l wavelengths to complete a rotation. Therefore, the lower the value of l, the higher the rotation speed. Finally, negative values of l will result in right hand spirals while positive values of l will result in left hand spirals.

From the shape of the wavefront 9.5 and knowing that the electric and magnetic fields are orthogonal to each other and contained in the wavefront, we deduce, using Eq. (9.20b), that the linear momentum acquires an azimutal component p_ϕ. Using Eq. (9.32), it follows that these fields will carry an OAM in the z direction. We will carry the exact computation of the OAM in a next section.

In order to further compare the different beams, we normalize them to the same power. Using Eq. (9.21), the power of the beam is given by the integral, on a surface orthogonal to the direction of propagation, of the average Poynting vector $\int d^2x_\perp \langle \Pi_z \rangle$, where

$$\langle \Pi_z \rangle = \frac{1}{2}\left[(E_x^* B_y - E_y^* B_x) + (E_x B_y^* - E_y B_x^*)\right] \sim |u|^2. \tag{9.39}$$

Choosing $C_{pl}^{\mathrm{LG}} = \sqrt{\frac{p!}{(p+|l|)!}}$ and using the following identity

$$\int_0^{+\infty} dx\, x^{|l|} \left[L_p^{|l|}(x)\right]^2 e^{-x} = \frac{p!}{(p+|l|)!}, \tag{9.40}$$

yields

$$\int d^2x_\perp \langle \Pi_z \rangle = 1, \tag{9.41}$$

for any (p, l). For convenience, we normalize all the distance to the wavelength λ of the field in the rest of this work. Equation (9.37) then takes the form

$$u_{lp}^{\mathrm{LG}}(\rho, \varphi, \zeta) = C_{p,l} \frac{w_0}{w(\zeta)} \left(\frac{\sqrt{2}\rho}{w(\zeta)}\right)^{|l|} L_p^{|l|}\left(\frac{2\rho^2}{w(\zeta)^2}\right) \times \exp\left[-\frac{\rho^2}{w(\zeta)^2}\right] \exp\left[-i\psi_{pl}(\zeta) + il\varphi + i\frac{\zeta\rho^2}{w(\zeta)^2}\right], \tag{9.42}$$

with

$$C_{pl}^{\mathrm{LG}} = \sqrt{\frac{p!}{(p+|l|)!}}, \tag{9.43a}$$

$$w(\zeta) = w_0\sqrt{1+\zeta^2}, \tag{9.43b}$$

$$\zeta = \frac{z}{z_R}, \tag{9.43c}$$

$$z_R = \pi w_0^2, \tag{9.43d}$$

$$\varphi = \arctan\left(\frac{y}{x}\right), \tag{9.43e}$$

$$\psi_{pl}(z) = (2p + |l| + 1)\arctan(\zeta). \tag{9.43f}$$

In the same way as for the gaussian beam, we use Eq. (9.19) to compute the corresponding electric and magnetic fields in linear and right-handed and left-handed circular polarizations ($\sigma = 0$, 1 and -1 resp.) for the $LG_{(0,1)}$ mode. The results are plotted in Figs. 9.6, 9.7 and 9.8. First, we notice that in these three cases, the field at the center $(x, y) = (0, 0)$ is equal to zero. This can be explained by the fact that the phase term $i\,l\arctan(y/x)$ is not defined for $x = 0$ and the amplitude thus has to go to zero. We have what we call a *phase singularity*.

In the case of the linearly polarized $LG_{(0,1)}$ field, the direction of the field remains along the y axis (as for the case of the gaussian field 9.3). However, at different times, there is an axis rotating around the center and across which the field changes sign.

Concerning the case of the right-handed circularly polarized $LG_{(0,1)}$ beam, the electric field does not point to the same direction depending on the position (x, y) in the transverse plane, contrarily to the case of the circularly polarized gaussian beam 9.4. However, since the polarization is circular, at each position, the field only rotates (clockwise since the polarization is right-handed) at frequency ω exactly like in 9.4. The particular shape of the vector field 9.7 is due to the differences in phase due to the phase term $i\arctan(y/x)$ (or equivalently to the rotating wavefront). As time changes the electric field goes from purely radial and oriented toward the center, to counterclockwise purely orthoradial, to purely radial oriented toward the outside and to clockwise purely orthoradial etc. periodically with period $2\pi\omega^{-1}$. Figure 9.7(a) shows an intermediate situation between the purely radial and purely orthoradial cases while Fig. 9.7(b) represents the field at a time close to the time at which it is purely radial and oriented toward the center.

Let us now look at the left-handed circularly polarized $LG_{(0,1)}$ beam 9.8. Similarly to the case of right-handed circularly polarized $LG_{(0,1)}$ beam and of the circularly polarized gaussian beam, the field rotates at each point at frequency ω (counter-clockwise here since the polarization is left-handed), while the global shape of the vector field 9.8 is due to the difference in phase between the points (x, y) due to the phase term $i\arctan(y/x)$. Here the field is never purely radial nor purely orthoradial. Instead, there is an axis along which the field points toward the center and an other axis, orthogonal to the previous one, along which the field points toward the outside. Because of the rotation of the field, these two axis are displaced and appear as rotating around the center.

If we consider a LG mode with $l = 0$ and $p \neq 0$, the phase of the field is that of a gaussian beam. The field therefore does not present any singularity in $(x, y) = (0, 0)$. Let us, for example, consider a $LG_{(1,0)}$ beam. It is represented on Fig. 9.9 in the case of a linearly polarized (LP) field and on Fig. 9.10 for a circularly polarized (CP) field. In both cases, the field presents a zero for $\rho \neq 0$. Moreover, it points in both cases in the same direction and its amplitude oscillates in the LP case and rotates at constant amplitude in the CP case exactly like the gaussian case (since as mentioned previously, the phase is that of a gaussian beam).

Finally, in the case where neither l nor p are zero, the situation is much more complicated as can be seen on Fig. 9.11 for the case of a $LG_{(2,3)}$ beam.

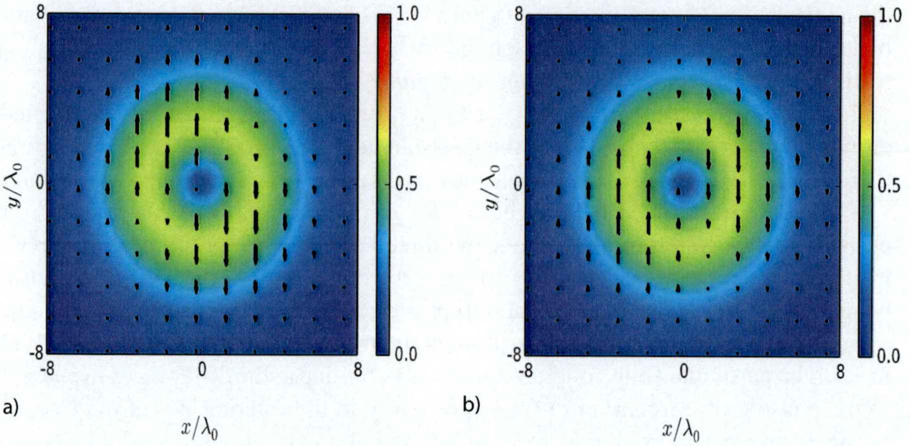

Fig. 9.6 Electric field of a linearly polarized $LG_{(0,1)}$ beam at focus ($z = 0$) at two different times (corresponding to (a) and (b)). The colormap represents the amplitude (9.33) of the field while the vector field represents the electric field as given by (9.19)

9.6 Angular Momentum of LG Beams

As stated before, we expect the LG modes with $l \neq 0$ to possess orbital angular momentum because of their spiral wavefront. Let us compute their linear and angular momenta exactly (still within the paraxial approximation). Inserting Eq. (9.42) into Eq. (9.31) yields

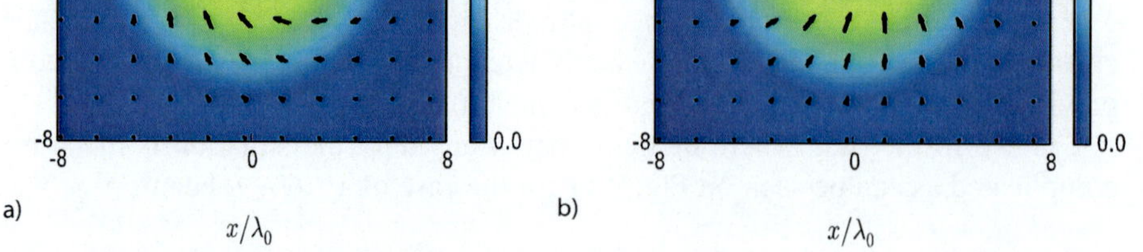

Fig. 9.7 Electric field of a right-handed ($\sigma = +1$) circularly polarized $LG_{(0,1)}$ beam at focus ($z = 0$) at two different times (corresponding to (a) and (b)). The colormap represents the amplitude (9.33) of the field while the vector field represents the electric field as given by (9.19)

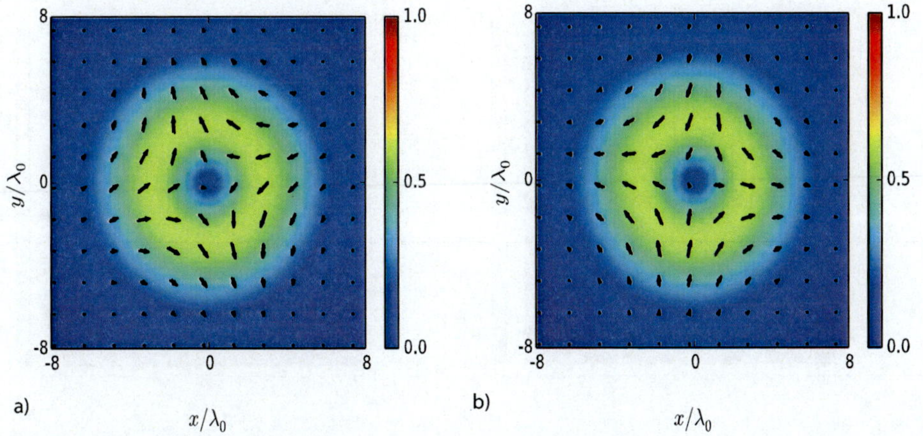

Fig. 9.8 Electric field of a left-handed ($\sigma = -1$) circularly polarized $LG_{(0,1)}$ beam at focus ($z = 0$) at two different times (corresponding to (a) and (b)). The colormap represents the amplitude (9.33) of the field while the vector field represents the electric field as given by (9.19)

$$\langle \mathbf{p} \rangle = \epsilon_0\, \omega \left[\frac{2\zeta\rho}{w(\zeta)^2}\, |u|^2\, \hat{\boldsymbol{\rho}} + \left(\frac{l}{\rho}\, |u|^2 - \frac{\sigma}{2}\, \partial_\rho |u|^2 \right) \hat{\boldsymbol{\varphi}} + k|u|^2\, \hat{\mathbf{z}} \right] \tag{9.44}$$

where $\hat{\rho}$, $\hat{\varphi}$ and $\hat{z}$ are unit vectors. The radial component of the linear momentum relates to the spread of the beam and its azimuthal to the angular momentum along z (as shown in Eq. (9.32)) [3]. The cycle averaged angular momentum is then deduced by $\langle \mathbf{j} \rangle = \mathbf{r} \times \langle \mathbf{p} \rangle$ and yields

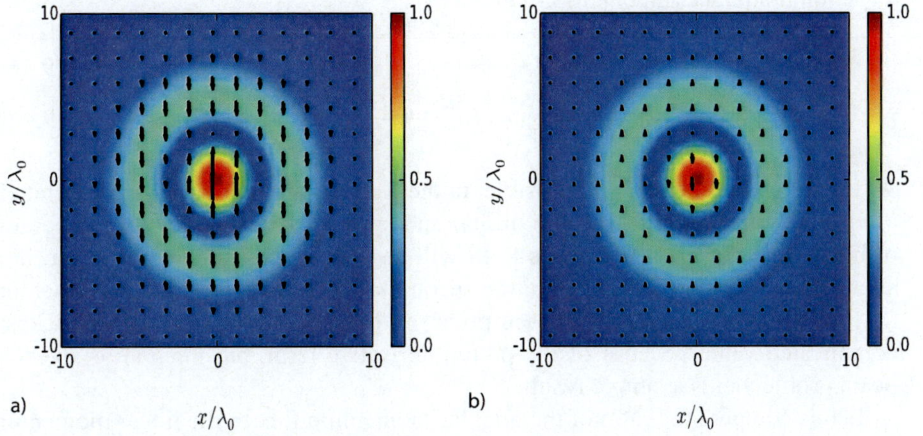

Fig. 9.9 Electric field of a linearly polarized $LG_{(1,0)}$ beam at focus ($z = 0$) at two different times (corresponding to (a) and (b)). The colormap represents the amplitude (9.33) of the field while the vector field represents the electric field as given by (9.19)

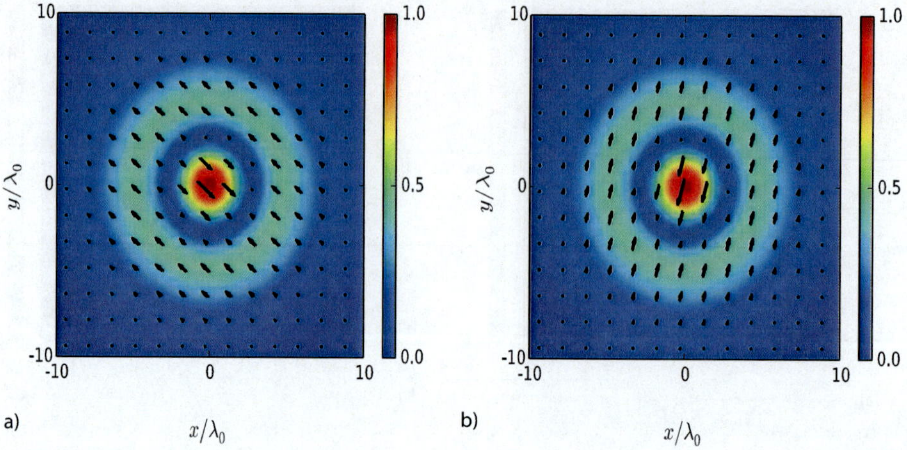

Fig. 9.10 Electric field of a circularly polarized $LG_{(1,0)}$ beam at focus ($z = 0$) at two different times (corresponding to (a) and (b)). The colormap represents the amplitude (9.33) of the field while the vector field represents the electric field as given by (9.19)

$$\boxed{\begin{aligned}\langle \mathbf{j} \rangle = \epsilon_0\, \omega \Bigg[-z \left(\frac{l}{\rho} |u|^2 - \frac{\sigma}{2} \partial_\rho |u|^2 \right) \hat{\boldsymbol{\rho}} + k\rho \left(\frac{2\zeta\rho}{kw(\zeta)^2} - 1 \right) |u|^2\, \hat{\boldsymbol{\varphi}} \\ + \left(l|u|^2 - \frac{\sigma\rho}{2} \partial_\rho |u|^2 \right) \hat{\mathbf{z}} \Bigg]\end{aligned}} \tag{9.45}$$

The longitudinal components are given by

$$\langle p_z \rangle = k\epsilon_0\omega|u|^2, \tag{9.46a}$$

$$\langle j_z \rangle = \epsilon_0\omega \left(l|u|^2 - \frac{\sigma\rho}{2} \partial_\rho |u|^2 \right). \tag{9.46b}$$

As explained previously, it is possible, in the paraxial approximation, to separate in a physically meaningful way the orbital and spin parts of the angular momentum of light. Here the first term of the RHS will be interpreted as the density of orbital angular momentum (OAM) while the second term as the spin angular momentum (SAM). It can be noted that at each point on the beam the density of spin takes a complicated value because of the partial derivative term, but integration over the beam profile yields a simple result.

Let us compute the ratio of the angular momentum flux to the linear momentum flux through a surface orthogonal to the direction of propagation. It is given by

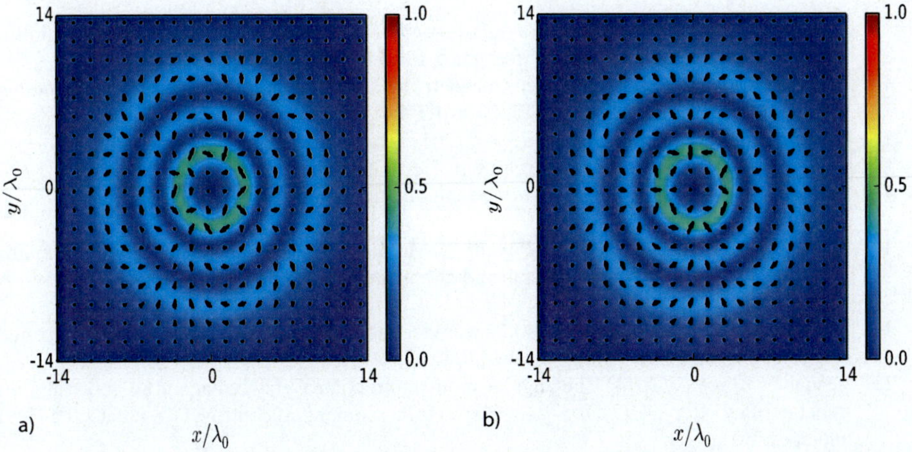

Fig. 9.11 Electric field of a circularly polarized $LG_{(2,3)}$ beam at focus ($z = 0$) at two different times (corresponding to (a) and (b)). The colormap represents the amplitude (9.33) of the field while the vector field represents the electric field as given by (9.19)

$$\frac{\iint \rho\, d\rho\, d\varphi\, c\langle j_z\rangle}{\iint \rho d\rho d\varphi c^2\langle p_z\rangle} = \frac{2\pi c\epsilon_0\omega\left[\int_0^{+\infty} d\rho\, \rho^2|u|^2 - \sigma\int_0^{+\infty} d\rho\, \frac{\rho^2}{2}\,\partial_\rho|u|^2\right]}{2\pi c^2 k\omega\epsilon_0\int_0^{+\infty} d\rho\rho|u|^2}$$
$$= \frac{l+\sigma}{\omega}. \tag{9.47}$$

Multiplying the numerator and denominator by $\hbar$ allows us to interpret this ratio as the fact that each photon has an energy of $\hbar\omega$ and an angular momentum of $\hbar(l + \sigma)$. In other words, Laguerre-Gauss beams have a well defined OAM.

References

1. Poynting J (1909) The wave-motion of a revolving shaft, and a suggestion as to the angular momentum in a beam of circularly-polarized light. Proc Roy Soc Lond A 82:560–567
2. Beth RA (1936) Mechanical detection and measurement of the angular momentum of light. Phys Rev 50:115–125
3. Allen L (1992) Orbital angular momentum of light and the transformation of Laguerre-Gaussian laser modes. Phys Rev A 45(11):8185
4. Lax M, Louisell WH, McKnight WM (1975) From Maxwell to paraxial wave optics. Phys Rev A 11:1365–1370
5. Davis L (1979) Theory of electromagnetic beams. Phys Rev A 19:1177–1179
6. Barnett SM, Allen L (1994) Orbital angular momentum and nonparaxial light beams. Opt Commun 110(5):670–678
7. Enk SV, Nienhuis G (1994) Commutation rules and eigenvalues of spin and orbital angular momentum of radiation fields. J Mod Opt 41(5):963–977
8. Berry MV (2009) Optical currents. J Opt A Pure Appl Opt 11(9):094001

9. Bliokh KY, Alonso MA, Ostrovskaya EA, Aiello A (2010) Angular momenta and spin-orbit interaction of nonparaxial light in free space. Phys Rev A 82:063825
10. Barnett SM (2010) Rotation of electromagnetic fields and the nature of optical angular momentum. J Mod Opt 57(14–15):1339–1343. PMID: 24808629
11. Bialynicki-Birula I, Bialynicka-Birula Z (2011) Canonical separation of angular momentum of light into its orbital and spin parts. J Opt 13(6):064014
12. Ornigotti M, Aiello A (2014) Surface angular momentum of light beams. Opt Express 22(6):6586–6596
13. Fernandez-Corbaton I, Zambrana-Puyalto X, Molina-Terriza G (2014) On the transformations generated by the electromagnetic spin and orbital angular momentum operators. J Opt Soc Am B 31(9):2136–2141
14. Bliokh KY, Dressel J, Nori F (2014) Conservation of the spin and orbital angular momenta in electromagnetism. New J Phys 16(9):093037
15. Leader E, Lorcé C (2014) The angular momentum controversy: what's it all about and does it matter? Phys Rep 541(3):163–248. The angular momentum controversy: What's it all about and does it matter?
16. Siegman AE (1986) Lasers, revised edn. University Science Books, Melville

Chapter 10
Soft Pair Showers in the Collision of Gamma Rays with Laguerre-Gauss Beams

10.1 Outline

In addition to the study of radiation-reaction (which has been extensively described in the first two parts of this work), one of the main interest of the next generation of PW lasers [1, 2] will be the study of pair production by nonlinear Breit-Wheeler process. One of the possible configurations to trigger pair creation is that of two counter-propagating plane-waves (studied in the previous Chap. 8). In this case, for intensities [3] of the order $\sim 5 \times 10^{25}$ W/cm^2, the development of an electromagnetic cascade is expected. Cascades (or avalanches as they are sometimes called [4]) are qualified of self-sustained in that the field acts as an external source of energy that constantly reaccelerates the radiating particles. An other configuration of interest is that of the collision of a high-energy electron or photon beam with a counter-propagating laser beam. In this case, the particles are not reaccelerated by the external field but this later just provides a transversal acceleration. All the energy of the subsequently produced particles comes from the initial energy of the seed particles: we talk about showers [3]). This configuration has been used in particular in the Burke experiment [5] and we focus in this situation in this chapter. In particular, we will study the nonlinear Breit-Wheeler process in presence of a Laguerre-Gaussian external field.

At first we will consider the collision of a high-energy photon beam with a strong external LG beam, neglecting the dynamics and the radiation of the produced pairs. This will allow us to study the effects of the LG beam on the Breit-Wheeler cross process by itself.

The effects of a LG beam compared to a gaussian beam on the particular trajectory of a charged particles will be then included by considering self-consistent PIC simulation. This allows to study the electromagnetic shower produced by the collision of a high-energy gamma photons beam with an external LG beam, the acceleration of the produced pairs, and the emission of energetic photons and subsequent decay into further pairs.

As a preliminary step in the next section we will derive some characteristic quantities of the LG beams and compare them to the usual gaussian beam. This will

F. Niel, *Classical and Quantum Description of Plasma and Radiation in Strong Fields*, Springer Theses, https://doi.org/10.1007/978-3-030-73547-0_10

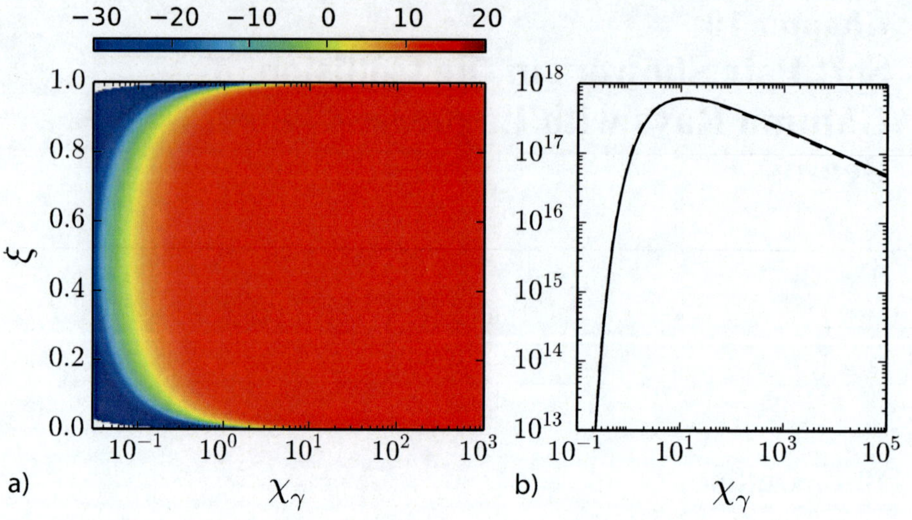

Fig. 10.1 **a** Time energy differential probability of pair production by nonlinear Breit-Wheeler process (10.1) as a function of the photon quantum parameter χ_γ and of the ratio between the electron quantum parameter and the photon quantum parameter $\xi' = \chi/\chi_\gamma$. **b** Time differential probability of pair production by nonlinear Breit-Wheeler process (10.3) as a function of the photon quantum parameter χ_γ

be useful in order to interpret the results obtained for the nonlinear Breit-Wheeler process with LG beams.

10.2 The Nonlinear Breit-Wheeler Process

Let us first recall here some of the main results about nonlinear Breit-Wheeler pair production (NBWPP) that will be useful in the rest of this Chapter. The differential probability of pair production is given by

$$\frac{d^2 N_{\mathrm{nBW}}}{dt d\chi} = A \frac{T_{\mathrm{nBW}}(\chi_\gamma, \chi)}{\chi_\gamma \gamma_\gamma}, \tag{10.1}$$

where

$$T_{\mathrm{nBW}}(\chi_\gamma, \xi') = \frac{\sqrt{3}}{2\pi}\left[-\int_{\nu'}^{+\infty} dy\, \mathrm{K}_{5/3}(y) + \frac{3}{2}\chi_\gamma \nu' \mathrm{K}_{2/3}(\nu')\right], \tag{10.2}$$

with $\nu' = \frac{2}{3}\frac{\chi_\gamma}{(\chi_\gamma - \chi)\chi}$, $\xi' = \chi/\chi_\gamma$ and $A = \frac{2}{3}\frac{\alpha^2}{\tau_e}$ (Fig. 10.1). We then define the production rate by

$$\frac{dN_{\text{nBW}}}{dt} = A\frac{W_{\text{nBW}}(\chi_\gamma)}{\gamma_\gamma}, \tag{10.3}$$

with

$$W_{\text{nBW}}(\chi_\gamma) = \frac{\sqrt{3}}{2\pi}\int_0^1 d\xi'\left[-\int_{\nu'}^{+\infty} dy\,\mathrm{K}_{5/3}(y) + \frac{3}{2}\chi_\gamma\nu'\,\mathrm{K}_{2/3}(\nu')\right]. \tag{10.4}$$

We can rewrite it like

$$\frac{d^2N_{\text{nBW}}}{dtd\xi'} = A\frac{\tilde{T}_{\text{nBW}}(\chi_\gamma, \xi')}{\gamma}, \tag{10.5}$$

with

$$\tilde{T}_{\text{nBW}}(\chi_\gamma, \xi') = \frac{\sqrt{3}}{2\pi}\xi'\left[-\int_{\nu'}^{+\infty} dy\,\mathrm{K}_{5/3}(y) + \frac{3}{2}\chi_\gamma\nu'\,\mathrm{K}_{2/3}(\nu')\right], \tag{10.6}$$

where $\nu' = \frac{2}{3}\frac{1}{\chi_\gamma(1-\xi')\xi'}$. The following asymptotic forms will be useful in the rest of this chapter

$$\frac{dN_{\text{nBW}}}{dt} \sim A\frac{\chi_\gamma}{\gamma_\gamma}\exp\left(-\frac{8}{3\chi_\gamma}\right), \text{ for } \chi_\gamma \ll 1, \tag{10.7a}$$

$$\frac{dN_{\text{nBW}}}{dt} \sim A\frac{\chi_\gamma^{2/3}}{\gamma_\gamma}, \text{ for } \chi_\gamma \gg 1. \tag{10.7b}$$

10.3 Characteristics of the LG Beams

One of the quantities that will play a role in the total yield of pairs by NBWPP is the effective radius of the external beam. As already noticed in the previous chapter, in particular on Figs. 9.6, 9.7, 9.8, 9.9, 9.10 and 9.11, the effective size of the different LG modes increases with p and l, while the amplitude of these fields seems to decrease. Let us analyze more precisely these effects. The LG modes given by Eq. (9.42) all have the same energy, while their amplitude is given by

$$C_{pl}^{\text{LG}} = \sqrt{\frac{p!}{(p+|l|)!}}. \tag{10.8}$$

We therefore see that, for $p = 0$, the maximum amplitude of the fields decrease as $1/\sqrt{|l|!}$, while for $l = 0$, the maximum amplitude of the LG mode ($p \neq 0$) is the same as for the gaussian mode. More precisely, Eq. (10.8) is plotted on Fig. 10.2. We see that the amplitude of the LG beam decreases very quickly with l at fixed p,

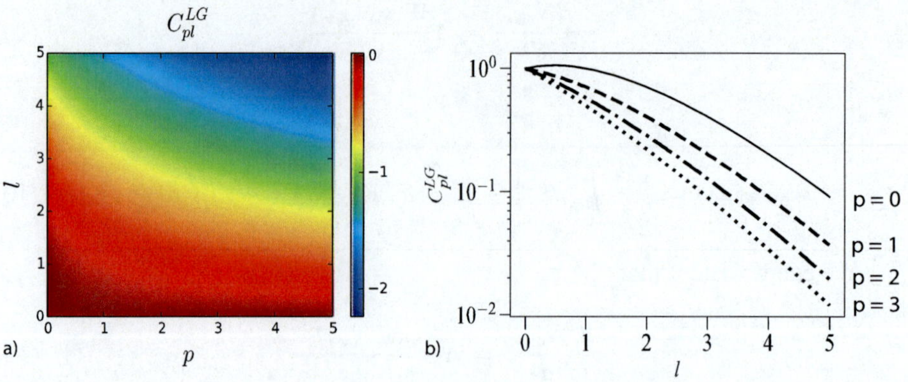

Fig. 10.2 Coefficient C_{pl}^{LG} for the LG modes (9.37) **a** for l and p going from 0 to 5 and **b** for $p = 0, 1, 2, 3$ in terms of l. Note that, in practice, l and p can only take integer values. We represent them here as taking continuous values for representation purposes.

while it decreases slowly with p at fixed l. In particular, the amplitude of the $LG_{(0,5)}$ mode is approximately 10 times smaller than the amplitude of the gaussian mode, while the amplitude of the $LG_{(3,5)}$ mode is approximately 100 times less than the gaussian mode. In this work, we only consider pretty small values for l and p. Indeed, the focus being here on strong-fields, it is impossible with the current technology to produce fields having at the same time a very large a_0 and a very large value[1] of l and/or p.

Contrarily to the case of the gaussian mode, for LG modes with $p \neq 0$, the amplitude does not fall off monotonically to 0 but possesses p zeros at different radiuses $\rho \neq 0$. We can therefore not define the radius of the beam as the distance to the axis at which the amplitude falls off of a factor $1/e$. We will instead define an effective radius as

$$\rho_{\text{eff}}^2 = \frac{\sqrt{2}}{\pi} \iint d^2x_\perp \, \rho^2 \, |u|^2. \tag{10.9}$$

Using the following identity

$$\int_0^{+\infty} dx \, x^{|l|+1} \left[L_p^{|l|}(x) \right]^2 e^{-x} = \frac{(|l| + p)!}{p!} (|l| + p + 1), \tag{10.10}$$

we get

$$\rho_{\text{eff}} = w_0 \sqrt{1 + \zeta^2} \sqrt{|l| + 2p + 1}, \tag{10.11}$$

[1] Outside of strong fields, it should be noted that it is currently possible to produce fields with values of l as large as 10000 [6].

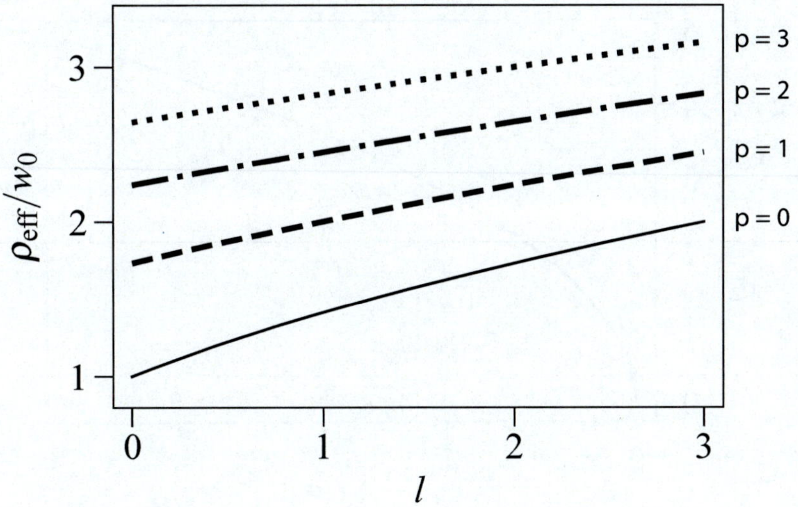

Fig. 10.3 Effective radius (10.11) (normalized to the waist w_0) in terms of l for $p = 0, 1, 2, 3$. In practice, l can only take integer values and it is represented here with a continuous axis for representation purposes.

where the factor $\sqrt{2}/\pi$ in Eq. (10.9) allowed us to find $\rho_{\text{eff}} = w_0$ for $\zeta = 0$ and $l = p = 0$, which is what we would expect for a gaussian beam at focus.

The result is plotted on Fig. 10.3. The effective radius increase with the square root of l and p, that is, much more slowly than the coefficients C_{pl}^{LG} (10.8) were decreasing with l.

Another quantity of interest is the position of the maximum of the field. As shown on Figs. 9.6, 9.7, 9.8, 9.9, 9.10 and 9.11, the maximum of the field is no longer at $(x, y) = (0, 0)$ contrarily to the case of the gaussian beam. We obtain this radius by solving $\partial_\rho |u|^2 = 0$. This yields

$$\frac{|l|}{\rho} - \frac{4\rho}{w(\zeta)^2} \frac{L_{p-1}^{|l|+1}\left(\frac{2\rho^2}{w(\zeta)^2}\right)}{L_p^{|l|}\left(\frac{2\rho^2}{w(\zeta)^2}\right)} - \frac{2\rho}{w(\zeta)^2} = 0, \tag{10.12}$$

which for $p = 0$ (and $l \neq 0$) gives

$$\rho_{\max} = w(\zeta)\sqrt{\frac{|l|}{2}}. \tag{10.13}$$

For $p \neq 0$, Eq. (10.12) cannot be solved analytically. We solve it numerically and display the results on Fig. 10.4. We see that any p, the radius of the maximum of the field increases with l. In particular, for $p = 0$, it increases like the square root of l, while for $p \neq 0$ it increases more slowly with l. At a given l, the radius of

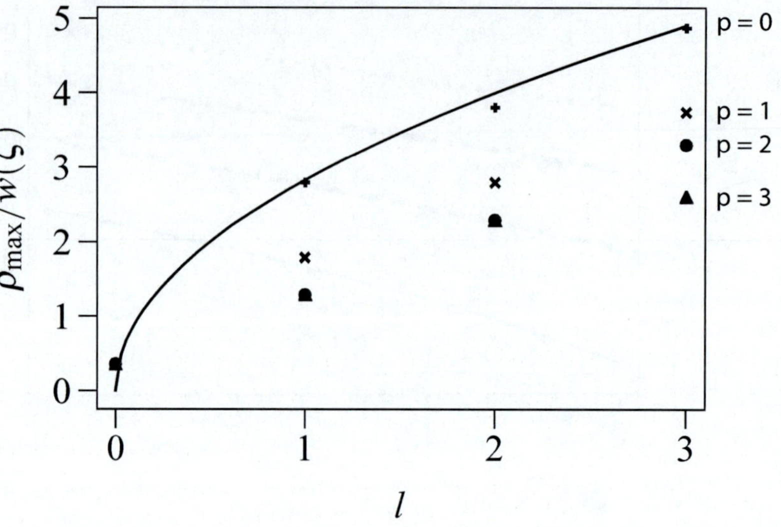

Fig. 10.4 Radius of the maximum of the field as given by (10.12) [normalized to the waist $w(\zeta)$] in terms of l for $p = 0, 1, 2, 3$. For $p = 0$ Eq. (10.12) can be solved analytically and is plotted in continuous line (10.13) while for $p \neq 0$ we plot the integer values for $l = 0, 1, 2, 3$ after numerical resolution of (10.12).

the maximum of the field decreases with p. This can be explained by the fact that $LG_{(p,l)}$ modes have p zeros for $\rho \neq 0$ while their effective size increases like the square root of p (10.11). The number of zeros therefore increases more quickly (linearly) than the effective size of the beam with p. Moreover, the maximum of the field being between the center (0, 0) and the first zero [because of the gaussian term in Eq. (9.37)] has to get closer and closer to the center (i.e. ρ_{max} decreases with p).

Finally, let us compute the value of the maximum of the amplitude u. This is easily done by inserting the value of the radius of the maximum as given by (10.12) into (9.37). Here again it can be done analytically for $p \neq 0$ using (10.13), while for $p \neq 0$ we solve (10.12) numerically and insert the value into (9.37). For $p = 0$ we then get the following equation for the envelop

$$|u_{lp}^{\mathrm{LG}}(\rho_{max}, \varphi, \zeta)| = \frac{C_{pl}^{\mathrm{LG}}}{\sqrt{1+\zeta^2}} |l|^{|l|/2} e^{-|l|/2}. \tag{10.14}$$

The results are plotted on Fig. 10.5. The maximum of the amplitude decreases with increasing l. This is due to the fact that the fields are computed so that they have the same energy. Knowing that their radius increases (10.13) with l, the maximum amplitude $u(\rho_{max})$ can only decrease to keep (9.41) the same. Moreover, $u(\rho_{max})$ almost does not depend on p for a given l.

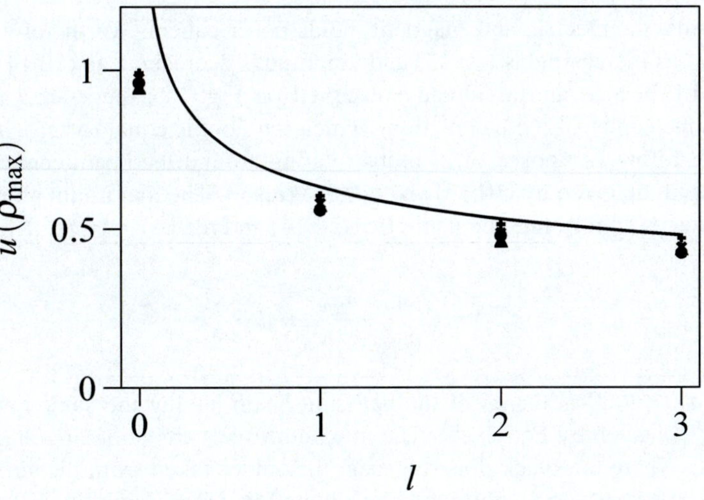

Fig. 10.5 Maximum of the field $u(\rho_{\max})$ for $p = 0, 1, 2, 3$ as a function of l.

10.4 Nonlinear Breit-Wheeler Pair Production in an External LG Beam

We now consider the collision of an external LG beam (with different values of l) with a high-energy photon beam. In a first time, we freeze the pairs created by the photon conversion, that is to say, we prevent them to move and to radiate. This will allow us to isolate the pure effect of the external field on the photon pair-conversion from the dynamical effect of this field on the created pairs.

In this case (and because we are in the LCFA), the total pair production rate is just the sum over the beam profile of the production rate at each point as given by (10.4)

$$W_{\mathrm{nBW}}^{\mathrm{tot}} = \int d^3x \, W_{\mathrm{nBW}} \left[\chi_\gamma(\mathbf{x}) \right], \tag{10.15}$$

where

$$\begin{aligned} \chi_\gamma(\mathbf{x}) &= \frac{\gamma_\gamma}{E_S} \left| \mathbf{E}_\perp(\mathbf{x}) + \mathbf{c} \times \mathbf{B}_\perp(\mathbf{x}) \right| \\ &= \frac{\gamma_\gamma}{E_S} \sqrt{\left[E_x(\mathbf{x}) + cB_y(\mathbf{x}) \right]^2 + \left[E_y(\mathbf{x}) - cB_x(\mathbf{x}) \right]^2}, \end{aligned} \tag{10.16}$$

with $\mathbf{c} = -c\hat{\mathbf{z}}$. The photon quantum parameter χ_γ therefore just corresponds to the modulus of the envelop (9.42) u. For example, in the case of a linearly polarized (along y) beam, $E_y = -cB_x$ while $E_x = B_y = 0$. We therefore have $\chi_\gamma = 2\gamma_\gamma |u|/E_s$. At a given position z, in the transverse plane, photons have therefore a quantum parameter of the form 10.6(a) (for a LP beam) and 10.7(b) (for a CP beam). In the case of the

CP beams, the electric and magnetic fields never cancel. We therefore have rings with radius increasing as (10.13) and amplitudes decreasing as (10.14). In the case of the LP beams, and as already observed on Fig. 9.6, there are l axis rotating around the center of the beam, along which the field is equal to zero. The resulting χ_γ is therefore composed of $2l$ blobs rotating around the beam center with radius and amplitude given by (10.13) and (10.14) (resp.). The maximum photon quantum parameter $\chi_\gamma^{\max}$ is computed using Eq. (10.14) and reads

$$\chi_\gamma^{\max} = 2 \frac{a_0}{a_{0S}} \frac{C_{pl}^{\mathrm{LG}}}{\sqrt{1+\zeta^2}} |l|^{|l|/2} e^{-|l|/2}, \tag{10.17}$$

where $a_0 = 2000$ is the a_0 of the gaussian beam having the same energy and w_0 and C_{pl}^{LG} is given by Eq. (10.8). The maximum $\chi_\gamma^{\max}$ are summarized (at $\zeta = 0$) on Fig. 10.7 where the black dots represent the values taken from the simulation (that we describe just below) while the black dotted line represents Eq. (10.17). The fact that the wavefront is helically shaped and that the beam carries OAM has therefore no influence on the photon quantum parameter. The change in the total number of created pairs can therefore only come from geometrical properties of the LG beams (for frozen pairs).

To verify this, we realize three-dimensional Particle-In-Cell (PIC) simulations with the PIC code SMILEI. In these simulations, a γ-ray flash (slab of γ-photons with normalized energy $\gamma_\gamma = 400$, corresponding to an energy of ~ 200 MeV) collides head-on with a LG light pulse ($l = 0-4$, $p = 0$). For all considered values of l, the LG pulse has a waist $w_0 = 3\,\lambda_0$ [as defined by Eq. (9.43)], the pulse total energy is kept the same as that of the Gaussian pulse ($l = 0$) with field strength $a_0 = 2000$, corresponding to a laser intensity of 2.8×10^{24} W/cm^2 (as we have considered the laser wavelength $\lambda_0 = 1\ \mu$m for these simulations). The LG pulse has a sin^2 intensity profile with $6\lambda_0/c$ full-width-at-half-maximum and the two light pulses are tuned to meet at the best focus of the LG pulse when the latter reaches its maximum intensity. For these simulations, a spatial resolution $\Delta x = \Delta y = \Delta z = \lambda_0/16$ was chosen and the timestep was $c\Delta t \simeq 0.55\,\Delta x$, corresponding to 95% of the Courant-Friedrich-Lewy condition for the electromagnetic solver employed in SMILEI. A quick convergence test was made: running the simulation with a resolution increased by a factor 2 showed similar results.

Two series of simulations were performed. In the first series, all electron-positron pairs created by the interaction of the γ-ray flash with the LG beams were created *frozen*, i.e. they did not further impact the simulation and only their creation was recorded. This first series was used to explore the total cross section of NBWPP driven by a LG beam. In the second series, all electron-positron pairs created were consistently evolving in the field, i.e. they could produce additional γ-photons through inverse nonlinear Compton scattering, which themselves could produce additional pairs through NBWPP.

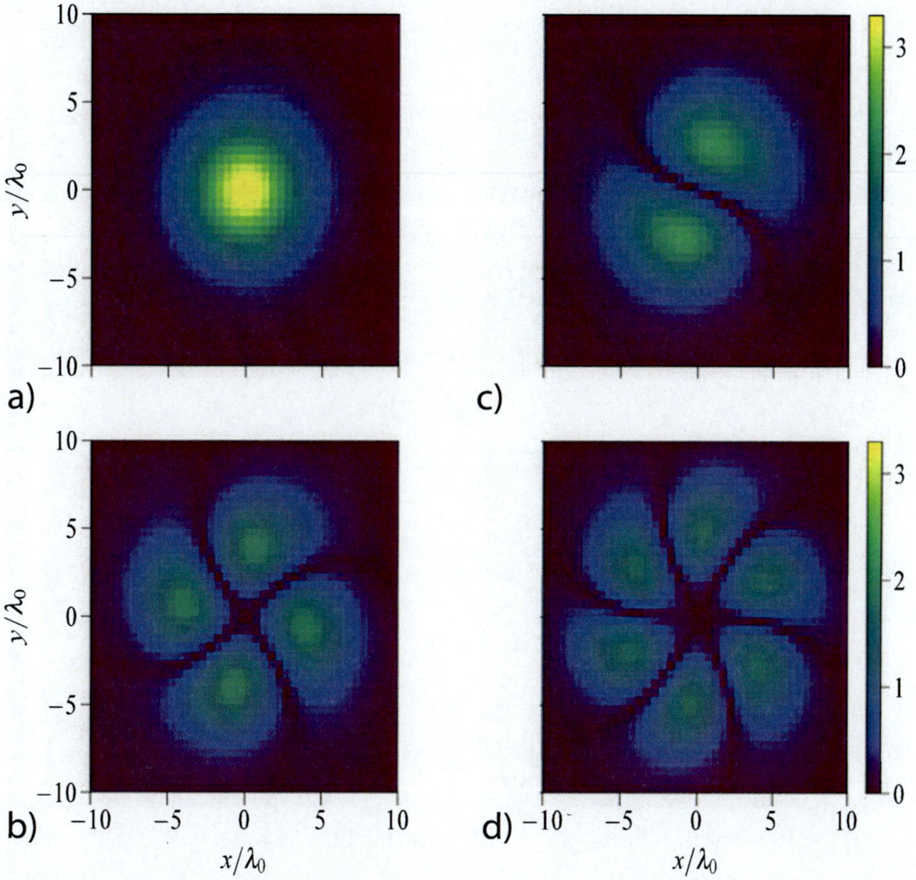

Fig. 10.6 Photon quantum parameter χ_γ in the transverse plane at focus ($z = 0$) for a photon of energy $\gamma_\gamma = 400$ and different linearly-polarized LG beams with $a_0 = 2000$, $p = 0$ and a) $l = 0$, b) $l = 1$, c) $l = 2$, d) $l = 3$.

Figure 10.9 summarizes the results extracted from the first series of simulations (no feedback from the produced pairs). It shows (blue dots), as a function of the LG beam parameter l ranging from 0 to 4, the total number of produced pairs (normalized to that obtained considering the Gaussian beam $l = 0$). We see that the number of produced pairs in a LG beam increases with l compared to the gaussian case. However, as mentioned previously, this effect is purely geometric. To verify this, we ran a simulation with a Gaussian beam (l=0) with a reduced field strength $a_0^* = 0.606 \times a_0$ corresponding to the maximum field strength obtained for the $LG_{(0,1)}$ beam, but with an increased beam waist $w_0^* = w_0/0.606$ in order to keep the beam energy constant. This case is represented as a red dot on Fig. 10.9 (and the corresponding $\chi_\gamma^{\max}$ as a red dot on Fig. 10.8). We therefore conclude that the increased number of created pairs is due to a geometric effect because of the increment in the beam effective size.

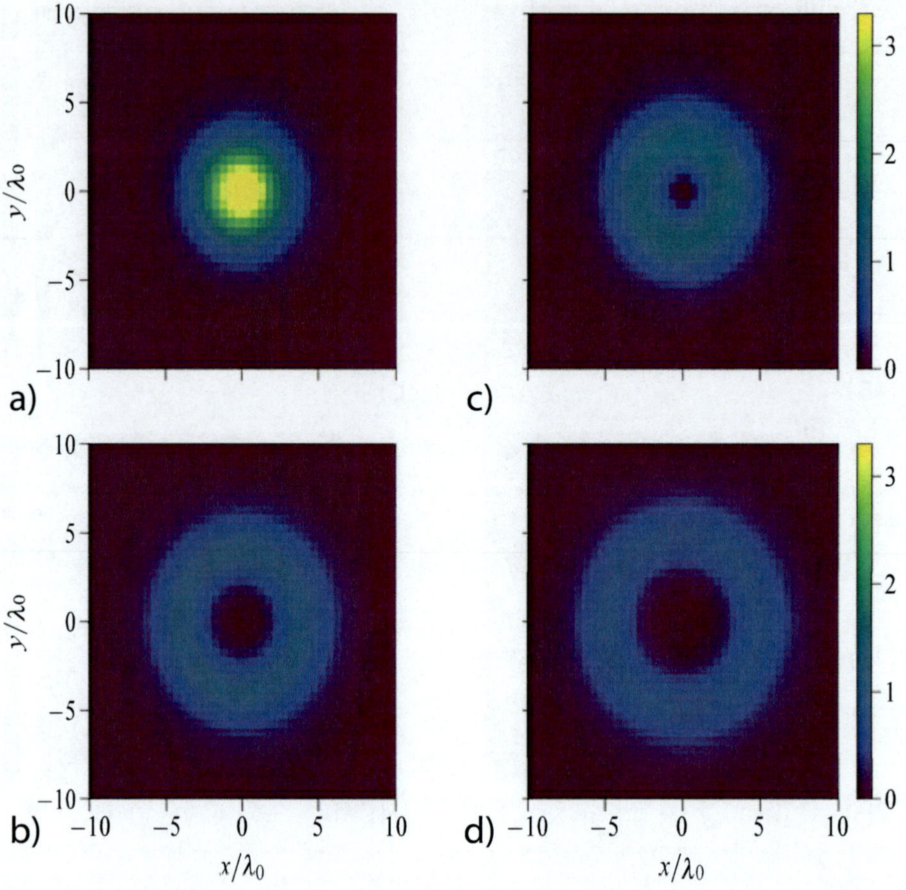

Fig. 10.7 Photon quantum parameter χ_γ in the transverse plane at focus ($z = 0$) for a photon of energy $\gamma_\gamma = 400$ and different circularly-polarized LG beams with $a_0 = 2000$, $p = 0$ and a) $l = 0$, b) $l = 1$, c) $l = 2$, d) $l = 3$.

As it turns out, using a LG beam with a given effective radius is less efficient than a direct increase of the waist of a gaussian beam ($l = 0$) to the same value as the effective radius, and with the same total energy.

To qualitatively understand the origin of the rise in the number of created pairs, we use Eq. (10.15) with the asymptotic form for $W_{\text{nBW}}\left[\chi_\gamma(\mathbf{x})\right]$ for $\chi_\gamma \gg 1$ as given by Eq. (10.7a). We get

$$\frac{N_{\text{pair}}(l)}{N_{\text{pair}}(l=0)} \simeq \frac{\int d^2x_\perp \left[\chi_\gamma^{LG}(\mathbf{x})\right]^{2/3}}{\int d^2x_\perp \left[\chi_\gamma^{G}(\mathbf{x})\right]^{2/3}}, \tag{10.18}$$

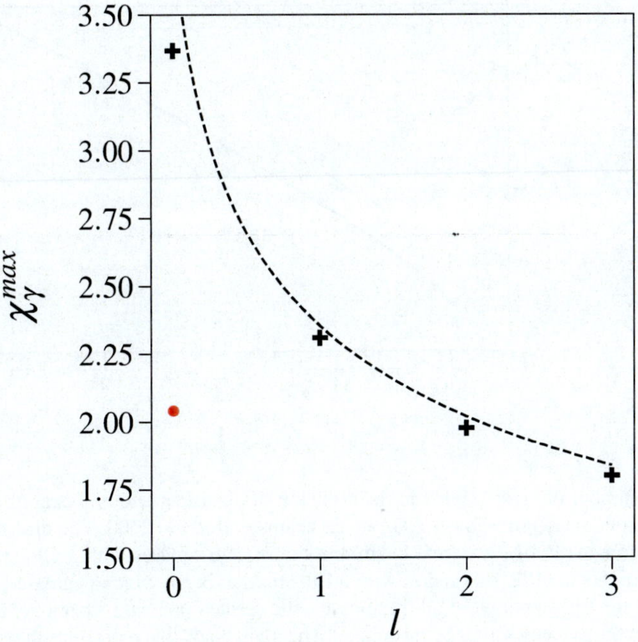

Fig. 10.8 Maximum photon quantum parameter $\chi_\gamma^{\max}$ as given by the PIC simulation (black dots) and by Eq. (10.17) (black dotted line) as a function of l.

where $\chi_\gamma^{LG}(\mathbf{x})$ is the photon quantum parameter computed with the external LG beam while $\chi_\gamma^{G}(\mathbf{x})$ is the photon quantum parameter computed with the usual gaussian beam. To get an analytical estimate, we approximate $\chi_\gamma^{LG}(\mathbf{x})$ by its maximum $\chi_\gamma^{\max}$ as given by (10.17) and the surface integral by the surface $\rho_{\max}^2$ with $\rho_{\max}$ given by (10.13) for the LG beams and to $w(\zeta)$ for the gaussian beam. We obtain

$$\frac{N_{\text{pair}}(l)}{N_{\text{pair}}(l=0)} \simeq \frac{l}{2(l!)^{1/3}} |l|^{|l|/3} e^{-|l|/3}. \tag{10.19}$$

This estimation is plotted in black line on Fig. 10.9. It is in good agreement with the value given by the PIC simulation. We therefore have the counter-intuitive conclusion that the number of created pairs increases with decreasing a_0. This is easily understood thanks to (10.18) as the fact that the beam surface increase more quickly than the production rate decreases with decreasing a_0. Of course, this argument only works for big enough χ_γ so that the cross section of pair creation is large enough, and it makes sense to approximate $W_{\text{nBW}}\left[\chi_\gamma(\mathbf{x})\right]$ by its asymptotic form. However, we see that this condition is not very stringent since the approximate formula works for $\chi_\gamma^{\max} \sim 3$ only (see Fig. 10.8). For $l = 4$, $\chi_\gamma^{\max} \sim 1.8$ and we see that our approximation (10.19) overestimates the simulation value (Fig. 10.9). We interpret that as due to the fact that for such a χ_γ, the production rate starts to be closer to (10.7b)

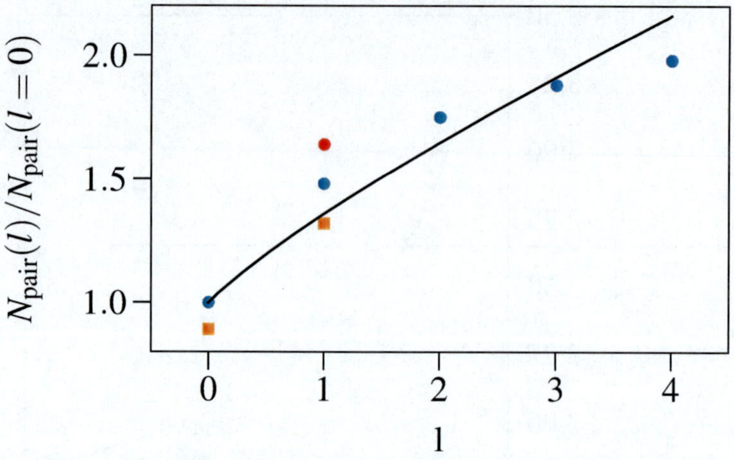

Fig. 10.9 Number of created pairs in the collision of a gamma photon beam of energy $\gamma_\gamma = 400$ with different counter-propagative $LG_{0,l=1-4}$ beams with $a_0 = 2000$. The blue dots represent the PIC simulations with a LP external beam and the orange squares with a CP external beam. The red dot represents the PIC simulation with a LP gaussian beam with a decreased a_0 and increased w_0, keeping the same energy. In all these cases, the created pairs are frozen i.e. they don't further participate in the dynamics after being created. The plain black line represents the estimation (10.19)

than to (10.7a) and so it is no longer beneficial to decrease χ_γ while increasing ρ_{max}. In addition, we report for the cases $l = 0$ and $l = 1$ (orange square on Fig. 10.9) the results of simulations for which the LG beam had a circular (both left- and right-handed case led to the same result) polarization. It is important to stress here that these simulations consider once more that the energy carried by the LG beams is the same than that considered for all the other cases. Doing so, we find that using a circularly polarized light pulse is not beneficial for efficient pair production as, at constant energy, this requires a reduction of the effective laser field strength, while keeping the same waist.

Of course, in practice, created pairs are not frozen but can move in the external field, radiate, and this subsequent radiation can in turn create new pairs. We therefore run a second set of PIC simulations, in the same conditions as before, but with the full dynamics. Figure 10.10 summarizes these results. In this figure we plot the increment (%) of produced pairs for the different LG beam configurations as in Fig. 10.9. Instead of normalizing the number of produced pairs by that obtain in the Gaussian case ($l = 0$), we here normalize the number of produced pair by that obtained in the very same configuration (i.e. same parameter l and polarization) but in the first series.

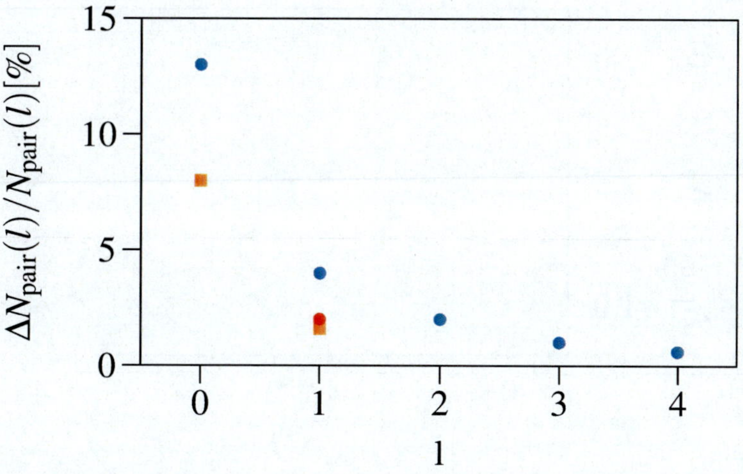

Fig. 10.10 Pourcentage of additional pairs created by subsequent radiation and pair production by the first pair conversion of the initial gamma photon beam, as a function of l. The parameters are the same as on Fig. 10.9, that is to say, the incoming gamma photons have an energy $\gamma_\gamma = 400$ and the counter-propagative $LG_{0,l=1-4}$ beams a normalized amplitude $a_0 = 2000$. The blue dots represent the PIC simulations with a LP external beam and the orange squares with a CP external beam. The red dot represents the PIC simulation with a LP gaussian beam with a decreased a_0 and increased w_0, keeping the same energy. The plain black line represents the estimation (10.19).

We see that the increase in the number of created pairs due to the subsequent radiation goes from +13% to +0.6% (for $l = 0$ to 4) compared to the case with frozen pairs. We qualify this situation of *soft shower* since most of the created pairs come from direct conversion of the incoming photon beam. The $LG_{(0,1)}$ beam is slightly beneficial compared to the gaussian beam, for the same transverse size (see blue vs red dot on Fig. 10.10). However, this difference is marginal and the overall behavior still follows the conclusion given for the first series of simulation. Moreover, we see that as l increases, the dynamics of the electron in the LG beam does not seem to be beneficial since the number of additional created pairs decreases strongly. As in the previous series of simulation, the CP beams produce less additional pairs than the LP beams (for the same reason that the CP beams have a smaller amplitude at given energy).

Finally, we plot the total number of created pairs in the full auto-consistent PIC simulation as a function of l on Fig. 10.11 (where we followed the same representation as on Fig. 10.9). We see the same tendency as in the case where the first created pairs were frozen.

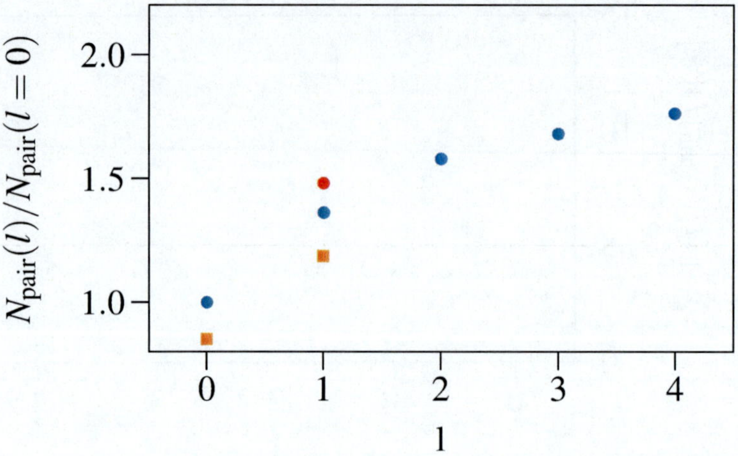

Fig. 10.11 Number of created pairs in the collision of a gamma photon beam of energy $\gamma_\gamma = 400$ with different counter-propagative $LG_{0,l=1-4}$ beams with $a_0 = 2000$. The blue dots represent the PIC simulations with a LP external beam and the orange squares with a CP external beam. The red dot represents the PIC simulation with a LP gaussian beam with a decreased a_0 and increased w_0, keeping the same energy. In all these cases, the created pairs are frozen i.e. they don't further participate in the dynamics after being created. The plain black line represents the estimation (10.19).

10.5 Conclusion

In this chapter, we studied soft showers in the collisions of a gamma ray beam with Laguerre-Gauss beams with parameters $l = 0 - 4$. We find that the number of created pairs increases with increasing l. This effect is purely geometrical and due to the fact that the total production rate increases more quickly with the effective surface coming into play in the process than the decrease of the production rate with the photon quantum parameter. We verify this claim by simulating the same situation with a usual gaussian beam with increased size and decreased amplitude. We therefore have the counter-intuitive conclusion that the number of created pair in the collision of a gamma ray beam with an external laser field increases with decreasing intensity. Of course, this effect is valid only for large enough photon quantum parameter. In practice, this condition is not so stringent since in this study we used photon quantum parameter going from 3.3 to 1.8 approximately. Moreover, this effect is stronger by defocusing a gaussian beam than by increasing its size by using higher order modes such as $LG_{(0,l)}$ beams. These findings are interesting in the context of the upcoming experiments on the new PW lasers such as ELI or Apollon in that they contradict the common belief that one has to use the highest possible intensities in order to observe nonlinear Breit-Wheeler pair production.

References

1. Cros B, Paradkar B, Davoine X, Chancé A, Desforges F, Dobosz-Dufrénoy S, Delerue N, Ju J, Audet T, Maynard G, Lobet M, Gremillet L, Mora P, Schwindling J, Delferrière O, Bruni, C, Rimbault C, Vinatier T, Piazza AD, Grech M, Riconda C, Marquès J, Beck A, Specka A, Martin P, Monot P, Normand D, Mathieu F, Audebert P, Amiranoff F (2014b) Laser plasma acceleration of electrons with multi-PW laser beams in the frame of CILEX. Nucl Instrum Methods Phys Res Sect A 740:27–33. Proceedings of the first European advanced accelerator concepts workshop 2013
2. Extreme light infrastructure (ELI). https://eli-laser.eu/
3. Fedotov AM (2016) Quantum regime of laser-matter interactions at extreme intensities. arXiv:1612.02038v1
4. Fedotov AM, Narozhny NB, Mourou G, Korn G (2010) Limitations on the attainable intensity of high power lasers. Phys. Rev. Lett. 105:080402
5. Burke DL, Field RC, Horton-Smith G, Spencer JE, Walz D, Berridge SC, Bugg WM, Shmakov K, Weidemann AW, Bula C, McDonald KT, Prebys EJ, Bamber C, Boege SJ, Koffas T, Kotseroglou T, Melissinos AC, Meyerhofer DD, Reis DA, Ragg W (1997) Positron production in multiphoton light-by-light scattering. Phys. Rev. Lett. 79:1626–1629
6. Courtial J, Dholakia K, Allen L, Padgett M (1997) Gaussian beams with very high orbital angular momentum. Opt Commun 144(4):210–213

Chapter 11
Conclusion and Perspectives

In the near future, the new class of petawatt lasers is expected to reach peak intensities above 10^{23} W/cm^2. In particular, in France, the laser Apollon is expected to reach, by 2020, a peak power of 10 PW. More long-term projects, such as the upgrade of ELI, aim at reaching powers above several hundreds of PW. At such intensities, the electron dynamic is expected to change drastically.

In this thesis, we studied radiation reaction in the classical and quantum regime using a kinetic description for the evolution of the electron distribution function. In Chap. 5, we start from a general linear Boltzmann (lB) equation describing RR in the quantum regime and perform an expansion for small ratios of the energy of the emitted photon over the electron's energy. Stopping this expansion to second order allows to simplify the linear Boltzmann (lB) equation to a Fokker-Planck (FP) equation. This equation is equivalent to a stochastic differential equation for the electron momentum. The first term of this equation is a purely deterministic term corresponding to the leading term of the LL equation, corrected by the quantum Gaunt factor. We refer to this equation as the corrected Landau-Lifshitz (cLL) equation. The second term corresponds to a stochastic process, representing the quantum effects in a regime of intermediate electron quantum parameters χ. We refer to this equation as the FP model. This "effective" stochastic differential equation is interesting from both a fundamental and practical point of view. From a fundamental point of view, this allows to study the characteristics of RR in a regime of radiation beyond the classical/deterministic limit, and including the lowest order quantum effects. From a practical point of view, this yields an effective stochastic pusher that can be very easily implemented in a numerical code like a PIC code.

We then derive the equation of evolution of the successive moments of the electron distribution function as described by the full linear Boltzmann (lB) equation, the second order expansion [referred to as the Fokker-Planck (FP) model] and the first order expansion [referred to as the corrected Landau-Lifshitz (cLL) model]. We show that the equation of evolution of the mean electron energy is formally the same in the three models (lB, FP and cLL). The difference between the average energy in the lB and cLL models is computed and shown to be less than a few % for $\chi \lesssim 1$. We find that the equation of evolution of the electron energy spread is formally the same

F. Niel, *Classical and Quantum Description of Plasma and Radiation in Strong Fields*, Springer Theses, https://doi.org/10.1007/978-3-030-73547-0_11

in the lB and FP descriptions but different from the cLL description. In the classical case (cLL), the electron distribution function can only cool down due to RR while in the FP and lB models, depending on the initial electron energy spread and the initial average electron quantum parameter, RR can either lead to heating or cooling of the electron distribution function. We compute analytically the limit value of energy spread between the heating and cooling regimes in terms of the electron average χ parameter (called $\hat{\sigma}_{\max}$). Typically narrow beams will present heating, while if the initial energy spread is very large, there will be cooling. When heating is present, it only lasts for a finite amount of time, (denoted t_{heat}), until the maximum energy spread $\hat{\sigma}_{\max}$ is reached. After that, the system starts to cool down. The maximum $\hat{\sigma}_{\max}$ coincides with the threshold between heating and cooling calculated theoretically. It is shown that even for χ of order one, the calculation allows to predict the value of the spread of the distribution function that separates heating from cooling. The equation of evolution of the third order moment in energy of the electron distribution function is then investigated. We find it is formally different for the three descriptions (cLL, FP and lB). This third order moment μ_3 can either increase or decrease in the lB model and the threshold between these two domains is computed analytically. We identify the decrease of this moment yielding to a negative μ_3 as a signature of quantum RR. Another interesting consequence of this study is that, for physical problems in which only the average electron energy is important, the cLL description of RR is sufficient. If one needs only the energy average and spread, the FP description is enough. If finally the exact shape of the electron distribution function is important, one needs to use the full lB description.

In Chap. 6, we study the domains of validity of the three models, and perform numerical simulations in order to confirm the theoretical findings of the previous chapter. Contrarily to the common belief that the classical description is valid for small values of χ and that a quantum description is needed when χ approaches unity, the domains of validity we derive are statistical ones that depend on the electron distribution itself. In particular, if the electron energy spread is large enough, a classical description can be sufficient, even for relatively high values of χ. On the contrary, very narrow electron beams may require a quantum treatment even for relatively small values of χ. This has fundamental but also practical implications, for example for the implementation of RR in PIC codes, for which it allows to better choose the threshold between the classical and the quantum modules for RR, depending on the particular physical situation under investigation. We support these claims with numerical simulations of the three models with initially narrow electron beams as well as broad Maxwell-Jüttner distributions, interacting with linearly or circularly polarized plane-waves and constant uniform magnetic fields. When stochastic heating is important the electron distribution function displays, not only an increasing σ_γ but also a decreasing third order moment μ_3 (skewness). In particular for a narrow beam, this reflects the fact that the heating of the electron distribution function happens in a non symmetric way, making it acquire a long tail toward the low energies. We quantify this asymmetry, find a theoretical condition to have increasing or decreasing μ_3 depending on the initial value of the average χ, the energy spread $\hat{\sigma}_\gamma$ and the initial skewness μ_3. In the case of a narrow beam, the initial skewness decreases leading to a negative value, eventually stops and μ_3 goes to zero before changing sign. The typical time scale for this inversion is denoted in the manuscript as t^* and it is found to be of the same order as t_{heat} in the examined case. We can link the asymmetry of the electron distribution function for times shorter than t^* to the so-called "quenching" of radiation reaction, that is the fact that, even if the average energy of the electrons decreases with time, energetic particles have reduced radiation losses for short time scales (less than t*). This gives an interesting way of controlling radiation in laser-plasma interaction schemes by appropriate tuning of the laser pulse duration.

All the results that we derived here are very general and apply to a wide range of configurations and geometries of the statistical ensemble and electromagnetic field. The configurations considered in this thesis are i) head-on collision of a linearly polarized plane-wave with a narrow high energy electron beam ii) head-on collision of a circularly polarized plane-wave with a narrow high energy electron beam iii) collision of a narrow high energy electron beam with a constant-uniform magnetic field iv) interaction of a 3D Maxwell-Jüttner distribution with zero drift and high temperature with a constant-uniform magnetic field.

In Chap. 7, we turn our attention to the radiation spectrum. In particular, we link its properties to the electron distribution function described in the previous chapters. The instantaneous and time integrated spectra are investigated using numerical simulations of the three models. We consider both initially narrow electron beams and broad Maxwell-Jüttner distributions, interacting with linearly or circularly polarized plane-waves and constant uniform magnetic fields. We find that the predictions of the FP and lB models are in good agreement even for χ close to one and different from the cLL model. In particular, the lB and FP models exhibit a hardening of the photon spectrum. The fact that the lB and FP models exhibit such close spectra shows that the photon distribution is especially sensitive to the overall broadening of the electron distribution function (energy spread), but not so much to the details of the asymmetries of this electron distribution function. Moreover, this implies that the most difference is observed around t_{heat}, where the broadening of the electron distribution function is the maximum in the FP and lB models. Finally, we see that the discrepancies between the three models are much smaller than on the electron distribution functions which makes us conclude that the electron distribution function is a better candidate to observe the transition between the classical and quantum regime.

In Chap. 8, we consider the interaction of an electron distribution with the superposition of two circularly polarized plane waves. This configuration is of outmost importance for the creation of the so-called QED cascades. In the general case, the motion of an electron in such a stationary wave is chaotic. To avoid this difficulty, it is common to consider the configuration in the magnetic nodes, which reduces to a pure rotating electric field. In this case, the motion is 2D and the electrons remain infinitely in the magnetic node plane. Moreover, contrarily to the external fields considered in the previous chapters, this configuration increases the mean energy of the electron distribution. We find that after a few rotations of the electric field, the electron distribution function reaches a stationary state. All of its moments are therefore constant. We then apply the previous kinetic equations, taking into account the new source term, that comes from the fact that the external field now brings energy to the particles. The average asymptotic energy is computed numerically in the cLL and lB models and the discrepancy between these two values is deduced analytically from the kinetic equations. Similarly to Chap. 5, this discrepancy is smaller than a few %, while it can reach several tenth of % for $\chi > 1$, but in this case, the analytical equation allows to compute exactly the difference between the mean energy as given by the cLL and lB models.

We then consider the equation of evolution of the energy spread. In this case, the source term is negligible for $\chi \lesssim 1$ and the asymptotic energy spread in this configuration is therefore the same as the threshold of the energy spread computed in Chap. 5. We verify these findings with numerical simulations, considering both an initially at rest electron distribution, and an initially hot Maxwell-Jüttner (MJ) distribution. In the case of the initially cold distribution, and for $\chi \lesssim 1$, the average energy, energy spread and third order moment μ_3 oscillates strongly in a first phase. After a few rotations of the electric field, they then stabilize at their asymptotic values. The average energy stabilizes to approximately a_0, the energy spread to the theoretical value $\hat{\sigma}_{\max}$, and the third order moment μ_3 to approximately 0. In the case of the initially broad MJ distribution, its initial average energy, energy spread, and third order moment are bigger than their asymptotic values. They therefore all decrease to reach, similarly to the case of the initially cold distribution, a_0 for the average energy, $\hat{\sigma}_{\max}$ for the energy spread, and 0 for μ_3. In both cases, the asymptotic electron distribution function is a gaussian in energy. As χ increases, the time it takes for the electron distribution to reach its asymptotic stationary state decreases, and the number of oscillations as well. We then consider the case where $\chi > 1$. We run a set of simulations, first without taking into account pair production, in order to compare with the previous case. In this situation, the average energy remains close to that of the cLL model, while the energy spread is no longer given by the theoretical $\hat{\sigma}_{\max}$. The third order moment is no longer equal to 0. For large values of χ, pair production [by nonlinear Breit-Wheeler pair production (NBWPP)] has to be taken into account. We find that the asymptotic average energy is close to the case without pairs, while the asymptotic energy spread is larger than the case without pairs.

In Chap. 9 we focus on electromagnetic field configurations that can be produced in the laboratory, in particular, having a finite size and duration. We then turn our study to the so-called Laguerre-Gaussian (LG) beams. These beams have recently received a great attention from the physics community. They indeed carry orbital angular momentum (OAM) which is believed to play a role in some astrophysical phenomena, such as the radiation from the accretion disk around Kerr black holes or the radiation by rotating pulsars or quasars. In this chapter, we derive them from the paraxial equation and review their main properties such as their helical phase and orbital angular momentum. We examine the electromagnetic field of LG beams as well as their phase and see how it differs from simple gaussian beams.

Finally, in Chap. 10, we examine the possibility of producing pairs by colliding LG beams with high-energy photon beams. Such a configuration produces a so-called electromagnetic shower. We first analyze in depth the properties of the LG beams, and compute in particular their effective transverse size, the radius of the maximum of the field, and the value of this maximum amplitude, for a beam fixed energy. We then estimate analytically the total rate of pairs production by nonlinear Breit-Wheeler (nBW) for LG beams and compare it to that of a pure gaussian beam, both for linear and circular polarization. We find that the nBW production rate increases with the orbital angular momentum of the LG beam. This effect is purely geometrical and due to the increase of the transverse effective size of the field. We confirm this effect by comparing the pair yields of gaussian beams with different transverse sizes. We then perform full 3D PIC simulations in order to verify our results. We find the number of pairs increases with increasing OAM, but less quickly than by directly increasing the waist of a gaussian beam (therefore decreasing its amplitude). We therefore obtain the counter-intuitive conclusion that the number of created pairs increases with decreasing intensity at fixed energy (as long as the photon quantum parameter remains larger than a threshold value).

Appendix A
Classical Electrodynamics

A.1 Bessel Functions

Bessel functions are solutions of the differential equation

$$x^2\, y'' + x\, y' + (x^2 - \alpha^2)\, y = 0\,, \tag{A.1}$$

where the prime denotes the differentiation with respect to x and where α is any complex value. This equation has many solutions which will be called differently depending on their behavior.

A.1.1. Bessel Functions of the First Kind

Bessel functions of the first kind, denoted $J_\alpha(x)$, are solutions of Eq. (A.1) that are finite at $x = 0$ for integer or positive values of α and diverge for $x \to 0$ for negative non-integer values of α. In this work, we will mainly use their integral representation for $\alpha = n$ where n is an integer

$$J_n(x) = \frac{1}{2\pi} \int_{-\pi}^{\pi} e^{i(n\tau - x\sin\tau)} d\tau\,. \tag{A.2}$$

From this representation, we can derive the so-called *Jacobi-Anger expansion*

$$e^{iz\cos\theta} \equiv \sum_{n=-\infty}^{+\infty} i^n\, J_n(z) e^{in\theta}\,. \tag{A.3}$$

F. Niel, *Classical and Quantum Description of Plasma and Radiation in Strong Fields*,
Springer Theses, https://doi.org/10.1007/978-3-030-73547-0

Substituting θ by $\theta - \pi/2$, we get

$$e^{iz\sin\theta} \equiv \sum_{n=-\infty}^{+\infty} J_n(z)e^{in\theta} , \tag{A.4}$$

and in particular, we will use

$$e^{i[\psi+l\phi]} = e^{i\psi_0} \sum_{m,n=-\infty}^{+\infty} J_m(\psi_3)\, J_{n-2m+l}(\psi_2)\, e^{i[\psi_1-n]\phi} , \tag{A.5a}$$

$$\psi(\phi) = \psi_0 + \psi_1\phi - \psi_2 \sin\phi - \psi_3 \sin 2\phi . \tag{A.5b}$$

and

$$e^{i[\psi+l(\phi-\varphi)]} = e^{i\psi_0} \sum_{n=-\infty}^{+\infty} J_{n+l}(\psi_2)\, e^{i[\psi_1-n]\phi} e^{in\varphi} , \tag{A.6a}$$

$$\psi(\phi) = \psi_0 + \psi_1\phi - \psi_2 \sin(\phi - \varphi) . \tag{A.6b}$$

Finally, the following recurrence relations will be useful

$$2\, J_n'(x) = J_{n-1}(x) - J_{n+1}(x) , \tag{A.7a}$$

$$\frac{2n}{x} J_n(x) = J_{n+1} + J_{n-1}(x) . \tag{A.7b}$$

Let us note that J_n satisfies

$$J_n(-x) = J_{-n}(x) = (-1)^n J_n(x) . \tag{A.8}$$

A.1.2 Modified Bessel Functions of the Second Kind

In the case where the argument x is purely imaginary, solutions of Eq. (A.1) are called *modified Bessel functions*. In particular, we consider here the modified Bessel functions of the second kind, denoted $K_\alpha(x)$. In this work, modified Bessel functions of the second kind will mainly be used to obtain analytical approximate forms of Bessel functions of the first kind. In particular, for $l \gg 1$

$$J_l(lz) \simeq \frac{\sqrt{x}}{\pi} \frac{1}{(1-z^2)^{1/4}} K_{1/3}(lx) , \tag{A.9a}$$

$$J_l'(lz) \simeq -\frac{\sqrt{x}}{\pi z}(1-z^2)^{1/4} K_{2/3}(lx) , \tag{A.9b}$$

where $x = \ln\left[1+\sqrt{1-z^2}\right] - \ln z - \sqrt{1-z^2}$. In particular, for $\sqrt{1-z^2} \ll 1$, we have at first order $x \simeq \frac{1}{3}(1-z^2)^{3/2}$. For $lx \gg 1$, we have

$$K_{1/3}(lx) \simeq K_{2/3}(lx) \simeq \frac{\pi}{2lx}\exp(-lx)\,. \tag{A.10}$$

A.2 Airy Functions

Airy functions are solutions of the differential equation

$$y'' - x\,y = 0\,, \tag{A.11}$$

that satisfy $y \to 0$ as $x \to +\infty$, denoted Ai(x) (the prime denotes the differentiation with respect to x). Throughout this work, we will mainly use the integral representation

$$\mathrm{Ai}(x) = \frac{1}{\pi}\int_0^{+\infty} dt\,\cos\left(xt+\frac{t^3}{3}\right) = \frac{1}{2\pi}\int_{-\infty}^{+\infty} dt\,\exp\left[-i\left(xt+\frac{t^3}{3}\right)\right]. \tag{A.12}$$

Its derivative is given by

$$\mathrm{Ai}'(x) = \frac{1}{\pi}\int_0^{+\infty} dt\,t\,\sin\left(xt+\frac{t^3}{3}\right) = \frac{1}{2i\pi}\int_{-\infty}^{+\infty} dt\,t\,\exp\left[-i\left(xt+\frac{t^3}{3}\right)\right]. \tag{A.13}$$

In Sect. B.5, we will use instead the following integral representations

$$\frac{1}{\sqrt[3]{3a}}\mathrm{Ai}\left(\pm\frac{x}{3a}\right) = \frac{1}{2\pi}\int_{-\infty}^{+\infty} dt\,\exp\left[i\left(at^3 \pm xt\right)\right] \tag{A.14}$$

It will be useful to link Airy functions to modified Bessel functions of the second kind, in particular, we have

$$\mathrm{Ai}(x) = \frac{1}{\pi}\sqrt{\frac{x}{3}}\,K_{1/3}\left(\frac{2}{3}x^{3/2}\right), \tag{A.15a}$$

$$\mathrm{Ai}'(x) = -\frac{x}{\pi\sqrt{3}}\,K_{2/3}\left(\frac{2}{3}x^{3/2}\right). \tag{A.15b}$$

A.3 Tensor Identities

A.3.1 Levi-Civita Symbols

$$\epsilon^{\alpha\beta\gamma\delta}\epsilon_{\alpha\beta\gamma\delta} = -4!\,, \tag{A.16a}$$

$$\epsilon^{\alpha\beta\gamma\mu}\epsilon_{\alpha\beta\gamma\nu} = -3!\,\delta^\mu_\nu\,, \tag{A.16b}$$

$$\epsilon^{\alpha\beta\mu\rho}\epsilon_{\alpha\beta\nu\sigma} = -2!\left(\delta^\mu_\nu\delta^\rho_\sigma - \delta^\mu_\sigma\delta^\rho_\nu\right), \tag{A.16c}$$

$$\begin{aligned}\epsilon^{\mu\nu\rho\sigma}\epsilon_{\alpha\beta\gamma\sigma} = -\Big(&\delta^\mu_\alpha\delta^\nu_\beta\delta^\rho_\gamma - \delta^\mu_\alpha\delta^\nu_\gamma\delta^\rho_\beta + \delta^\mu_\gamma\delta^\nu_\alpha\delta^\rho_\beta \\ &- \delta^\mu_\gamma\delta^\nu_\beta\delta^\rho_\alpha + \delta^\mu_\beta\delta^\nu_\gamma\delta^\rho_\alpha - \delta^\mu_\beta\delta^\nu_\alpha\delta^\rho_\gamma\Big)\,,\end{aligned} \tag{A.16d}$$

$$\epsilon^{\mu\nu\rho\sigma}\epsilon_{\alpha\beta\gamma\delta} = -\det\begin{pmatrix}\delta^\mu_\alpha & \delta^\mu_\beta & \delta^\mu_\gamma & \delta^\mu_\delta \\ \delta^\nu_\alpha & \delta^\nu_\beta & \delta^\nu_\gamma & \delta^\nu_\delta \\ \delta^\rho_\alpha & \delta^\rho_\beta & \delta^\rho_\gamma & \delta^\rho_\delta \\ \delta^\sigma_\alpha & \delta^\sigma_\beta & \delta^\sigma_\gamma & \delta^\sigma_\delta\end{pmatrix}, \tag{A.16e}$$

$$\epsilon^{ijk}\epsilon_{ijk} = 3!\,, \tag{A.16f}$$

$$\epsilon^{imn}\epsilon_{jmn} = 2!\,\delta^j_i\,, \tag{A.16g}$$

$$\epsilon^{ijk}\epsilon_{imn} = \delta^m_j\delta^n_k - \delta^n_j\delta^m_k\,. \tag{A.16h}$$

A.3.2 Constant Field Tensor Identities

The constant field tensor is defined as $\mathfrak{f}_i^{\mu\nu} = k^\mu a_i^\nu - k^\nu a_i^\mu$ and its dual as $\mathfrak{f}_i^{*\mu\nu} = \frac{1}{2}\epsilon^{\mu\nu\rho\sigma}\mathfrak{f}_{i,\rho\sigma}$. Then, the following results hold

$$\mathfrak{f}_i^{\mu\rho}\mathfrak{f}^\nu_{j,\rho} = \mathfrak{f}_i^{*\mu\rho}\mathfrak{f}^{*\nu}_{j,\rho} = -k^\mu k^\nu a_i^2\delta_{ij}\,, \tag{A.17a}$$

$$k_\mu\mathfrak{f}_i^{\mu\nu} = k_\mu\mathfrak{f}_i^{*\mu\nu} = 0\,, \tag{A.17b}$$

$$\mathfrak{f}_i^{2\mu\nu} = \mathfrak{f}_i^{\mu\rho}\mathfrak{f}^\nu_{i,\rho} = -k^\mu k^\nu a_i^2\,, \tag{A.17c}$$

$$\mathfrak{f}_i^{\mu\nu}a_{j,\nu} = \delta_{ij}a_i^2 k^\mu\,, \tag{A.17d}$$

$$\mathfrak{f}_i^{\mu\nu}u_\nu = (a_i u)\,k^\mu - (ku)a_i^\mu\,, \tag{A.17e}$$

$$\mathfrak{f}_i^{\mu\nu}\mathfrak{f}_{j,\nu\eta}u^\eta = -(ku)\delta_{ij}a_i^2 k^\mu\,, \tag{A.17f}$$

$$(\mathfrak{f}_i^{\alpha\nu}u_\nu)(\mathfrak{f}_i^{\alpha\lambda}u_\lambda) = (ku)^2 a_i^2\,. \tag{A.17g}$$

Defining $f^{\mu\nu} = \sum_{i=1,2} \mathfrak{f}_i^{\mu\nu} \psi_i'(\phi)$, this implies that

$$f_{\eta\nu}\, u^\eta = \sum_{i=1,2} \psi_i' \left[(ku) a_{i,\nu} - k_\nu (a_i u) \right] , \tag{A.18a}$$

$$f^{2\mu\nu} = - \sum_{i=1,2} a_i^2 \psi_i'^2\, k^\mu k^\nu , \tag{A.18b}$$

$$f^{\mu\nu}\, f_{\eta\nu}\, u^\eta = (ku) \sum_{i=1,2} a_i^2 \psi_i'^2 k^\mu , \tag{A.18c}$$

$$(f^{\nu\eta}\, u_\eta)\, (f_{\nu\alpha}\, u^\alpha) = (ku)^2 \sum_{i=1,2} a_i^2 \psi_i'^2 , \tag{A.18d}$$

$$f^{\mu\nu}(\phi)\, f_{\nu\eta}(\phi') = - \sum_{i=1,2} \psi_i'(\phi) \psi_i'(\phi') a_i^2 k^\mu k_\eta , \tag{A.18e}$$

$$f^{\mu\nu}(\phi)\, f_{\nu\eta}(\phi')\, f_{\mu\sigma}(\phi'') = f^{\mu\nu}(\phi)\, f_{\nu\eta}(\phi')\, f^{\eta\sigma}(\phi'') = 0 , \tag{A.18f}$$

Finally, writing,

$$\mathcal{F}^{\mu\nu}(\phi, \phi_0) \equiv \int_{\phi_0}^{\phi} d\phi' f^{\mu\nu}(\phi') , \tag{A.19}$$

$$\mathcal{F}^{\mu\nu}(\phi) \equiv \mathcal{F}^{\mu\nu}(\phi, -\infty) , \tag{A.20}$$

we have

$$\begin{aligned} \mathcal{F}^{\mu\nu}(\phi, \phi_0) &= k^\mu \left[a^\nu(\phi) - a^\nu(\phi_0) \right] - k^\nu \left[a^\mu(\phi) - a^\mu(\phi_0) \right] \\ &= \sum_{i=1,2} \mathfrak{f}_i^{\mu\nu} \left[\psi_i(\phi) - \psi_i(\phi_0) \right] , \end{aligned} \tag{A.21a}$$

$$\begin{aligned} \mathcal{F}^{\mu\nu}(\phi) &= k^\mu a^\nu(\phi) - k^\nu a^\mu(\phi) \\ &= \sum_{i=1,2} \mathfrak{f}_i^{\mu\nu} \psi_i(\phi) , \end{aligned} \tag{A.21b}$$

$$\mathcal{F}^{2\mu\nu}(\phi, \phi_0) = - \sum_{i=1,2} a_i^2 \left[\psi_i(\phi) - \psi_i(\phi_0) \right]^2 k^\mu k^\nu , \tag{A.21c}$$

$$\mathcal{F}^{2\mu\nu}(\phi) = - \sum_{i=1,2} a_i^2 \psi_i^2(\phi)\, k^\mu k^\nu . \tag{A.21d}$$

A.4 Light-Cone Coordinates

The light-cone (or also called light-front) coordinates of any 4-vector a^μ are defined by

$$a_- \equiv na , \qquad a_+ \equiv \bar{n}a , \tag{A.22}$$

where $n^\mu = (1, \boldsymbol{n}) = k^\mu/\omega_0$ and $\bar{n}^\mu = (1, -\boldsymbol{n})$. In the case of a plane-wave propagating in the $\hat{z}$ direction, $\boldsymbol{n} = \hat{z}$ and we have

$$a^- = a^0 - a^3 , \qquad a^+ = a^0 + a^3 , \tag{A.23}$$

while the perpendicular components $a^\perp = (a^1, a^2)$ are unchanged. The usual coordinates of a are therefore given by

$$a^0 = \frac{1}{2}(x^+ + x^-) , \qquad a^3 = \frac{1}{2}(x^+ - x^-) , \tag{A.24}$$

from which we deduce the metric tensor in light-cone coordinates

$$(\hat{\eta}_{\mu\nu}) = \begin{pmatrix} 0 & 0 & 0 & \frac{1}{2} \\ 0 & -1 & 0 & 0 \\ 0 & 0 & -1 & 0 \\ \frac{1}{2} & 0 & 0 & 0 \end{pmatrix} , \quad (\hat{\eta}^{\mu\nu}) = \begin{pmatrix} 0 & 0 & 0 & 2 \\ 0 & -1 & 0 & 0 \\ 0 & 0 & -1 & 0 \\ 2 & 0 & 0 & 0 \end{pmatrix} , \tag{A.25}$$

The covariant components of a vector in light-cone coordinates are therefore related to its contravariant components as

$$a_- = \frac{1}{2}a^+ , \quad a_+ = \frac{1}{2}a^- , \quad \boldsymbol{a}_\perp = -\boldsymbol{a}^\perp , \tag{A.26}$$

while the Minkowski scalar product reads

$$xy = x^+ y_+ + x^- y_- + \boldsymbol{x}_\perp \cdot \boldsymbol{y}^\perp = \frac{1}{2}x^+ y^- + \frac{1}{2}x^- y^+ - \boldsymbol{x}_\perp \cdot \boldsymbol{y}_\perp . \tag{A.27}$$

The integration measure is then

$$d^4x = dx_- \, dx_+ \, d^2\boldsymbol{x}_\perp . \tag{A.28}$$

A.5 Solution of the Equation of Motion of a Charged Particle in a Plane-Wave Electromagnetic Field Without RR

We start from Newton-Lorentz equation (2.30) with the expression (2.74) for the electromagnetic strength tensor. Changing the derivation with respect to the proper time by a derivation with respect to the phase of the field ϕ, we get

$$\frac{dp^\mu}{d\phi} = \frac{q}{(kp_0)} \sum_{i=1,2} \psi_i'(\phi)(a_i p)\, k^\mu - q \sum_{i=1,2} \psi_i'(\phi)\, a_i^\mu . \tag{A.29}$$

We now need to find an expression for $(a_i p)$ in terms of ϕ. This is done by multiplying (A.29) by a_i and integrating over ϕ. This yields

$$a_i p = a_i p_0 - q a_i^2 \left[\psi_i(\phi) - \psi_i(\phi_0)\right] . \tag{A.30}$$

Inserting this expression into (A.29) and integrating over ϕ gives the desired solution

$$p^\mu(\phi) = p_0^\mu - q A^\mu(\phi) + \frac{q\,(A(\phi)p_0)}{kp_0} k^\mu - \frac{q^2 A(\phi)^2}{2kp_0} k^\mu . \tag{A.31}$$

Writing

$$a^\mu(\phi) = a(0, \delta\cos(\phi), \sqrt{1-\delta^2}\sin(\phi), 0) , \quad u^\mu = \gamma(1, \boldsymbol{\beta}) , \quad u_0^\mu = \gamma_0(1, \boldsymbol{\beta}_0) ,$$
$$k^\mu = \omega_0(1, \hat{\boldsymbol{k}}) , \quad x^\mu = (t, \boldsymbol{x}) , \quad \phi = \omega(t - \hat{\boldsymbol{k}}\cdot\boldsymbol{x}) , \quad ku_0 = \gamma_0\omega_0(1 - \hat{\boldsymbol{k}}\cdot\boldsymbol{\beta}_0) ,$$

where $a_0 = eA/m$, we find the 3-vector components of (A.31)

$$\gamma(\phi) = \gamma_0 - \frac{\mathbf{a}(\phi)\cdot\boldsymbol{\beta}_0}{1 - \hat{\boldsymbol{k}}\cdot\boldsymbol{\beta}_0} + \frac{\mathbf{a}(\phi)^2}{2\gamma_0(1 - \hat{\boldsymbol{k}}\cdot\boldsymbol{\beta}_0)} , \tag{A.32a}$$

$$\mathbf{u}(\phi) = \mathbf{u}_0 - \mathbf{a}(\phi) - \frac{\mathbf{a}(\phi)\cdot\boldsymbol{\beta}_0}{1 - \hat{\boldsymbol{k}}\cdot\boldsymbol{\beta}_0}\hat{\boldsymbol{k}} + \frac{\mathbf{a}(\phi)^2}{2\gamma_0(1 - \hat{\boldsymbol{k}}\cdot\boldsymbol{\beta}_0)}\hat{\boldsymbol{k}} , \tag{A.32b}$$

from which we deduce

$$\begin{aligned}\mathbf{x}(\phi) = \mathbf{x}_0 &+ \int_{\phi_0}^{\phi} d\phi' \frac{\mathbf{u}_0 - \mathbf{a}(\phi')}{\gamma_0\omega_0(1 - \hat{\boldsymbol{k}}\cdot\boldsymbol{\beta}_0)} \\ &- \int_{\phi_0}^{\phi} d\phi' \left[\frac{\mathbf{a}(\phi')\cdot\boldsymbol{\beta}_0}{\gamma_0\omega_0(1 - \hat{\boldsymbol{k}}\cdot\boldsymbol{\beta}_0)^2} - \frac{\mathbf{a}(\phi')^2}{2\gamma_0^2\omega_0(1 - \hat{\boldsymbol{k}}\cdot\boldsymbol{\beta}_0)^2}\right]\hat{\boldsymbol{k}} \quad .\end{aligned} \tag{A.33}$$

More explicitly, we parametrize $\boldsymbol{\beta}_0$ as $\boldsymbol{\beta}_0 = \beta_0(\sin\theta_0\cos\varphi_0, \sin\theta_0\sin\varphi_0, \cos\theta_0)$, which allows us to rewrite in 3-vector notation

$$\gamma(\phi) = a_\gamma + b_\gamma\cos\phi + c_\gamma\sin\phi + d_\gamma\cos 2\phi , \tag{A.34a}$$

$$\boldsymbol{u}(\phi) = \boldsymbol{a}_u + \boldsymbol{b}_u\cos\phi + \boldsymbol{c}_u\sin\phi + \boldsymbol{d}_u\cos 2\phi , \tag{A.34b}$$

where

$$a_\gamma = \gamma_0 + \frac{a_0^2}{4\gamma_0(1-\beta_0\cos\theta_0)}\,, \tag{A.35a}$$

$$b_\gamma = -\frac{a_0\beta_0\delta\cos\varphi_0\sin\theta_0}{1-\beta_0\cos\theta_0}\,, \tag{A.35b}$$

$$c_\gamma = -\frac{a_0\beta_0\sqrt{1-\delta^2}\sin\varphi_0\sin\theta_0}{1-\beta_0\cos\theta_0}\,, \tag{A.35c}$$

$$d_\gamma = \frac{a_0^2(\delta^2-1/2)}{2\gamma_0(1-\beta_0\cos\theta_0)}\,, \tag{A.35d}$$

and

$$\boldsymbol{a}_u = \gamma_0\boldsymbol{\beta}_0 + \frac{a_0^2}{4\gamma_0(1-\beta_0\cos\theta_0)}\,\hat{\boldsymbol{k}}\,, \tag{A.36a}$$

$$\boldsymbol{b}_u = -a_0\,\delta\,\boldsymbol{a}_1 - \frac{a_0\beta_0\delta\cos\varphi_0\sin\theta_0}{1-\beta_0\cos\theta_0}\,\hat{\boldsymbol{k}}\,, \tag{A.36b}$$

$$\boldsymbol{c}_u = -a_0\sqrt{1-\delta^2}\,\boldsymbol{a}_2 - \frac{a_0\beta_0\sqrt{1-\delta^2}\sin\varphi_0\sin\theta_0}{1-\beta_0\cos\theta_0}\,\hat{\boldsymbol{k}}\,, \tag{A.36c}$$

$$\boldsymbol{d}_u = \frac{a_0^2(\delta^2-1/2)}{2\gamma_0(1-\beta_0\cos\theta_0)}\,\hat{\boldsymbol{k}}\,. \tag{A.36d}$$

Writing the integration constant $\mathbf{x_0}$, we get

$$\boldsymbol{x}(\phi) = \mathbf{x_0} + \frac{1}{ku_0}\left[\boldsymbol{c}_u + \boldsymbol{a}_u\phi - \boldsymbol{c}_u\cos\phi + \boldsymbol{b}_u\sin\phi + \frac{\boldsymbol{d}_u}{2}\sin 2\phi\right]. \tag{A.37}$$

A.6 Non-linear Thomson Scattering

A.6.1 Circularly Polarized Plane-Wave

From Eqs. (2.89a), we deduce

$$\frac{dx(\phi)}{d\phi} = x_2\cos\phi\,, \tag{A.38a}$$

$$\frac{dy(\phi)}{d\phi} = x_2\sin\phi\,, \tag{A.38b}$$

$$\frac{dz(\phi)}{d\phi} = z_1\,, \tag{A.38c}$$

and

$$\psi(\phi) = \psi_0 + \psi_1\phi - \psi_2 \sin(\phi - \varphi) , \tag{A.39}$$

with

$$\psi_0 = -\frac{\omega}{\omega_0} [x_0 \sin\theta \cos\varphi + y_0 \sin\theta \sin\varphi + z_0(1 + \cos\theta)] , \tag{A.40a}$$

$$\psi_1 = \frac{\omega}{\omega_0} [1 - z_1(1 + \cos\theta)] , \tag{A.40b}$$

$$\psi_2 = \frac{\omega}{\omega_0} x_2 \sin\theta . \tag{A.40c}$$

We then have

$$K_x = x_2 e^{i\psi_0} \int_{-\eta}^{\eta} d\phi \cos\phi \, e^{i[\psi_1\phi - \psi_2 \sin(\phi-\varphi)]} , \tag{A.41a}$$

$$K_y = x_2 e^{i\psi_0} \int_{-\eta}^{\eta} d\phi \sin\phi \, e^{i[\psi_1\phi - \psi_2 \sin(\phi-\varphi)]} , \tag{A.41b}$$

$$K_z = z_1 e^{i\psi_0} \int_{-\eta}^{\eta} d\phi \, e^{i[\psi_1\phi - \psi_2 \sin(\phi-\varphi)]} . \tag{A.41c}$$

We can then rewrite

$$\cos\varphi K_x + \sin\varphi K_y = x_2 e^{i\psi_0} \int_{-\eta}^{\eta} d\phi \cos(\phi - \varphi) \, e^{i[\psi_1\phi - \psi_2 \sin(\phi-\varphi)]} , \tag{A.42a}$$

$$\sin\varphi K_x - \cos\varphi K_y = -x_2 e^{i\psi_0} \int_{-\eta}^{\eta} d\phi \sin(\phi - \varphi) \, e^{i[\psi_1\phi - \psi_2 \sin(\phi-\varphi)]} . \tag{A.42b}$$

Using Eq. (A.6a), we get

$$\cos\varphi K_x + \sin\varphi K_y = 2 x_2 \eta \sum_{n=-\infty}^{+\infty} e^{i(\psi_0 + n\varphi)} \frac{n}{\psi_2} J_n(\psi_2) R(\omega, n\omega_0) , \tag{A.43a}$$

$$\sin\varphi K_x - \cos\varphi K_y = 2i x_2 \eta \sum_{n=-\infty}^{+\infty} e^{i(\psi_0 + n\varphi)} J_n'(\psi_2) R(\omega, n\omega_0) , \tag{A.43b}$$

$$K_z = 2\eta z_1 \sum_{n=-\infty}^{+\infty} J_n(\psi_2) e^{i(\psi_0 + n\varphi)} R(\omega, n\omega_0) , \tag{A.43c}$$

where $R(\omega, n\omega_0)$ is the so-called resonance function, defined by

$$R(\omega, n\omega_0) = \frac{1}{2\eta} \int_{-\eta}^{\eta} d\phi \, e^{i[\psi_1 - n]\phi} = \frac{\sin[(\psi_1 - n)\eta]}{(\psi_1 - n)\eta} . \tag{A.44}$$

Inserting these expression in Eq. (2.96) and using the fact that the resonance function $R(\omega, n\omega_0)$ is strongly peaked at $\psi_1 = n$, we get

$$\left.\frac{d^2 I}{d\omega\, d\Omega}\right|_{\rm CP} = \frac{e^2\omega^2\eta^2}{\pi^2}\sum_{n=1}^{+\infty} R(\omega, n\omega_0)^2 \left[\left(\frac{\cos\theta - z_1(1+\cos\theta)}{\sin\theta}\right)^2 J_n(\tilde{\psi}_2)^2 + x_2^2\, J_n'(\tilde{\psi}_2)^2\right],$$

where

$$\tilde{\psi}_2 = \frac{n\, x_2\, \sin\theta}{1 - z_1(1+\cos\theta)}\,. \tag{A.45}$$

A.6.2 Linearly Polarized Plane-Wave

From Eqs. (2.85), we deduce

$$\frac{dx(\phi)}{d\phi} = x_1 \cos\phi\,, \tag{A.46a}$$

$$\frac{dy(\phi)}{d\phi} = 0\,, \tag{A.46b}$$

$$\frac{dz(\phi)}{d\phi} = z_1 + 2\, z_2 \cos 2\phi\,, \tag{A.46c}$$

and

$$\psi(\phi) = \psi_0 + \psi_1\phi - \psi_2 \sin\phi - \psi_3 \sin 2\phi\,, \tag{A.47}$$

with

$$\psi_0 = -\frac{\omega}{\omega_0}\left[x_0 \sin\theta\cos\varphi + y_0 \sin\theta\sin\varphi + z_0(1+\cos\theta)\right]\,, \tag{A.48a}$$

$$\psi_1 = \frac{\omega}{\omega_0}\left[1 - z_1(1+\cos\theta)\right]\,, \tag{A.48b}$$

$$\psi_2 = \frac{\omega}{\omega_0} x_1 \sin\theta\cos\varphi\,, \tag{A.48c}$$

$$\psi_3 = \frac{\omega}{\omega_0} z_2(1+\cos\theta)\,. \tag{A.48d}$$

We then have

$$K_x = x_1\, e^{i\psi_0} \int_{-\eta}^{\eta} d\phi\, \cos\phi\, e^{i[\psi_1\phi-\psi_2 \sin\phi-\psi_3 \sin 2\phi]}\,, \tag{A.49a}$$

$$K_y = 0\,, \tag{A.49b}$$

$$K_z = z_1\, e^{i\psi_0} \int_{-\eta}^{\eta} d\phi\, e^{i[\psi_1\phi-\psi_2 \sin\phi-\psi_3 \sin 2\phi]}$$

$$+\, 2z_2\, e^{i\psi_0} \int_{-\eta}^{\eta} d\phi\, \cos 2\phi\, e^{i[\psi_1\phi-\psi_2 \sin\phi-\psi_3 \sin 2\phi]}\,. \tag{A.49c}$$

Using Eq. (A.5a), we get

$$K_x = x_1\eta\, e^{i\psi_0} \sum_{m,n=-\infty}^{+\infty} R(\omega, n\omega_0)\, J_m(\psi_3)[J_{n-2m-1}(\psi_2) + J_{n-2m+1}(\psi_2)]\,, \tag{A.50a}$$

$$K_z = 2\, z_1\eta\, e^{i\psi_0} \sum_{m,n=-\infty}^{+\infty} R(\omega, n\omega_0)\, J_m(\psi_3) J_{n-2m}(\psi_2) \tag{A.50b}$$

$$+\, 2z_2\eta\, e^{i\psi_0} \sum_{m,n=-\infty}^{+\infty} R(\omega, n\omega_0)\, J_m(\psi_3)\left[J_{n-2m-2}(\psi_2) + J_{n-2m+2}(\psi_2)\right]\,,$$

where $R(\omega, n\omega_0)$ is the so-called resonance function, defined by

$$R(\omega, n\omega_0) = \frac{1}{2\eta}\int_{-\eta}^{\eta} d\phi\, e^{i[\psi_1-n]\phi} = \frac{\sin[(\psi_1-n)\eta]}{(\psi_1-n)\eta}\,. \tag{A.51}$$

The resonance function $R(\omega, n\omega_0)$ is strongly peaked in $\psi_1 = n$, so in the following, we make the approximation $\psi_1 \simeq n$

$$K_x = x_1\eta\, e^{i\psi_0} \sum_{m,n=-\infty}^{+\infty} R(\omega, n\omega_0)\,(-1)^m\, J_m(\tilde{\psi}_3)[J_{n-2m-1}(\tilde{\psi}_2) + J_{n-2m+1}(\tilde{\psi}_2)]\,, \tag{A.52a}$$

$$K_z = 2\, z_1\eta\, e^{i\psi_0} \sum_{m,n=-\infty}^{+\infty} R(\omega, n\omega_0)\,(-1)^m\, J_m(\tilde{\psi}_3) J_{n-2m}(\tilde{\psi}_2) \tag{A.52b}$$

$$+\, 2\, z_2\eta\, e^{i\psi_0} \sum_{m,n=-\infty}^{+\infty} R(\omega, n\omega_0)\,(-1)^m\, J_m(\tilde{\psi}_3)\left[J_{n-2m-2}(\tilde{\psi}_2) + J_{n-2m+2}(\tilde{\psi}_2)\right]\,,$$

where

$$\tilde{\psi}_2 = \frac{n x_1 \sin\theta \cos\varphi}{1 - z_1(1+\cos\theta)}\,, \tag{A.53a}$$

$$\tilde{\psi}_3 = \frac{-n z_2 (1+\cos\theta)}{1 - z_1(1+\cos\theta)}\,. \tag{A.53b}$$

Inserting Eq. (A.52) into Eq. (2.96) and knowing that the resonance functions are well separated for $n \neq n'$, we get

$$\frac{d^2 I}{d\omega\, d\Omega} = \frac{e^2\omega^2}{4\pi^2} \sum_{n=1}^{+\infty} R(\omega, n\omega_0)^2 \left[(1 - \sin^2\theta \cos^2\varphi)\tilde{K}_x^2 - \sin^2\theta \cos\varphi\, \tilde{K}_x \tilde{K}_z + \sin^2\theta\, \tilde{K}_z^2\right],$$

where

$$\tilde{K}_x = x_1\eta \sum_{m=-\infty}^{+\infty} (-1)^m J_m(\tilde{\psi}_3)\,[J_{n-2m-1}(\tilde{\psi}_2) + J_{n-2m+1}(\tilde{\psi}_2)], \tag{A.54}$$

$$\tilde{K}_z = 2\eta \sum_{m=-\infty}^{+\infty} (-1)^m J_m(\tilde{\psi}_3)\left[z_1 J_{n-2m}(\tilde{\psi}_2) + z_2\left(J_{n-2m-2}(\tilde{\psi}_2) + J_{n-2m+2}(\tilde{\psi}_2)\right)\right]. \tag{A.55}$$

For $\varphi = \pi/2$, and in the ultra-relativistic regime ($a_0 \gg 1$ and therefore $n \gg 1$ and $\theta \ll 1$), we have the following approximations

$$\tilde{K}_x^2 \simeq x_1^2\eta^2 J_l'^2(l\hat{\psi}_3), \tag{A.56}$$

$$\tilde{K}_z^2 \simeq \eta^2 J_l^2(l\hat{\psi}_3), \tag{A.57}$$

with $l = (n-1)/2$ and $\hat{\psi}_3 = \tilde{\psi}_3/l$. We have

$$\frac{d^2 I}{d\omega\, d\Omega} \simeq \frac{e^2\omega^2}{4\pi^2} \sum_{n=1}^{+\infty} R(\omega, n\omega_0)^2 \left[\tilde{K}_x^2 + \theta^2\,\tilde{K}_z^2\right]. \tag{A.58}$$

We have $1 - 2z_1 = (1 + a_0^2/2)/h_0^2$ and $-4z_2/(1 - 2z_1) \simeq 1 - 2/a_0^2$ and so

$$\hat{\psi}_3 \simeq 1 - \frac{2}{a_0^2}\left(1 + \frac{h_0^2}{4}\theta^2\right), \tag{A.59}$$

Using (A.9a), we get

$$J_l^2(l\hat{\psi}_3) \simeq \frac{1}{3\pi^2}\left(1 - \hat{\psi}_3^{\,2}\right) K_{1/3}^2(\xi), \tag{A.60}$$

$$J_l'^2(l\hat{\psi}_3) \simeq \frac{1}{3\pi^2}\frac{\left(1 - \hat{\psi}_3^{\,2}\right)^2}{\hat{\psi}_3^{\,2}} K_{2/3}^2(\xi), \tag{A.61}$$

where $\xi = \frac{n}{n_c}\left(1 + \frac{h_0^2}{4}\theta^2\right)$ and $n_c = 3a_0^3/4$.

A.7 Regularized Self-field

We will compute the regularized self-potential and field defined as

$$A^\mu_\epsilon(x) = \frac{e}{2\pi}\int_{-\infty}^{+\infty} d\tau\,\theta(x_0 - x'_0)\,\delta\left[(x-x')^2 - \epsilon^2\right] u^\mu(\tau)\,, \tag{A.62}$$

and $F^{\mu\nu}_\epsilon(x) = 2\partial^{[\mu}A^{\nu]}_\epsilon(x)$. Choosing the origin of coordinates such as $x^\mu = r^\mu(0)$, we have

$$r^\mu(\tau) = x^\mu + u^\mu_0\tau + a^\mu_0\frac{\tau^2}{2} + O(\tau^3)\,, \tag{A.63}$$

$$(x - r(\tau))^2 = \tau^2 + O(\tau^3)\,. \tag{A.64}$$

Inserting these expression into Eq. (A.62) and computing the integral yields

$$A^\mu_\epsilon[r(0)] = \frac{e}{4\pi}\frac{u^\mu(-\epsilon)}{\epsilon} = \frac{e}{4\pi}\left(\frac{u^\mu(0)}{\epsilon} - a(0)^\mu\right)\,. \tag{A.65}$$

Generalized to any proper time τ, we get

$$A^\mu_\epsilon[r(\tau)] = \frac{e}{4\pi}\left(\frac{u^\mu(\tau)}{\epsilon} - a^\mu(\tau)\right) + O(\epsilon)\,, \tag{A.66}$$

from which we deduce

$$\partial^\mu A^\nu_\epsilon[r(\tau)] = \frac{e}{2\pi}\int_{-\infty}^{0} d\tau\,\delta[(x - r(\tau))^2 - \epsilon^2]\,N^{\mu\nu}(\tau)\,, \tag{A.67}$$

where $N^{\mu\nu}(\tau) = \frac{dL^{\mu\nu}(\tau)}{d\tau}$ and $L^{\mu\nu}(\tau) = \frac{[x^\mu - r^\mu(\tau)]v^\nu(\tau)}{[x - r(\tau)]v(\tau)}$. Taking the antisymmetric part of Eq. (A.67) and performing the integration, we get

$$F^{\mu\nu}_\epsilon[r(\tau)] = \frac{e}{2\pi}\frac{N^{[\mu\nu]}(-\epsilon)}{\epsilon}\,. \tag{A.68}$$

Expanding $r^\mu(\tau)$ and $v^\mu(\tau)$ around $\tau = 0$ yields

$$-\left[x^\mu - r^\mu(\tau)\right]v^\nu(\tau) = A^{\mu\nu}\tau + B^{\mu\nu}\tau^2 + C^{\mu\nu}\tau^3 + D^{\mu\nu}\tau^4 + O(\tau^5)\,, \tag{A.69}$$

with

$$A^{\mu\nu} = u_0^\mu u_0^\nu\,, \tag{A.70a}$$

$$B^{\mu\nu} = u_0^\mu a_0^\nu + \frac{1}{2} a_0^\mu u_0^\nu\,, \tag{A.70b}$$

$$C^{\mu\nu} = \frac{1}{2} u_0^\mu \dot{a}_0^\nu + \frac{1}{2} a_0^\mu a_0^\nu + \frac{1}{6} \dot{a}_0^\mu u_0^\nu\,, \tag{A.70c}$$

$$D^{\mu\nu} = \frac{1}{6} u_0^\mu \ddot{a}_0^\nu + \frac{1}{4} a_0^\mu \dot{a}_0^\nu + \frac{1}{6} \dot{a}_0^\mu a_0^\nu + \frac{1}{24} \ddot{a}_0^\mu u_0^\nu\,. \tag{A.70d}$$

where $\dot{}$ denotes the derivation with respect to the proper time τ. Contracting the two indices yields [and using Eq. (2.32a) and (2.32b)], we get

$$-\left[x - r(\tau)\right] v(\tau) = \tau + \frac{u_0 \dot{a}_0}{6}\tau^3 - \frac{5\dot{a}_0^2}{24}\tau^4 + O(\tau^5)\,. \tag{A.71}$$

We can therefore deduce $L^{\mu\nu}(\tau)$,

$$L^{\mu\nu}(\tau) = A^{\mu\nu} + B^{\mu\nu}\tau + \tau^2\left[C^{\mu\nu} - \frac{u_0 \dot{a}_0}{6} A^{\mu\nu}\right] + O(\tau^3)\,, \tag{A.72}$$

which after differentiation yields

$$N^{\mu\nu}(\tau) = B^{\mu\nu} + 2\tau\left[C^{\mu\nu} - \frac{u_0 \dot{a}_0}{6} A^{\mu\nu}\right] + O(\tau^2)\,, \tag{A.73}$$

whose anti-symmetric part is

$$N^{[\mu\nu]}(\tau) = \frac{1}{4} u_0^{[\mu} a_0^{\nu]} + \frac{\tau}{3} u_0^{[\mu} \dot{a}_0^{\nu]} + O(\epsilon^2)\,. \tag{A.74}$$

The self-field is finally given by

$$F^{\mu\nu}[r(\tau)] = \frac{e}{4\pi}\left[-\frac{1}{2\epsilon} a^{[\mu} u^{\nu]} + \frac{2}{3} \dot{a}^{[\mu} u^{\nu]}\right] + O(\epsilon)\,, \tag{A.75}$$

where we dropped the "0" subscript.

A.8 Solution of the Equation of Motion of a Charged Particle in a Plane-Wave Electromagnetic Field with RR

We consider the LL equation

$$a^\mu = f^{\mu\nu} u_\nu + \tau_0 \left[\partial_\eta f^{\mu\nu} u_\nu u^\eta - f^{\mu\nu} f_{\eta\nu} u^\eta + (f^{\nu\eta} u_\eta)(f_{\nu\alpha} u^\alpha) u^\mu\right].$$

We follow the same steps as for the situation without RR and look for ku. Using (2.75), we get

$$\frac{d(ku)}{d\tau} = \tau_0 \sum_{i=1,2} a_i^2 \psi_i'^2(\phi)\,(ku)^3\,, \tag{A.76}$$

which, using $d\tau/d\phi = (ku)^{-1}$, yields

$$-\frac{1}{(ku)^2}\frac{d(ku)}{d\phi} = -\tau_0 \sum_{i=1,2} a_i^2 \psi_i'^2(\phi)\,. \tag{A.77}$$

This readily integrates into $ku(\phi) = \frac{ku_0}{h(\phi,\phi_0)}$ with

$$h(\phi,\phi_0) = 1 - \tau_0\,\rho_0 \sum_{i=1,2} \int_{\phi_0}^{\phi} d\varphi\, a_i^2 \psi_i'^2(\varphi)\,, \tag{A.78}$$

and where $\rho_0 = ku_0$. We define the "reduced" 4-velocity as $\tilde{u}^\mu(\phi) \equiv h(\phi,\phi_0)u^\mu(\phi)$. Its equation of evolution is then given by $\tilde{u}'(\phi) = h'(\phi,\phi_0)u^\mu(\phi) + h(\phi,\phi_0)u'^\mu(\phi)$ with

$$\begin{aligned} hu'^\mu &= \frac{1}{\rho_0} h\, f^{\mu\nu}\tilde{u}_\nu + \tau_0 f'^{\mu\nu}\tilde{u}_\nu \\ &\quad - h\,\tau_0 \sum_{i=1,2} a_i^2 \psi_i'^2 k^\mu + \rho_0\tau_0 \sum_{i=1,2} a_i^2 \psi_i'^2 u^\mu\,, \end{aligned} \tag{A.79}$$

$$h'u^\mu = -\rho_0\tau_0 \sum_{i=1,2} a_i^2 \psi_i'^2 u^\mu\,, \tag{A.80}$$

where we dropped the dependencies in ϕ for clarity. We therefore get

$$\frac{d\tilde{u}^\mu}{d\phi} = \frac{1}{\rho_0}\left[h\, f^{\mu\nu} + \tau_0\rho_0\, f'^{\mu\nu}\right]\tilde{u}_\nu - h\,\tau_0 \sum_{i=1,2} a_i^2 \psi_i'^2 k^\mu\,, \tag{A.81}$$

where the prime denotes the derivation with respect to ϕ. Using Picard's method and the orthogonality properties of $f^{\mu\nu}$, we get after two iterations the exact solution

$$
\begin{aligned}
\tilde{u}^{\mu}(\phi) &= u_0^{\mu} + \frac{1}{\rho_0}\int_{\phi_0}^{\phi} d\varphi \left[h(\varphi,\phi_0) f^{\mu\nu}(\varphi) + \tau_0\rho_0 f'^{\mu\nu}(\varphi)\right] u_{0,\nu} \\
&+ \frac{1}{\rho_0}\int_{\phi_0}^{\phi} d\varphi \left[h(\varphi,\phi_0) f^{\mu\nu}(\varphi) + \tau_0\rho_0 f'^{\mu\nu}(\varphi)\right] \\
&\times \frac{1}{\rho_0}\int_{\varphi_0}^{\varphi} d\varphi' \left[h(\varphi',\phi_0) f^{\mu\nu}(\varphi') + \tau_0\rho_0 f'^{\mu\nu}(\varphi')\right] u_0^{\eta} \\
&- \int_{\phi_0}^{\phi} d\varphi\, h(\varphi,\phi_0) \sum_{i=1,2} a_i^2 \psi_i'^2(\varphi) k^{\mu} ,
\end{aligned}
\tag{A.82}
$$

which simplifies to

$$
\begin{aligned}
u^{\mu}(\phi) &= \frac{1}{h(\phi,\phi_0)}\Bigg\{ u_0^{\mu} + \frac{1}{2\rho_0}\left[h^2(\phi,\phi_0) - 1\right] k^{\mu} \\
&+ \frac{1}{\rho_0}\sum_{i=1,2}\left[I_i(\phi,\phi_0) f_i^{\mu\nu} u_{0,\nu} - \frac{1}{2} a_i^2 I_i^2(\phi,\phi_0) k^{\mu}\right]\Bigg\} ,
\end{aligned}
\tag{A.83}
$$

with

$$
\begin{aligned}
I_i(\phi,\phi_0) &\equiv \int_{\phi_0}^{\phi} d\varphi \left[h(\varphi,\phi_0)\psi_i'(\varphi) + \tau_0\rho_0\psi_i''(\varphi)\right] \\
&= h(\phi,\phi_0)\psi_i(\phi) - \psi_i(\phi_0) + \tau_0\rho_0\left[\psi_i'(\phi) - \psi_i'(\phi_0)\right] \\
&+ \tau_0\rho_0 \int_{\phi_0}^{\phi} d\varphi\, \psi_i(\varphi) \sum_{j=1,2} a_j^2 \psi_j'^2(\varphi) .
\end{aligned}
\tag{A.84}
$$

To make the comparaison with the case without RR simpler, we can rewrite the solution as

$$
\begin{aligned}
h(\phi,\phi_0) u^{\mu}(\phi) &= u_0^{\mu} + \frac{\mathcal{F}_{\mathrm{RR}}^{\mu\nu}(\phi,\phi_0) u_{0,\nu}}{k u_0} + \frac{\mathcal{F}_{\mathrm{RR}}^{2\mu\nu}(\phi,\phi_0) u_{0,\nu}}{2(k u_0)^2} \\
&+ \frac{1}{2(k u_0)}\left[h^2(\phi,\phi_0) - 1\right] k^{\mu} ,
\end{aligned}
\tag{A.85}
$$

with

$$
\mathcal{F}_{\mathrm{RR}}^{\mu\nu}(\phi,\phi_0) = \mathcal{I}_i(\phi,\phi_0)\mathcal{F}^{\mu\nu}(\phi,\phi_0) , \tag{A.86}
$$

$$
\mathcal{I}_i(\phi,\phi_0) = \frac{1}{\left[\psi_i(\phi) - \psi_i(\phi_0)\right]} I_i(\phi,\phi_0) . \tag{A.87}
$$

In 3-vector (with the same notations as for the case without RR)

$$a^\mu(\phi) = a(0, \delta\cos(\phi), \sqrt{1-\delta^2}\sin(\phi), 0)\,, \quad u^\mu = \gamma(1, \boldsymbol{\beta})\,, \quad u_0^\mu = \gamma_0(1, \boldsymbol{\beta}_0)\,,$$
$$k^\mu = \omega_0(1, \hat{\boldsymbol{k}})\,, \quad x^\mu = (t, \boldsymbol{x})\,, \quad \phi = \omega(t - \hat{\boldsymbol{k}}\cdot\boldsymbol{x})\,, \quad ku_0 = \gamma_0\omega_0(1 - \hat{\boldsymbol{k}}\cdot\boldsymbol{\beta}_0)\,,$$

this reads

$$\gamma(\phi) = \frac{1}{h(\phi,\phi_0)}\left[\gamma_0 + \frac{1}{2h_0}\left[h^2(\phi,\phi_0) - 1\right] - \frac{\gamma_0}{h_0}\sum_{i=1,2} I_i(\phi,\phi_0)\,\boldsymbol{a}_i\cdot\boldsymbol{\beta}_0 + \frac{1}{2h_0}\sum_{i=1,2}\boldsymbol{a}_i^2 I_i^2(\phi,\phi_0)\right], \tag{A.88a}$$

$$u_x(\phi) = \frac{1}{h(\phi,\phi_0)}\left(\gamma_0\beta_{0,x} - \sum_{i=1,2}\mathcal{I}_i(\phi,\phi_0)a_{i,x}\right), \tag{A.88b}$$

$$u_y(\phi) = \frac{1}{h(\phi,\phi_0)}\left(\gamma_0\beta_{0,y} - \sum_{i=1,2}\mathcal{I}_i(\phi,\phi_0)a_{i,y}\right), \tag{A.88c}$$

$$u_z(\phi) = \frac{1}{h(\phi,\phi_0)}\left(\gamma_0\beta_{0,z} + \frac{1}{2h_0}\left[h^2(\phi,\phi_0) - 1\right]\hat{k}_z - \frac{\gamma_0}{h_0}\sum_{i=1,2} I_i(\phi,\phi_0)\,\boldsymbol{a}_i\cdot\boldsymbol{\beta}_0\,\hat{k}_z + \frac{1}{2h_0}\sum_{i=1,2}\boldsymbol{a}_i^2 I_i^2(\phi,\phi_0)\,\hat{k}_z\right). \tag{A.88d}$$

More explicitly, we have

$$h(\phi,\phi_0) = 1 - \frac{1}{2}\tau_0\rho_0 a^2\left(\delta^2 - \frac{1}{2}\right)[\sin 2\phi - \sin 2\phi_0] - \frac{1}{2}\tau_0\rho_0 a^2(\phi - \phi_0)\,, \tag{A.89a}$$

$$I_1(\phi,\phi_0) = h(\phi,\phi_0)\delta\sin\phi - \delta\sin\phi_0 + \tau_0\rho_0\delta\left[1 + \frac{1}{2}a^2\left(\delta^2 - \frac{3}{2}\right)\right][\cos\phi - \cos\phi_0] - \frac{1}{6}\tau_0\rho_0 a^2\delta\left(\delta^2 - \frac{1}{2}\right)[\cos 3\phi - \cos 3\phi_0]\,, \tag{A.89b}$$

$$I_2(\phi,\phi_0) = -h(\phi,\phi_0)\sqrt{1-\delta^2}\cos\phi + \sqrt{1-\delta^2}\cos\phi_0 + \tau_0\rho_0\sqrt{1-\delta^2}\left[1 - \frac{1}{2}a^2\left(\delta^2 - \frac{1}{2}\right)\right][\sin\phi - \sin\phi_0] - \frac{1}{6}\tau_0\rho_0 a^2\sqrt{1-\delta^2}\left(\delta^2 - \frac{1}{2}\right)[\sin 3\phi - \sin 3\phi_0]\,. \tag{A.89c}$$

Appendix B
Quantum Electrodynamics

B.1 Gamma Matrix Algebra

The gamma matrices γ^μ generate a matrix representation of the Clifford algebra $\mathrm{Cl}_{1,3}(\mathcal{R})$ and are defined by the anticommutation relation

$$\{\gamma^\mu, \gamma^\nu\} = 2\eta^{\mu\nu}\,. \tag{B.1}$$

In this thesis, we use the so-called Weyl (or chiral) representation of the gamma matrices

$$\gamma^0 = \begin{pmatrix} 0 & 1 \\ 1 & 0 \end{pmatrix}, \quad \gamma^i = \begin{pmatrix} 0 & \sigma^i \\ -\sigma^i & 0 \end{pmatrix}, \tag{B.2}$$

where $(\sigma^i)_{i=1..3}$ are the Pauli matrices

$$\sigma^1 = \begin{pmatrix} 0 & 1 \\ 1 & 0 \end{pmatrix}, \quad \sigma^2 = \begin{pmatrix} 0 & -i \\ i & 0 \end{pmatrix}, \quad \sigma^3 = \begin{pmatrix} 1 & 0 \\ 0 & -1 \end{pmatrix}. \tag{B.3}$$

They satisfy the algebraic identity

$$\sigma_i \sigma_j = \delta^{ij} + i\epsilon^{ijk}\sigma^k\,. \tag{B.4}$$

Similarly, the gamma matrices satisfy

$$\gamma^\mu \gamma^\nu = \eta^{\mu\nu} + \sigma^{\mu\nu}\,, \tag{B.5}$$

where $\sigma^{\mu\nu} = \frac{1}{2}\left[\gamma^\mu, \gamma^\nu\right]$.

F. Niel, *Classical and Quantum Description of Plasma and Radiation in Strong Fields*, Springer Theses, https://doi.org/10.1007/978-3-030-73547-0

Product of Gamma Matrices

Using (B.1), we can show the following identities

$$\gamma^\mu \gamma_\mu = 4\,, \tag{B.6a}$$
$$\gamma^\mu \gamma^\nu \gamma_\mu = -2\gamma^\nu\,, \tag{B.6b}$$
$$\gamma^\mu \gamma^\nu \gamma^\rho \gamma_\mu = 4\eta^{\nu\rho}\,, \tag{B.6c}$$
$$\gamma^\mu \gamma^\nu \gamma^\rho \gamma^\sigma \gamma_\mu = -2\gamma^\sigma \gamma^\rho \gamma^\nu\,. \tag{B.6d}$$

Trace of Gamma Matrices

Using the cyclic property of the trace, one can simply calculate the trace of any product of gamma matrices by simply permuting the first gamma matrix to the last position and using (B.1)

$$\frac{1}{4}\operatorname{Tr}\gamma^\mu \gamma^\nu = \eta^{\mu\nu}\,, \tag{B.7a}$$
$$\frac{1}{4}\operatorname{Tr}\gamma^\mu \gamma^\nu \gamma^\rho \gamma^\sigma = \eta^{\mu\sigma}\eta^{\nu\rho} - \eta^{\mu\rho}\eta^{\nu\sigma} + \eta^{\mu\nu}\eta^{\rho\sigma}\,, \tag{B.7b}$$
$$\frac{1}{4}\operatorname{Tr}\sigma^{\mu\nu}\gamma^\rho \gamma^\sigma = \eta^{\mu\sigma}\eta^{\nu\rho} - \eta^{\mu\rho}\eta^{\nu\sigma}\,, \tag{B.7c}$$
$$\frac{1}{4}\operatorname{Tr}\gamma^\mu \gamma^\nu \gamma^\rho \gamma^\sigma \gamma^5 = i\epsilon^{\mu\nu\rho\sigma}\,. \tag{B.7d}$$

Finally, it is easy to convince oneself that the trace of any product of an odd number of gamma matrices is zero.

B.2 Useful Identities for QED

The computation of cross-sections in QED will require extensive use of the gamma matrices. In this optic, we summarize here some of the most used identities.

Trace Identities

We will be lead to compute products of the form

$$\sum_{s,s'} M'^{\dagger}_{p,s;p',s'} M_{p,s;p',s'}\,, \tag{B.8}$$

with

$$M'_{p,s;p',s'} = \bar{u}_{p',s'}\Gamma' u_{p,s} \,, \tag{B.9a}$$

$$M_{p,s;p',s'} = \bar{u}_{p',s'}\Gamma u_{p,s} \,. \tag{B.9b}$$

These can be rewritten as

$$\begin{aligned}
\sum_{s,s'} M'^{\dagger}_{p,s;p',s'} M_{p,s;p',s'} &= \sum_{s,s'} u^{\dagger}_{p,s}\,\Gamma'^{\dagger}\,\bar{u}^{\dagger}_{p',s'}\,\bar{u}_{p',s'}\,\Gamma u_{p,s} = \sum_{s,s'} \bar{u}_{p,s}\,\bar{\Gamma}' u_{p',s'}\bar{u}_{p',s'}\Gamma u_{p,s} \,, \\
&= \sum_{s} \bar{u}_{p,s}\left[\Gamma'(\not{p}' + m)\Gamma\right] u_{p,s} \\
&= \sum_{\alpha,\beta}\sum_{s} \bar{u}_{p,s,\alpha}\left[\Gamma'(\not{p}' + m)\Gamma\right]_{\alpha,\beta} u_{p,s,\beta} \\
&= \mathrm{Tr}\left[\Gamma'(\not{p}' + m)\Gamma(\not{p} + m)\right] .
\end{aligned} \tag{B.10}$$

The following identities will be extensively used in the derivation of QED cross-sections

$$\mathrm{Tr}\left[\not{a}\not{b}\right] = 4ab \,, \tag{B.11a}$$

$$-\frac{1}{4}\mathrm{Tr}\left[\not{a}\gamma^{\mu}\not{b}\gamma_{\mu}\right] = 2ab \,, \tag{B.11b}$$

$$-\frac{1}{4}\mathrm{Tr}\left[\not{a}\gamma^{\mu}\not{b}\not{n}\mathcal{A}(\phi)\gamma_{\mu}\right] = 2\left[b_{-}\left(a\mathcal{A}(\phi)\right) - a_{-}\left(b\mathcal{A}(\phi)\right)\right] , \tag{B.11c}$$

$$-\frac{1}{4}\mathrm{Tr}\left[\not{a}\gamma^{\mu}\not{b}\not{n}\mathcal{A}(\phi)\gamma_{\mu}\not{n}\mathcal{A}(\phi')\right] = -4\,a_{-}b_{-}\,\mathcal{A}(\phi)\mathcal{A}(\phi') \,, \tag{B.11d}$$

$$-\frac{1}{4}\mathrm{Tr}\left[\not{a}\gamma^{\mu}\not{n}\mathcal{A}(\phi)\not{b}\not{n}\mathcal{A}(\phi')\gamma_{\mu}\not{n}\mathcal{A}(\phi'')\right] = 0 \,. \tag{B.11e}$$

Slashed Quantities Identities

$$\not{a}\not{b} = 2ab - \not{b}\not{a} \,, \tag{B.12}$$

B.3 The Dirac Spinors

The solutions of the free Dirac equation $(i\not{\partial} - m)\Psi = 0$ read

$$\Psi(x) = u_s(\boldsymbol{p})\,e^{-ipx} \,, \tag{B.13a}$$

$$\Psi(x) = v_s(\boldsymbol{p})\,e^{ipx} \,, \tag{B.13b}$$

and are interpreted respectively as particles and anti-particles of energy $\epsilon_p = \sqrt{p^2 + m^2}$ and spin $s = \pm\frac{1}{2}$. The components $u_s(\boldsymbol{p})$ and $v_s(\boldsymbol{p})$ are called *Dirac spinors* (or *4-spinors*), and are solution of the following momentum space Dirac equations

$$(\not{p} - m)u = 0\,, \tag{B.14a}$$

$$(\not{p} + m)v = 0\,. \tag{B.14b}$$

They can be represented as

$$u_{+\frac{1}{2}}(\boldsymbol{p}) = \sqrt{\epsilon_p + m}\begin{pmatrix} \phi_1 \\ \frac{\boldsymbol{\sigma}\cdot\boldsymbol{p}}{\epsilon_p+m}\phi_1 \end{pmatrix}, \quad u_{-\frac{1}{2}}(\boldsymbol{p}) = \sqrt{\epsilon_p + m}\begin{pmatrix} \phi_2 \\ \frac{\boldsymbol{\sigma}\cdot\boldsymbol{p}}{\epsilon_p+m}\phi_2 \end{pmatrix}, \tag{B.15a}$$

$$v_{+\frac{1}{2}}(\boldsymbol{p}) = \sqrt{\epsilon_p + m}\begin{pmatrix} \frac{\boldsymbol{\sigma}\cdot\boldsymbol{p}}{\epsilon_p+m}\chi_1 \\ \chi_1 \end{pmatrix}, \quad v_{-\frac{1}{2}}(\boldsymbol{p}) = -\sqrt{\epsilon_p + m}\begin{pmatrix} \frac{\boldsymbol{\sigma}\cdot\boldsymbol{p}}{\epsilon_p+m}\chi_2 \\ \chi_2 \end{pmatrix}, \tag{B.15b}$$

where ϕ_i and χ_i are arbitrary constant bi-spinors normalized as $\chi_i^\dagger\chi_i = \phi_i^\dagger\phi_i = 1$. These states are normalized as

$$u_s^\dagger(\boldsymbol{p})u_s(\boldsymbol{p}) = 2\epsilon_p\,, \tag{B.16a}$$

$$v_s^\dagger(\boldsymbol{p})v_s(\boldsymbol{p}) = 2\epsilon_p\,. \tag{B.16b}$$

The states u and v are orthogonal

$$\bar{u}_i u_j = 2m\,\delta_{ij}\,, \tag{B.17a}$$

$$\bar{v}_i v_j = -2m\,\delta_{ij}\,, \tag{B.17b}$$

where $\bar{u} = u^\dagger\gamma^0$ and $\bar{v} = v^\dagger\gamma^0$ are the adjoint spinors associated to u and v which satisfy

$$\bar{u}(\not{p} - m) = 0\,, \tag{B.18a}$$

$$\bar{v}(\not{p} + m) = 0\,. \tag{B.18b}$$

Multiplying Eq. (B.18a) (with spin r) by $u_s(p')$ from the right (and similarly for v), we get the so-called Gordon identities

$$2m\,\bar{u}_r(p)\gamma^\mu u_s(p') = \bar{u}_r(p)\left[(p+p')^\mu + i\sigma^{\mu\nu}(p'-p)_\nu\right]u_s(p')\,, \tag{B.19a}$$

$$2m\,\bar{v}_r(p)\gamma^\mu v_s(p') = -\bar{v}_r(p)\left[(p+p')^\mu + i\sigma^{\mu\nu}(p'-p)_\nu\right]v_s(p')\,, \tag{B.19b}$$

which for $p = p'$ yield

$$\bar{u}_r(p)\gamma^\mu u_s(p) = 2p^\mu\delta_{rs}\,, \tag{B.20a}$$

$$\bar{v}_r(p)\gamma^\mu v_s(p) = 2p^\mu\delta_{rs}\,. \tag{B.20b}$$

Finally, the so-called completeness relation reads

$$\sum_{s=\pm\frac{1}{2}} u_s \bar{u}_s = \not{p} + m\,, \tag{B.21a}$$

$$\sum_{s=\pm\frac{1}{2}} v_s \bar{v}_s = \not{p} - m\,. \tag{B.21b}$$

B.4 The Dirac-Volkov States

We consider the Dirac equation with the external electromagnetic potential[1] $\mathcal{A}$

$$\left(\not{p} - e\not{\mathcal{A}} - m\right)\Psi = 0\,, \tag{B.22}$$

where we assume that $\mathcal{A}$ depends only on the phase $\phi = kx$. In Lorenz gauge, we therefore have

$$\partial\mathcal{A} = k\mathcal{A}' = 0 \implies k\mathcal{A} = const. = 0\,. \tag{B.23}$$

Multiplying Eq. (B.22) on the left by $\not{p} - e\not{\mathcal{A}} + m$, we get

$$\left[(p - e\mathcal{A})^2 - m^2 - \frac{i}{2}eF_{\mu\nu}\sigma^{\mu\nu}\right]\psi = 0\,, \tag{B.24}$$

which can be rewritten as

$$\left[-\partial^2 - 2ie(\mathcal{A}\partial) + e^2\mathcal{A}^2 - m^2 - ie\not{k}\not{\mathcal{A}}\right]\psi = 0\,. \tag{B.25}$$

We now look for a solution in the form $\psi(x) = e^{-ipx}F(\phi)$. We get

$$2i(kp)F' + \left[-2e(p\mathcal{A}) + e^2\mathcal{A}^2 - m^2 - ie\not{k}\not{\mathcal{A}}\right]F = 0\,, \tag{B.26}$$

which integrates into

$$F = \exp\left\{-i\int_{\phi_0}^{\phi} d\varphi\left[\frac{e(p\mathcal{A}(\varphi))}{(kp)} - \frac{e^2\mathcal{A}(\varphi)^2}{2(kp)}\right] + \frac{e\not{k}\not{\mathcal{A}}(\phi)}{2(kp)}\right\}u_r(\boldsymbol{p})\,, \tag{B.27}$$

---

[1]Here the electron charge is $q = e$ ($e < 0$).

where w_{ps} is an arbitrary constant bispinor (See Sect. 3.3.2). Using Eq. (B.12), we can show that $(\not{k}\not{\mathcal{A}})^2 = 0$, and therefore the solution can be rewritten as

$$\Psi^V_{p,r}(x) = \sqrt{\frac{1}{2\epsilon_p V}}\, E_p(x) u_r(\boldsymbol{p}) \,, \tag{B.28}$$

where $u_r(\boldsymbol{p})$ is the free Dirac spinor defined in Sect. 3.3.2, while

$$E_p(x) = \left[1 + \frac{e\not{k}\not{\mathcal{A}}(\phi)}{2(kp)}\right] e^{iS_p(x)} \,, \tag{B.29}$$

is the so-called *Ritus matrix*. The phase $S_p(x)$ is the classical Hamilton-Jacobi function for a charged particle in a potential $\mathcal{A}$ and is given by

$$S_p(x) = -px - \int_{\phi_0}^{\phi} d\varphi \left[\frac{e(p\mathcal{A}(\varphi))}{(kp)} - \frac{e^2\mathcal{A}(\varphi)^2}{2(kp)}\right] , \tag{B.30}$$

Finally, the constant $\sqrt{\frac{1}{2\epsilon_p V}}$ is chosen so that the states are normalized to one particle per volume V

$$\int_V \Psi^{V\dagger}_{p,r} \Psi^V_{p,r} = 1 \,. \tag{B.31}$$

These solutions are called the *Dirac-Volkov states* and will be used extensively to compute differential probability of processes in external laser fields. The adjoint of the Ritus matrix is given by

$$\bar{E}_p(x) = \left[1 - \frac{e\not{k}\not{\mathcal{A}}(\varphi)}{2(kp)}\right] e^{-iS_p(x)} \,, \tag{B.32}$$

since $k\mathcal{A} = 0$. The Ritus matrices form an orthogonal and complete set

$$\int \frac{d^4p}{(2\pi)^4}\, E_p(x)\bar{E}_p(x') = \delta^{(4)}(x - x') \,, \tag{B.33a}$$

$$\int \frac{d^4x}{(2\pi)^4}\, \bar{E}_{p'}(x) E_p(x) = \delta^{(4)}(p - p') \,. \tag{B.33b}$$

The Ritus matrices convert momentum operators into momentum variables

$$\left[i\not{\partial}_x - e\not{\mathcal{A}}(kx)\right] E_{p,x} = E_{p,x}\not{p} \,, \tag{B.34a}$$

$$\left[i\overleftarrow{\not{\partial}}_x + e\not{\mathcal{A}}(kx)\right] E_{p,x} = -\not{p}\bar{E}_{p,x} \,, \tag{B.34b}$$

B.5 Nonlinear Compton Scattering

We consider the Volkov states normalized to one particle per volume V on average, that is to say

$$\psi_{p,s}(x) = \left[1 + \frac{e \not{k} \not{A}(\phi)}{2(kp)}\right] e^{-i\left(px + \int_0^\phi d\varphi\left[\frac{ep\cdot A(\varphi)}{(kp)} - \frac{e^2 A^2(\varphi)}{2(kp)}\right]\right)} \frac{u_{p,s}}{\sqrt{2\epsilon_p V}}, \quad \text{(B.35)}$$

where k is the 4-momentum of the laser field and the state of the photon with 4-momentum k' is taken to be

$$A^\mu_{k',\lambda}(x) = \sqrt{\frac{1}{2\omega V}}\, \epsilon^\mu_{k',\lambda}\, e^{ik'x}, \quad \text{(B.36)}$$

where $\epsilon^\mu_{k',\lambda}$ is a real polarization vector with mode index λ and such that $(k'\,\epsilon_{k',\lambda}) = 0$. At first order, the S matrix is then given by

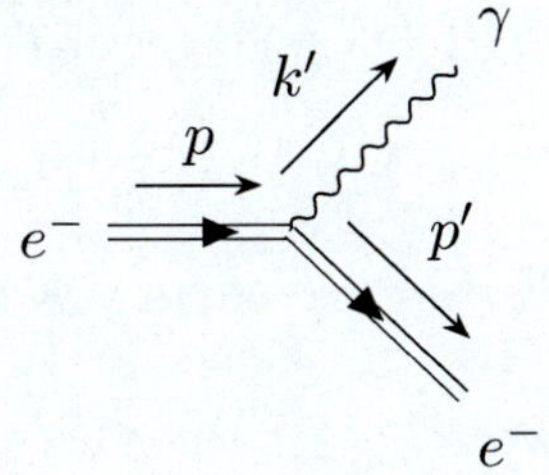

that is to say

$$S_{fi} = -ie \int d^4x\, \overline{\psi}_{p',s'}\, \not{A}^*_{k',\lambda}\, \psi_{p,s}, \quad \text{(B.37)}$$

which inserting Eq. (B.35) and (B.36) yields

$$S_{fi} = S_0 \int d^4x\, \mathfrak{M}_{fi}(p', s', k', \lambda, p, s)\, e^{-iqx} e^{i\Phi_{p,p'}(\phi)},$$

with

$$S_0 = -ie\sqrt{\frac{1}{2\omega V}}\sqrt{\frac{1}{2\epsilon V}}\sqrt{\frac{1}{2\epsilon' V}}\,, \tag{B.38a}$$

$$\mathfrak{M}_{fi}(p', s', k', \lambda, p, s; \phi) = \bar{u}_{p's'}\left[1 - \frac{e\,\not{k}\,\not{\mathcal{A}}(\phi)}{2(kp')}\right]\not{\epsilon}^*_{k',\lambda} \tag{B.38b}$$

$$\times\left[1 + \frac{e\,\not{k}\,\not{\mathcal{A}}(\phi)}{2(kp)}\right]u_{p,s}\,,$$

$$q^\mu = p^\mu - p'^\mu - k'^\mu\,, \tag{B.38c}$$

$$\tilde{\Phi}_{p,p'}(\phi, 0) = \int_0^\phi d\varphi\left[\frac{e\,p'\mathcal{A}(\varphi)}{(kp')} - \frac{e\,p\mathcal{A}(\varphi)}{(kp)}\right. \tag{B.38d}$$

$$\left. - e^2\frac{\mathcal{A}^2(\varphi)}{2}\left(\frac{1}{(kp')} - \frac{1}{(kp)}\right)\right].$$

Going to light-cone coordinates, we change (t, x, y, z) to $(\phi, T, \boldsymbol{x}_\perp)$ with $\phi = x_-$ and $T = x_+$ (See Appendix A.4). Integrating over T and $\boldsymbol{x}_\perp$, S_{fi} reduces to

$$S_{fi} = S_0\,(2\pi)^3\,\delta(p_- - p'_- - k'_-)\,\delta^{(2)}(\boldsymbol{p}_\perp - \boldsymbol{p}'_\perp - \boldsymbol{k}'_\perp) \tag{B.39}$$

$$\times\int d\phi\;\mathfrak{M}_{fi}(p', s', k', \lambda, p, s; \phi)\,e^{-iq_+\phi}e^{i\tilde{\Phi}_{p,p'}(\phi,0)}\,,$$

where $x_- = \phi$, $q_+ = p_+ - p'_+ - k'_+$ and where we used Eqs. (A.28) and (A.27).

What we are actually interested in is the transition probability for an incoming state to a end up in a volume $d\Pi$ of the phase space of outcoming particles. In our case, the outcoming state has two particles and this infinitesimal volume of phase space is given by

$$d\Pi = V\frac{d^3p'}{(2\pi)^3}V\frac{d^3k'}{(2\pi)^3}\,. \tag{B.40}$$

and the transition probability by

$$dP = V^2\frac{d^3k'}{(2\pi)^3}\int\frac{d^3p'}{(2\pi)^3}\frac{1}{2}\sum_{s,s',\lambda}|S_{fi}|^2\,, \tag{B.41}$$

since we look here at *unpolarized probabilities*, that is to say we average over all incoming electron spins and sum over all final electron spins and photon polarizations. Moreover, we integrate over all final electron energies. Using Eq. (B.10), we get

$$\begin{aligned}
&\tfrac{1}{2}\sum_{s,s',\lambda} \mathfrak{M}^{\dagger}_{fi}(p',s',k',\lambda,p,s;\phi)\,\mathfrak{M}_{fi}(p',s',k',\lambda,p,s;\phi') \\
&= -\frac{1}{2}\,\mathrm{Tr}\left\{(\not p' + m)\left[1 - \frac{e\not n \not A(\phi)}{2p'_-}\right]\gamma^{\mu}\left[1 + \frac{e\not n \not A(\phi)}{2p_-}\right](\not p + m)\right. \\
&\quad \left.\times \left[1 - \frac{e\not n \not A(\phi')}{2p_-}\right]\gamma_{\mu}\left[1 + \frac{e\not n \not A(\phi')}{2p'_-}\right]\right\} = 2\mathcal{T}\,.
\end{aligned} \tag{B.42}$$

After lengthy calculations, and extensive use of the trace identities derived in Appendix B.2, we obtain

$$\begin{aligned}
\mathcal{T} &= m^2\left(\frac{p'_-}{p_-} + \frac{p_-}{p'_-} - 4\right) + \frac{p'_-}{p_-}\boldsymbol{p}_{\perp}^2 - 2\boldsymbol{p}_{\perp}\cdot\boldsymbol{p}'_{\perp} + \frac{p_-}{p'_-}\boldsymbol{p}'^2_{\perp} \\
&+ ek_-\left(\frac{\boldsymbol{p}_{\perp}}{p_-} - \frac{\boldsymbol{p}'_{\perp}}{p'_-}\right)\cdot\left[\boldsymbol{A}_{\perp}(\phi) + \boldsymbol{A}_{\perp}(\phi')\right] \\
&- e^2\left[\boldsymbol{A}^2_{\perp}(\phi) + \boldsymbol{A}^2_{\perp}(\phi') - \left(\frac{p'_-}{p_-} + \frac{p_-}{p'_-}\right)\boldsymbol{A}_{\perp}(\phi)\cdot\boldsymbol{A}_{\perp}(\phi')\right],
\end{aligned} \tag{B.43}$$

and

$$\begin{aligned}
\frac{1}{2}\sum_{s,s',\lambda}|S_{fi}|^2 &= S_0^2\,(2\pi)^6\left[\delta(p_- - p'_- - k'_-)\right]^2\left[\delta^{(2)}(\boldsymbol{p}_{\perp} - \boldsymbol{p}'_{\perp} - \boldsymbol{k}'_{\perp})\right]^2 \\
&\quad \times \int d\phi d\phi'\, 2\mathcal{T} e^{-iq_+(\phi-\phi')} e^{i\tilde{\Phi}_{p,p'}(\phi,\phi')}\,,
\end{aligned} \tag{B.44}$$

Using the following prescription for the delta function

$$(2\pi)^3\delta\,(q_- = 0)\,\delta^{(2)}\left(\boldsymbol{q}_{\perp} = \boldsymbol{0}\right) \to \frac{\epsilon}{p_-}V\,, \tag{B.45}$$

and changing the infinitesimal volume to light-cone coordinates

$$d^3p' = \frac{\epsilon'}{p'_-}dp'_-d^2p'_{\perp}\,, \tag{B.46a}$$

$$d^3k' = \frac{\omega'}{k'_-}dk'_-d^2k'_{\perp}\,, \tag{B.46b}$$

we get

$$dP = \frac{dk'_- d^2k'_\perp}{(2\pi)^3} \frac{e^2}{4\,k'_- p'_- p_-} \int d\phi d\phi' \, \mathcal{T} e^{-iq_+(\phi-\phi')} e^{i\tilde{\Phi}_{p,p'}(\phi,\phi')} \,, \tag{B.47}$$

where

$$p_- - p'_- = k'_- \,, \tag{B.48a}$$

$$\boldsymbol{p}_\perp - \boldsymbol{p}'_\perp = \boldsymbol{k}'_\perp \,, \tag{B.48b}$$

because of the delta functions. Writing $\boldsymbol{a}_\perp = \frac{e\boldsymbol{A}_\perp}{m}$, we can rewrite, after lengthy algebra, the phase $\Phi_{p,p'}(\phi,\phi') = -q_+(\phi-\phi') + \tilde{\Phi}_{p,p'}(\phi,\phi')$ as

$$\Phi_{p,p'}(\phi,\phi') = \frac{m^2 k'_-}{2p_- p'_-} \int_{\phi'}^{\phi} d\varphi \left[1 + \left(\frac{\boldsymbol{p}_\perp}{m} - \frac{p_-}{k_-}\frac{\boldsymbol{k}_\perp}{m} - \boldsymbol{a}_\perp(\varphi) \right)^2 \right] , \tag{B.49}$$

and the trace $\mathcal{T}$ as

$$\begin{aligned} \frac{\mathcal{T}}{m^2} = &\left(\frac{p'_-}{p_-} + \frac{p_-}{p'_-} - 4 \right) + \frac{p'_-}{p_-}\frac{\boldsymbol{p}_\perp^2}{m^2} - 2\frac{\boldsymbol{p}_\perp}{m} \cdot \frac{\boldsymbol{p}'_\perp}{m} + \frac{p_-}{p'_-}\boldsymbol{p}'^2_\perp \\ &+ ek_- \left(\frac{\boldsymbol{p}_\perp}{p_-} - \frac{\boldsymbol{p}'_\perp}{p'_-} \right) \cdot \left[\boldsymbol{A}_\perp(\phi) + \boldsymbol{A}_\perp(\phi') \right] \\ &- e^2 \left[\boldsymbol{A}^2_\perp(\phi) + \boldsymbol{A}^2_\perp(\phi') - \left(\frac{p'_-}{p_-} + \frac{p_-}{p'_-} \right) \boldsymbol{A}_\perp(\phi) \cdot \boldsymbol{A}_\perp(\phi') \right] . \end{aligned} \tag{B.50}$$

Appendix C
Electron Distribution Function

C.1 Derivation of the Master Equation (5.31)

We start from Eq. (5.28a) where $\frac{d}{dt}$ is the usual Vlasov operator

$$\frac{d}{dt} f_e = \frac{\partial}{\partial t} f_e(t, \mathbf{r}, \boldsymbol{p}) + \nabla_{\mathbf{r}} \cdot \left[\frac{\boldsymbol{p}}{m\gamma} f_e(t, \mathbf{r}, \boldsymbol{p}) \right] + \nabla_{\boldsymbol{p}} \cdot \left[q \left(\mathbf{E} + \frac{\boldsymbol{p}}{m\gamma} \times \mathbf{B} \right) f_e(t, \mathbf{r}, \boldsymbol{p}) \right]. \tag{C.1}$$

Let us now decompose $\boldsymbol{p}$ as $\boldsymbol{p} = p\boldsymbol{\Omega}$ and rewrite the Vlasov operator in accordance

$$\frac{d}{dt} f_e(t, \mathbf{r}, p, \boldsymbol{\Omega}) = \frac{\partial}{\partial t} f_e(t, \mathbf{r}, p, \boldsymbol{\Omega}) + \nabla_{\mathbf{r}} \cdot \left[\frac{p\boldsymbol{\Omega}}{m\gamma} f_e(t, \mathbf{r}, p, \boldsymbol{\Omega}) \right] + \nabla_{\boldsymbol{p}} \cdot \left[q \left(\mathbf{E} + \frac{p\boldsymbol{\Omega}}{m\gamma} \times \mathbf{B} \right) f_e(t, \mathbf{r}, p, \boldsymbol{\Omega}) \right].$$

Using the weak formulation and going to spherical coordinates, we obtain

$$\begin{aligned} \int_{S^2} d\boldsymbol{\Omega} \int_0^\infty dp\, p^2 \frac{d}{dt} f_e(t, \mathbf{r}, p, \boldsymbol{\Omega}) \phi_1(\gamma) \phi_2(\boldsymbol{\Omega}) = \\ \frac{\partial}{\partial t} \int_{S^2} d\boldsymbol{\Omega} \int_0^\infty dp\, p^2 f_e(t, \mathbf{r}, p, \boldsymbol{\Omega}) \phi_1(\gamma) \phi_2(\boldsymbol{\Omega}) \\ + \int_{S^2} d\boldsymbol{\Omega} \int_0^\infty dp\, p^2 \nabla_{\mathbf{r}} \cdot \left[\frac{p\boldsymbol{\Omega}}{m\gamma} f_e(t, \mathbf{r}, p, \boldsymbol{\Omega}) \right] \phi_1(\gamma) \phi_2(\boldsymbol{\Omega}) \\ - \int_{S^2} d\boldsymbol{\Omega} \int_0^\infty dp\, p^2 \left[q(\mathbf{E} + \frac{p}{m\gamma} \boldsymbol{\Omega} \times \mathbf{B}) f_e(t, \mathbf{r}, p, \boldsymbol{\Omega}) \right] \cdot \\ \left[\phi_1'(\gamma) \phi_2(\boldsymbol{\Omega}) \nabla_{\boldsymbol{p}} \gamma + \phi_1(\gamma) \nabla_{\Omega} \phi_2(\boldsymbol{\Omega}) \cdot \nabla_{\boldsymbol{p}} \boldsymbol{\Omega} \right]. \end{aligned}$$

F. Niel, *Classical and Quantum Description of Plasma and Radiation in Strong Fields*,
Springer Theses, https://doi.org/10.1007/978-3-030-73547-0

where ϕ_1 and ϕ_2 are two test functions and where we performed an integral by parts in the third integral. After some algebra, we get the following equation

$$
\begin{aligned}
mc^2 \int_{S^2} d\mathbf{\Omega} \int_0^\infty d\gamma \frac{d}{dt} \widetilde{f}_e(t, \mathbf{r}, p, \mathbf{\Omega})\phi_1(\gamma)\phi_2(\mathbf{\Omega}) = \\
mc^2 \frac{\partial}{\partial t} \int_{S^2} d\mathbf{\Omega} \int_0^\infty d\gamma\, \widetilde{f}_e(t, \mathbf{r}, p, \mathbf{\Omega})\phi_1(\gamma)\phi_2(\mathbf{\Omega}) \\
+ mc^2 \int_{S^2} d\mathbf{\Omega} \int_0^\infty d\gamma \nabla_{\mathbf{r}} \cdot \left[\frac{p\mathbf{\Omega}}{m\gamma} \widetilde{f}_e(t, \mathbf{r}, p, \mathbf{\Omega}) \right] \phi_1(\gamma)\phi_2(\mathbf{\Omega}) \\
- \int_{S^2} d\mathbf{\Omega} \int_0^\infty d\gamma \left[q(\mathbf{E} + \frac{p}{m\gamma} \mathbf{\Omega} \times \mathbf{B}) \widetilde{f}_e(t, \mathbf{r}, p, \mathbf{\Omega}) \right] \cdot \\
\left[\phi_1'(\gamma)\phi_2(\mathbf{\Omega}) \frac{p}{m\gamma} \mathbf{\Omega} + \phi_1(\gamma)\nabla_{\mathbf{\Omega}}\phi_2(\mathbf{\Omega}) \cdot mc^2 \nabla_p \Omega \right],
\end{aligned}
$$

where we have defined a new distribution function $\widetilde{f}_e(t, \mathbf{r}, p, \mathbf{\Omega}) \equiv m\gamma p f_e(t, \mathbf{r}, p, \mathbf{\Omega})$. Finally, performing an other integration by parts in the last integral we get

$$
\begin{aligned}
mc^2 \int_{S^2} d\mathbf{\Omega} \int_0^\infty d\gamma \frac{d}{dt} \widetilde{f}_e(t, \mathbf{r}, p, \mathbf{\Omega})\phi_1(\gamma)\phi_2(\mathbf{\Omega}) = \\
mc^2 \frac{\partial}{\partial t} \int_{S^2} d\mathbf{\Omega} \int_0^\infty d\gamma\, \widetilde{f}_e(t, \mathbf{r}, p, \mathbf{\Omega})\phi_1(\gamma)\phi_2(\mathbf{\Omega}) \\
+ mc^2 \int_{S^2} d\mathbf{\Omega} \int_0^\infty d\gamma \nabla_{\mathbf{r}} \cdot \left[\frac{p\mathbf{\Omega}}{m\gamma} \widetilde{f}_e(t, \mathbf{r}, p, \mathbf{\Omega}) \right] \phi_1(\gamma)\phi_2(\mathbf{\Omega}) \\
+ mc^2 \int_{S^2} d\mathbf{\Omega} \int_0^\infty d\gamma \frac{q\mathbf{E}}{mc^2} \cdot \frac{p\mathbf{\Omega}}{m\gamma} \widetilde{f}_e(t, \mathbf{r}, p, \mathbf{\Omega})\phi_1(\gamma)\phi_2(\mathbf{\Omega}) \\
+ mc^2 \int_{S^2} d\mathbf{\Omega} \int_0^\infty d\gamma \nabla_{\mathbf{\Omega}} \cdot \left[\left[\frac{1}{p} (\mathbb{1} - \mathbf{\Omega} \otimes \mathbf{\Omega}) \cdot \right] \cdot \right. \\
\left. q \left(\mathbf{E} + \frac{p}{m\gamma} \mathbf{\Omega} \times \mathbf{B} \right) \widetilde{f}_e(t, \mathbf{r}, p, \mathbf{\Omega}) \right] \phi_1(\gamma)\phi_2(\mathbf{\Omega}).
\end{aligned}
$$

We can perform the same algebra on the collision operator to get

$$
\begin{aligned}
\mathscr{C}[f_e] = mc^2 \int_0^\infty d\gamma_\gamma \int_{S^2} d\mathbf{\Omega} \int_0^\infty d\gamma\, w_\chi^{\mathrm{nCs}}(\gamma + \gamma_\gamma, \gamma_\gamma) \\
\widetilde{f}_e(t, \mathbf{r}, p + p_\gamma, \mathbf{\Omega})\phi_1(\gamma)\phi_2(\mathbf{\Omega}) \\
- mc^2 \int_0^\infty d\gamma_\gamma \int_{S^2} d\mathbf{\Omega} \int_0^\infty d\gamma\, w_\chi^{\mathrm{nCs}}(\gamma, \gamma_\gamma) \\
\widetilde{f}_e(t, \mathbf{r}, p, \mathbf{\Omega})\phi_1(\gamma)\phi_2(\mathbf{\Omega}),
\end{aligned}
$$

Now equating both equations and knowing that the equality is true for any test function ϕ_1 and ϕ_2, we get

$$\frac{\partial}{\partial t} f_e'(t, \mathbf{r}, \gamma, \boldsymbol{\Omega}) + \nabla_{\mathbf{r}} \cdot \left[\frac{p\boldsymbol{\Omega}}{m\gamma} f_e'(t, \mathbf{r}, \gamma, \boldsymbol{\Omega})\right]$$
$$+ \frac{\partial}{\partial \gamma}\left[\frac{1}{mc^2} q\mathbf{E} \cdot \frac{p}{m\gamma}\boldsymbol{\Omega}\, f_e'(t, \mathbf{r}, \gamma, \boldsymbol{\Omega})\right]$$
$$+\nabla_{\boldsymbol{\Omega}} \cdot \left[\left[\frac{1}{p}(\mathbb{1} - \boldsymbol{\Omega} \otimes \boldsymbol{\Omega})\right] \cdot q\left(\mathbf{E} + \frac{p}{m\gamma}\boldsymbol{\Omega} \times \mathbf{B}\right) f_e'(t, \mathbf{r}, \gamma, \boldsymbol{\Omega})\right]$$
$$= \int_0^{\infty} d\gamma_\gamma\, w_\chi^{\mathrm{nCs}}(\gamma + \gamma_\gamma, \gamma_\gamma)\, f_e'(t, \mathbf{r}, \gamma + \gamma_\gamma, \boldsymbol{\Omega})$$
$$- \int_0^{+\infty} d\gamma_\gamma\, w_\chi^{\mathrm{nCs}}(\gamma, \gamma_\gamma)\, f_e'(t, \mathbf{r}, \gamma, \boldsymbol{\Omega}),$$

where we introduced $f_e'(t, \mathbf{r}, \gamma, \boldsymbol{\Omega})$ which displays explicitly the dependence on γ.

C.2 Exact and Approximate Expressions of $a_n(\chi)$ Functions

In this Section, we detail the analytical computation of the functions $a_n(\chi)$ appearing in the computation of the nth order moment of the nonlinear Compton scattering kernel $w_\chi^{nCs}(\gamma, \gamma_\gamma)$ (See Sect. 5.3.2). As a reminder, the successive moments of $w_\chi^{\mathrm{nCs}}(\gamma, \gamma_\gamma)$ are computed as

$$A_n(\chi, \gamma) = \int_0^{+\infty} d\gamma_\gamma\, \gamma_\gamma^n\, w_\chi(\gamma, \gamma_\gamma) = \frac{2}{3}\frac{\alpha^2}{\tau_e}\, \gamma^{n-1}\, a_n(\chi)\,, \tag{C.2}$$

where

$$a_n(\chi) = \int_0^1 d\xi\, \xi^{n-1}\, \widetilde{G}(\chi, \xi)\,. \tag{C.3}$$

Using Eq. (5.30), we obtain the explicit expression for $a_n(\chi)$

$$a_n(\chi) = \frac{\sqrt{3}}{2\pi} 3^{n+1} \chi^{n+1} \int_0^{+\infty} d\nu \left[\frac{9\chi^2 \nu^{n+2}}{(2 + 3\nu\chi)^{n+3}} K_{2/3}(\nu) + \frac{\nu^{n+1}}{(n+1)(2 + 3\nu\chi)^{n+1}} K_{5/3}(\nu)\right], \tag{C.4}$$

which admits the following asymptotic expansions

$$a_n(\chi) \simeq \begin{cases} \frac{\sqrt{3}}{4\pi(n+1)} 3^{n+1}\,\Gamma\left(\frac{n}{2}+\frac{1}{6}\right)\Gamma\left(\frac{n}{2}+\frac{11}{6}\right)\chi^{n+1}, & \chi \ll 1 \\ \frac{\sqrt{3}}{4\pi} 3^{2/3}\,\Gamma\left(\frac{2}{3}\right)\left[2\frac{\Gamma(5/3)\Gamma(n+1/3)}{(n+1)\Gamma(n+1)} + \frac{\Gamma(2/3)\Gamma(n+7/3)}{\Gamma(n+3)}\right]\chi^{2/3}. & \chi \gg 1 \end{cases}$$

The functions a_n are represented with their asymptotic expansions in Fig. C.1a).
The first derivative of a_n is given by

$$a_n'(\chi) = \frac{\sqrt{3}}{2\pi} 3^{n+1} \chi^n \int_0^{+\infty} d\nu \left[\frac{2\nu^{n+1}}{(2+3\nu\chi)^{n+2}} K_{5/3}(\nu) + \frac{18(n+3)\chi^2\nu^{n+2}}{(2+3\nu\chi)^{n+4}} K_{2/3}(\nu) \right], \tag{C.5}$$

with the following asymptotic expansions

$$a_n'(\chi) \simeq \begin{cases} \frac{\sqrt{3}}{4\pi} 3^{n+1}\,\Gamma\left(\frac{n}{2}+\frac{1}{6}\right)\Gamma\left(\frac{n}{2}+\frac{11}{6}\right)\chi^{n}, & \chi \ll 1 \\ \frac{\sqrt{3}}{6\pi} 3^{2/3}\,\Gamma\left(\frac{2}{3}\right)\left[2\frac{\Gamma(5/3)\Gamma(n+1/3)}{(n+1)\Gamma(n+1)} + \frac{\Gamma(2/3)\Gamma(n+7/3)}{\Gamma(n+3)}\right]\chi^{-1/3}. & \chi \gg 1 \end{cases}$$

The functions a_n' are represented with their asymptotic expansions in Fig. C.1b).
Finally, the second derivative a_n'' is given by

$$a_n''(\chi) = \frac{\sqrt{3}}{2\pi} 3^{n+1} \chi^{n-1} \int_0^{+\infty} d\nu \left[\frac{4\nu^{n+1}(n-3\nu\chi)}{(2+3\nu\chi)^{n+3}} K_{5/3}(\nu) + \frac{36(n+3)\chi^2\nu^{n+2}(-3\nu\chi+n+2)}{(2+3\nu\chi)^{n+5}} K_{2/3}(\nu) \right]. \tag{C.6}$$

The zeroth order moment $A_0(\chi,\gamma)$ corresponds to the production rate $W_{\rm nCs}(\chi,\gamma)$ which admits the following asymptotic expressions

$$W_{\rm nCs}(\chi,\gamma) \simeq \frac{2}{3}\frac{\alpha^2}{\tau_e} \begin{cases} \frac{5\sqrt{3}}{4}\frac{\chi}{\gamma}, & \chi \ll 1 \\ 2.19\frac{\chi^{2/3}}{\gamma}, & \chi \gg 1 \end{cases}$$

C.3 Exact and Approximate Expressions of $b_n(\chi)$ Functions

In this Section, we detail the analytical computation of the functions $b_n(\chi_\gamma)$ appearing in the computation of the nth order moment of the nonlinear Breit-Wheeler

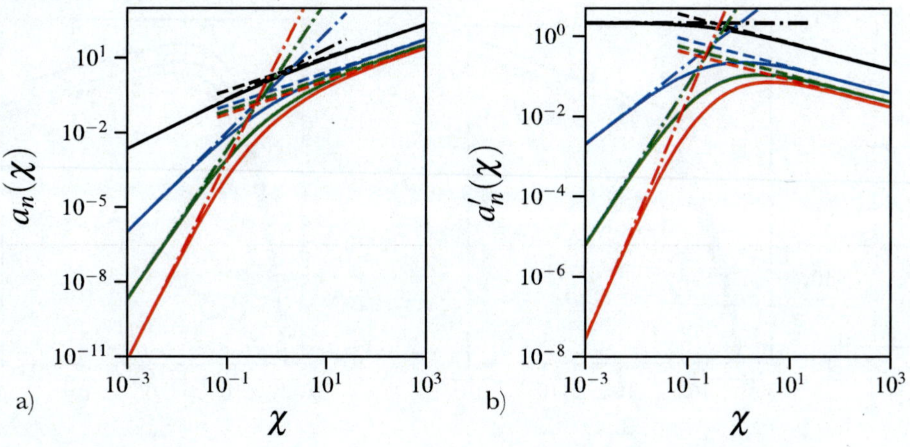

Fig. C.1 Plot of **a** $a_n(\chi)$ and **b** $a'_n(\chi)$, for $n = 0...3$ and their asymptotic expansion in dashed lines, for $\chi = 10^{-3}$ to 10^3

pair production kernel $w^{nBW}_{\chi_\gamma}(\gamma_\gamma, \gamma)$ (See Chap. 8). As a reminder, the successive moments of $w^{nBW}_{\chi_\gamma}(\gamma_\gamma, \gamma)$ are computed as

$$B_n(\chi, \gamma) = \int_1^{+\infty} d\gamma\, \gamma^n\, w^{\rm nBW}_{\chi_\gamma}(\gamma_\gamma, \gamma) = \frac{2}{3}\frac{\alpha^2}{\tau_e}\, \gamma_\gamma^{n-1}\, b_n(\chi_\gamma)\,, \tag{C.7}$$

where

$$b_n(\chi_\gamma) = \int_0^1 d\xi'\, \xi'^{n-1}\, \widetilde{T}_{\rm nBW}(\chi_\gamma, \xi')\,. \tag{C.8}$$

Using Eq. (3.92) yields

$$\begin{aligned} b_n(\chi_\gamma) = \frac{\sqrt{3}}{2\pi} \int_0^1 d\xi'\, \frac{\xi'^{n-1}}{1-\xi'} \Bigg[\frac{2}{3(n+1)\chi_\gamma} \frac{1-2\xi'}{1-\xi'} K_{5/3}\left(\nu'\right) \\ + K_{2/3}\left(\nu'\right)\Bigg]\,, \end{aligned} \tag{C.9}$$

which admits the following asymptotic expansions

$$b_n(\chi_\gamma) \simeq \begin{cases} \frac{9}{64}\frac{\sqrt{6}}{2^n}\chi_\gamma \exp\left(-\frac{8}{3\chi_\gamma}\right)\,, & \chi \ll 1 \\ \frac{\sqrt{3}}{4\pi}\, 3^{2/3}\, \Gamma\left(\frac{2}{3}\right) \left[2\frac{\Gamma(5/3)\Gamma(n+1/3)}{(n+1)\Gamma(n+1)} + \frac{\Gamma(2/3)\Gamma(n+7/3)}{\Gamma(n+3)}\right]\chi^{2/3}\,. & \chi \gg 1 \end{cases}$$

The functions b_n are represented with their asymptotic expansions in Fig. C.2a).

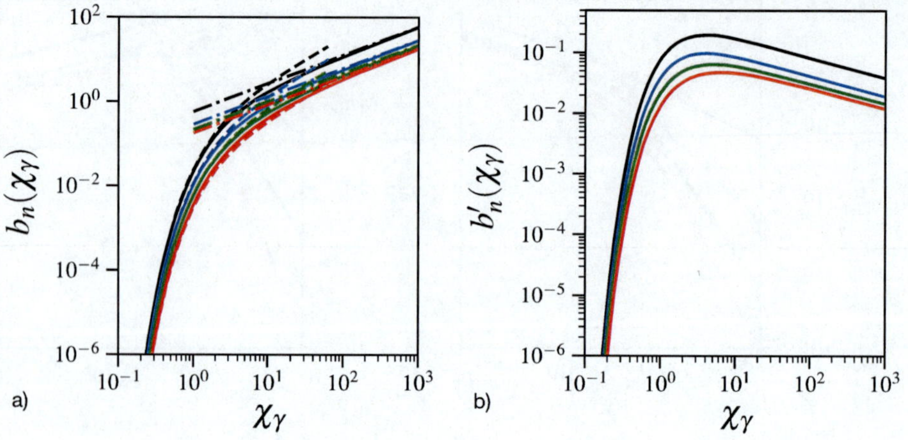

Fig. C.2 Plot of **a** $b_n(\chi_\gamma)$ and their asymptotic expansions in dashed lines, and **b** $b'_n(\chi_\gamma)$, for $n = 0...3$ and $\chi = 10^{-1}$ to 10^3

The first derivative of b_n is given by

$$b'_n(\chi_\gamma) = \frac{\sqrt{3}}{2\pi} \int_0^{+\infty} d\xi' \, \frac{\xi'^{\,n-1}}{1-\xi'} \frac{1}{\chi_\gamma} \left\{ \frac{2}{3(n+1)} \frac{1-2\xi'}{1-\xi'} \right. \times \frac{1}{\chi_\gamma} \left[\frac{\nu}{2} K_{2/3}(\nu) + \frac{\nu}{2} K_{8/3}(\nu) - K_{5/3}(\nu) \right] \left. + \frac{\nu}{2} \left[K_{-1/3}(\nu) + K_{5/3}(\nu) \right] \right\} . \tag{C.10}$$

The functions b_n are represented in Fig. C.2b).

The zeroth order moment $B_0(\chi, \gamma)$ corresponds to the production rate $W_{\mathrm{nBW}}(\chi_\gamma, \gamma_\gamma)$ which admits the following asymptotic expressions

$$W_{\mathrm{nBW}}(\chi_\gamma, \gamma_\gamma) \simeq \frac{2}{3} \frac{\alpha^2}{\tau_e} \begin{cases} 0.345 \, \frac{\chi_\gamma}{\gamma_\gamma} \exp\left[-\frac{8}{3\chi_\gamma} \right] , & \chi_\gamma \ll 1 \\ 0.57 \, \frac{\chi_\gamma^{2/3}}{\gamma_\gamma} , & \chi_\gamma \gg 1 \end{cases}$$